AF681475

HIGH VOLTAGE AND ELECTRICAL INSULATION ENGINEERING

IEEE Press
445 Hoes Lane
Piscataway, NJ 08854

IEEE Press Editorial Board
Lajos Hanzo, *Editor in Chief*

R. Abhari
J. Anderson
G. W. Arnold
F. Canavero
M. El-Hawary
B-M. Haemmerli
M. Lanzerotti
D. Jacobson
O. P. Malik
S. Nahavandi
T. Samad
G. Zobrist

Kenneth Moore, *Director of IEEE Book and Information Services (BIS)*

A complete list of titles in the IEEE Press Series on Power Engineering appears at the end of this book.

HIGH VOLTAGE AND ELECTRICAL INSULATION ENGINEERING

RAVINDRA ARORA
WOLFGANG MOSCH

IEEE PRESS SERIES ON POWER ENGINEERING

Mohamed E. El-Hawary, *Series Editor*

IEEE

IEEE PRESS

WILEY

A JOHN WILEY & SONS, INC., PUBLICATION

Copyright © 2011 by Institute of Electrical and Electronics Engineers. All rights reserved.

Published by John Wiley & Sons, Inc., Hoboken, New Jersey.
Published simultaneously in Canada.

No part of this publication may be reproduced, stored in a retrieval system, or transmitted in any form or by any means, electronic, mechanical, photocopying, recording, scanning, or otherwise, except as permitted under Section 107 or 108 of the 1976 United States Copyright Act, without either the prior written permission of the Publisher, or authorization through payment of the appropriate per-copy fee to the Copyright Clearance Center, Inc., 222 Rosewood Drive, Danvers, MA 01923, (978) 750-8400, fax (978) 750-4470, or on the web at www.copyright.com. Requests to the Publisher for permission should be addressed to the Permissions Department, John Wiley & Sons, Inc., 111 River Street, Hoboken, NJ 07030, (201) 748-6011, fax (201) 748-6008, or online at http://www.wiley.com/go/permission.

Limit of Liability/Disclaimer of Warranty: While the publisher and author have used their best efforts in preparing this book, they make no representations or warranties with respect to the accuracy or completeness of the contents of this book and specifically disclaim any implied warranties of merchantability or fitness for a particular purpose. No warranty may be created or extended by sales representatives or written sales materials. The advice and strategies contained herein may not be suitable for your situation. You should consult with a professional where appropriate. Neither the publisher nor author shall be liable for any loss of profit or any other commercial damages, including but not limited to special, incidental, consequential, or other damages.

For general information on our other products and services or for technical support, please contact our Customer Care Department within the United States at (800) 762-2974, outside the United States at (317) 572-3993 or fax (317) 572-4002.

Wiley also publishes its books in a variety of electronic formats. Some content that appears in print may not be available in electronic formats. For more information about Wiley products, visit our web site at www.wiley.com.

Library of Congress Cataloging-in-Publication Data:
Arora, Ravindra.
High voltage and electrical insulation engineering / Ravindra Arora, Wolfgang Mosch.
p. cm.
ISBN 978-0-470-60961-3 (hardback)
1. Electric insulators and insulation. 2. Electric power distribution–High tension. 3. Electric cables–Insulation. I. Mosch, Wolfgang. II. Title.
TK3401.A76 2010
621.319′37–dc22

2010030998

oBook ISBN: 978-0-470-94790-6
ePDF ISBN: 978-0-470-94789-0
ePub ISBN: 978-1-118-00896-6

Printed in Singapore.

10 9 8 7 6 5 4 3 2 1

CONTENTS

PREFACE

Earth has the unique characteristic of absorbing any amount of electricity (electric charge) and yet remaining neutral, that is, at zero potential. However, there could be no electricity without electrical insulation. The higher the potential, the greater the level of insulation required. The fundamentals of understanding high voltage engineering lie in the knowledge of the behavior of dielectrics, the electrical insulation, subjected to high potentials.

The insulation system is the basis of power systems. To create an optimally designed insulation system, that can provide long-lasting and satisfactory service, it is important to understand the behavior of dielectrics under electric stress. In a scientific subject, the fundamental knowledge and concepts evolve through continuous academic efforts supported by dedicated research work over decades, in some cases even centuries.

The contents of this book are derived from the lectures in High Voltage Engineering delivered by us for decades at Technical University (TU) Dresden, Germany and at Indian Institute of Technology, Kanpur, India, to the graduate and senior undergraduate students. Our first book in English on the subject was published in 1995 in India. Since then, much research and development work have been performed in our laboratories and elsewhere in the world. The innovative conceptual ideas, developed through discussions in the classrooms over the last two decades, have prompted us to write this book.

TU Dresden is one of the biggest and oldest technical universities in Europe. It celebrated its 150th anniversary in 1978. Germany is well known for its organized, systematic practical research in laboratories for the development of fundamental scientific approach and technology. The development in high voltage engineering at TU Dresden started more than a century ago, in early 1900. The research work in the field of gas discharge was initiated by the well-known persons in the field, Professors Teopler and Binder. Having had the opportunity to work in such a highly developed professional environment, the authors had full access to the fundamental concepts that evolved on the subject at TU Dresden.

A novel approach, the "field dependent behavior of the dielectrics", has been adopted throughout this book. In the classification of electric fields, a unique concept of "weakly nonuniform field" is introduced conceptually as well as analytically with the help of Schwaiger factor. It is an important tool for the design of high voltage equipment, especially for the Gas Insulated Systems (GIS).

For the preparation of this manuscript the authors have had the advantage of referring to the vast and rich literature available in German and in English. The advanced level of contents in this book is suitable for graduate and senior undergraduate engineering students. Research, design and practicing professionals will also find it useful for gaining in-depth knowledge and insights into the subject. For explaining a particular phenomenon, the actual measured curves, rather than schematic curves, have been provided throughout the book in order to make it more practice oriented.

In place of the hitherto commonly used term "Partial Discharge" (PD), a more appropriate term "Partial Breakdown" (PB) has been adopted for the first time in this book. In electrical engineering, the literal meaning of the word "discharge" is to get rid of a charge or electricity. Discharge is also described as the process of withdrawing or transference of an electric charge. At its initial stage, the electric discharge process between two electrodes leads to the mechanism of "conduction" of current through the dielectric. When the conduction is increased to the extent that the electric discharge current may lead to equalization of the difference of potential between two electrodes, the phenomenon is appropriately termed "breakdown", which is often mentioned as "discharge". Breakdown is the situation in which complete insulation failure takes place. Under extremely nonuniform field conditions, the electric breakdown process can confine locally to a region within the dielectric without affecting the total dielectric. Such a local breakdown process is appropriately termed as "Partial Breakdown", (PB). Stable Partial Breakdown process in any gaseous medium is known as "Corona". Stable PB process always precedes the complete breakdown in all the dielectrics working under extremely nonuniform field conditions.

The first chapter of the book, "Introduction", explains the real meaning of the relevant scientific terms commonly used in high voltage engineering. These terms have often been interpreted and adopted inappropriately. We saw the need to write this chapter through our involvement in teaching and interaction with our students. Discussions with graduate students while supervising their theses generated correct interpretations that have been incorporated in this text.

Chapter 2 on electric fields provides the base for understanding the field dependent behavior of dielectrics. The "electric field intensity" is the measure of "Electric Stress" a dielectric is subjected to and it depends upon the shape of the electrodes. Hence, the electric field intensity determines the overall performance of the dielectrics.

Chapter 3 in the book on gaseous dielectrics is the longest. The investigations made on free atmospheric air reveal the interesting conceptual developments in the breakdown process and the failure of insulating properties of dielectrics. Studying gaseous dielectrics is the best way of learning the behavior of all other types of dielectrics. The reader should find it interesting to learn how the breakdown strength of atmospheric air varies between very high magnitudes of the order of 90 kV/cm to an extremely low value of just 1 kV/cm under different field conditions. Distinction between the three types of Coronas, namely Star, Streamer and Leader Corona, and their peculiar characteristics are also described in this chapter. The performance of SF_6 gas and its mixtures is examined under different field conditions. Professionals involved with GIS will also find this part of the chapter useful for their specific interests.

The phenomenon of lightning, very closely related to the breakdown process in long air gaps, is presented in Chapter 4. Description of the rare phenomenon of

"Ball Lightning" should be interesting for all readers. The authors' experiences with the rare incidents of Ball Lightning due to man-made sources of charge are also described. Application of vacuum as a dielectric has increased considerably in the last three decades. Hence, it has been presented separately in Chapter 5.

Classification, properties and practical applications of liquid and solid dielectrics are presented in the Chapters 6 and 7. Their intrinsic and practical breakdown strengths are distinguished with respect to the processes, which affect the breakdown. Partial Breakdown in solid dielectrics is covered with special significance.

This book is our second joint venture. The first one was published in 1995 in India. We are always open to and would be grateful for suggestions from readers of our book.

Ravindra Arora
Wolfgang Mosch
June 2011

ACKNOWLEDGMENTS

The authors are grateful to the Technical University (TU) Dresden and Indian Institute of Technology (IIT), Kanpur, for providing us with facilities and encouragement in completing our second Indo-German joint book. After retirement from IIT Kanpur, one of the authors, Dr. Ravindra Arora, is working with Ansal Institute of Technology, Gurgaon, India. The last part of the manuscript was completed at this institute. We are grateful to AIT Gurgaon for providing us with wholehearted support and access to its excellent facilities.

We are indebted to our colleagues, Doz. Dr.-Ing. habil. Helmut Loebl and Dr.-Ing. Werner Keller of TU Dresden; Prof. Manjit Singh Kalra and Prof. Prem Kumar Kalra of IIT Kanpur and Prof. P.V. Dhamija of AIT Gurgaon for their help and advice in completing the manuscript.

The authors have no words to thank Prof. M. Ramamoorty, Rtd. Director General of Central Power Research Institute, Bangalore, who came forward so readily to review the manuscript and make positive and valuable comments at short notice.

Mr. Ajai Kumar Srivastava typed the complete text and the tables and Mr. Avijit Singh prepared all the figures and drawings in CorelDraw at IIT Kanpur. We are grateful to both of them for their commendable work and readiness to take up the job. We acknowledge their efficiency and perfection in completing the work. The assistance provided by Mr. Lekhraj Singh and Mr. S.V. Ghorpade, Technical Assistants in the HV Laboratory at IIT Kanpur, throughout the preparation of the text is invaluable.

The authors are indebted for the remarkable interest and gesture extended by two M.Tech. students, Major Vikram Mehndiratta and Mr. Sachin Garg, who were working on their theses in the HV Laboratory at IIT Kanpur. Their involvement in the subject and in the preparation of the manuscript is commendable.

We received great help from the staff of the Computer Center at AIT Gurgaon in processing the soft copy of the manuscript. We express our sincere gratitude to Mr. Pradeep Lal, Mr. Sanjay Singh, Mr. Narender Kumar Verma, Mr. Rakesh Kumar and Mr. Alok Singh.

It is difficult for us to express in words our gratefulness to Mrs. Kiran Bala Arora, Dr. Anubhav Arora and Mr. Pujak Arora, wife and sons of Dr. Ravindra Arora, for their encouragement and helpful contributions in giving shape to this book.

We must also express our sincere gratitude to Ms. Mary Mann, Ms. Melissa Yanuzzi and all others at Wiley-IEEE. They have shown tremendous patience and interest in coordinating the entire printing work.

While writing these words of acknowledgment, we realize that a venture like this requires teamwork by many dedicated professionals. We are grateful to all those whose names may have been left out by oversight.

Ravindra Arora
Wolfgang Mosch
June 2011

CHAPTER 1

INTRODUCTION

The subject, "High Voltage Engineering", is the knowledge of the behavior of dielectrics—electrical insulation when subjected to high voltage. Performance of dielectrics is electric field dependent. The electric field configuration to which a dielectric is subjected determines its life and function in the long run. It is always desirable to minimize the volume of the electrical insulation requirements yet a long and trouble-free life of all high voltage apparatus should be ensured. For an apparatus to be economically viable, its desirable life expectancy is thirty to forty years, depending upon the cost and technology of production involved.

The world has seen rapid advancement in the technology applied in high voltage apparatus in the second half of the twentieth century. Manufacturing of Gas Insulated Sub-stations (GIS), power transformers, cables and switchgears at the highest rated voltages up to 1100 kV involve the most sophisticated technologies. Such a development has taken place with dedicated efforts to understand the behavior of dielectrics, gaseous, solid, liquid, and vacuum.

The last half a century has also seen prominent advancement in the technology of dielectric finishes on equipment. To a limited extent, insulating materials with better dielectric properties and performance have been developed. Knowledge of electric field dependent behavior of dielectrics has led to better use of the insulating materials. Advancement in techniques of evaluating the quality of the finish of electrical insulation in an apparatus has contributed to producing quality power apparatus with more reliability up to the highest rated voltages. The non-destructive testing and condition monitoring techniques of equipment/insulation have improved considerably. The high voltage test apparatus and measuring instrumentation and their respective technologies have also made big advances. These have led to the production of more dependable and economical high voltage apparatus with sophisticated technologies.

The contents of this book were initially developed at the High Voltage Laboratory of Technische Universität Dresden, Germany, which is well known in the continent of Europe for its dedicated research and development work for more than one and a half centuries. These were published for the first time in English in our

High Voltage and Electrical Insulation Engineering, First Edition. Ravindra Arora, Wolfgang Mosch.
© 2011 Institute of Electrical and Electronics Engineers. Published 2011 by John Wiley & Sons, Inc.

earlier book, "High Voltage Insulation Engineering" in 1995. Advances in this subject, at TU Dresden, Germany and Indian Institute of Technology Kanpur (India) and in many other countries in the world are being incorporated into this second book.

While delivering the lectures based on our first book, interaction with the students revealed a number of lacunae in interpreting the basic concepts essential for understanding the behavior of dielectrics. Hence, some fundamental terminologies used commonly in this subject are explained in the following pages. Explanation of these terms has been mainly derived from various English-language dictionaries [1.1] to [1.4] that describe the same terminology in slightly different ways. Hence, a number of similar expressions available for a particular term are compiled. These descriptions are bulleted in the following text. A clear interpretation of these terms will help the reader to better understand the high voltage phenomena.

1.1 ELECTRIC CHARGE AND DISCHARGE

Electron:

- an elementary particle of negative charge found outside the nucleus of an atom
- negatively charged sub-atomic particle found in all atoms and acting as the primary carrier of electricity in solids

Proton:

- a subatomic particle with a positive electric charge occurring in all atomic nuclei-origin Greek, "first thing"
- a nuclear particle with positive charge equal and opposite to that of an electron negative charge

Ion:

- an electrified atom having either a positive or negative charge
- an electrified atom which has increased or decreased its number of electrons after electrolysis (ionisation)
- an atom or molecule with a net electric charge produced through loss or gain of electrons

Ionise:

- convert an atom, molecule or substance, into an ion or ions
- to convert into an ion form
- to convert wholly or partly into ions—to become ionized

Ionisation:

- the process of formation of ions

Electric Charge:

- the presence of an uncancelled excess of either positive subatomic particles (protons), or negative subatomic particles (electrons) in a substance
- free subatomic particles of a polarity, positive or negative

The behavior of electric charge can be explained with the following typical characteristics:

- ionisation is a process by which charges build up
- accumulation of charge (q) builds up potential ϕ
- concentration of like polarity charge (in dielectrics) is known as "**space charge**"
- when the positive and the negative charges are uniformly distributed in a dielectric, the volume charge density "ρ_v", is equal to zero
- on the contrary, when there is a concentration of any one polarity charge, ρ_v is not equal to zero
- the electric charge is at rest in dielectrics, however, it is restless in conductors
- the electric charge always acquires the least resistance path to flow
- flow of charge is electric current
- the electric charge finds its ultimate peace only inside the earth, the mother earth

Electric discharge:

- to get rid of a charge of electricity
- withdrawing or transference of an electric charge
- release or neutralise the electric charge
- a flow of electricity through the air or other gas
- a sudden movement of charge

The electric discharge process can be typically described by the following:

- ionization is the process by which electric charges—hence potential builds up; while discharge involves movement of charge—hence loss of potential
- ionization builds up potential on a body while discharge tends to lose it
- electric discharge leads to equalization of the difference of electric potential built by the charge between any two bodies/electrodes

1.2 ELECTRIC AND MAGNETIC FIELDS AND ELECTROMAGNETICS

Field is a quantity that is a function of space. The presence of a field is sensed by the force exerted on a particle or body. A wave can be defined as a function of both time and space [1.5, 1.6].

Electric Field:

- a quantitative description of the attraction or repulsion of one electric charge by another at any one point
- the ratio of the force exerted on a positive test charge, placed at that point, to the magnitude of the charge
- the source of electric field intensity is electric charge

Magnetic Field:

- the portion of space near a magnetic body or a current carrying body in which the forces from the body or current can be detected
- a region around a magnet within which the force of magnetism acts
- any space or region in which magnetic forces are present, as the space or region in or around a piece of magnetized steel, or in or around an electrical current

Electromagnetic:

- relating to the inter-relation of electric and magnetic fields
- pertaining to electromagnetism or an electromagnet

Electromagnetism:

- magnetism developed by a current of electricity
- branch of physical science that deals with the physical relations between electricity and magnetism
- the study of the relation between electric currents and magnetism
- magnetism caused by electric current

Electromagnetic Radiation:

- radiation in which electric and magnetic fields vary at the same time

Electromagnetic Wave:

- a wave whose characteristics are variations of electric and magnetic fields, such as a radio wave or a light wave
- one of the waves that are propagated by simultaneous periodic variations of electric and magnetic field intensity and that include radio wave, infrared, visible light, ultraviolet, X-rays and gamma rays

Electromagnetic waves can also be explained as follows:

- time varying magnetic field produces an electric field (Maxwell's equation)
- time varying electric field also produces a magnetic field, even in the absence of flow of electric current
- time varying electric and magnetic fields form electromagnetic waves that are characterized by their impedance, energy and velocity of propagation etc.

Electromagnetic Field:

- An electromagnetic field comprised of both electric and magnetic fields. The two fields are related to each other theoretically such that the Maxwell's equations are satisfied under the given boundary conditions. An electromagnetic field itself has no mathematical symbol and it is not a measurable quantity as such.

1.3 DIELECTRIC AND ELECTRICAL INSULATION

Electric:

- electricus produced from amber (a resin) by friction
- amber's substance that develops electricity under friction
- pertaining to, consisting of, or containing electric charge or electric current
- charged with or capable of developing electricity

Dielectric:

- archaic: a non conductor of electricity used to excite or accumulate electricity
- dia + electric: non conductor of direct electric current
- insulating (medium or substance), non-conductive, non-conductor, through which electricity is transmitted (without conduction).
- a non conducting or insulating material; a material which admits electrostatic and magnetic lines of force but resists passage of electric current.

However, there is no dielectric which does not have any conduction of charge or current. Conduction currents through dielectrics mainly depend upon their relative permittivity number ε_r and the type and amplitude of the voltage applied.

Before pico, nano or micro ampere of current magnitudes could not be detected or measured, the electrical insulating materials were considered to be totally non-conducting, hence called "dielectric".

Insulator:

- one that insulates; a material that is a poor conductor of electricity

Electrical insulant:

- an electrical insulating material, insulation, the material used for insulating

Insulate:

- to separate from conducting bodies by means of nonconductors so as to prevent transfer of electricity

The first and foremost enemy of an electrical insulator is water. It is the most **bitter** enemy of liquid and solid dielectrics.

1.4 ELECTRICAL BREAKDOWN

Failure of electrical insulation properties of insulating materials is known as "breakdown". The electrical breakdown of dielectrics can be distinguished between "Global" and "Local" breakdowns, described below.

1.4.1 Global Breakdown

The complete rupture or failure of the electrical insulation between two electrodes is described as "breakdown". It is generally termed as "electrical breakdown", or simply "breakdown".

1.4.2 Local Breakdown

The phenomenon of failure of insulating properties confined locally to a part of the total insulation system provided between two electrodes is known as local breakdown. Since it takes place partially, not globally, it is described as "Partial Breakdown" (PB) in an electrical insulation. The healthy part of the dielectric continues to provide electrical insulation between the two electrodes in spite of failure of insulating properties in some limited part. The terminology, used very widely so far, for describing this phenomenon has been "Partial Discharge" (PD) in the literature. Since the word discharge has several meanings, it is more appropriate to describe this phenomenon as "Partial Breakdown" (PB). This phenomenon can occur in any dielectric under adverse conditions. Like Breakdown, the Partial Breakdown phenomenon is injurious for the dielectrics. Hence it is most undesirable and should be prevented as much as possible.

1.5 CORONA, STREAMER AND AURORA

Corona:

- the gaseous envelope of the sun or star
- a small circle of light seen around the sun or moon
- origin Latin; crown, cornice, garland
- halo of white light seen around the black disc of moon in total eclipse of sun, Figure 1.1
- the brush discharge of electricity
- a circle of light made by the apparent convergence of the streamers of the aurora borealis
- a faint glow adjacent to the surface of an electrical conductor at high voltage
- a crown or garland, especially that bestowed upon the ancient Romans as a reward for distinguished services
- white or coloured circle of light seen around a luminous body, the sun or moon
- the thin, hot outer atmosphere of the sun that is shaped by solar magnetic fields

The stable Partial Breakdown (PB) phenomenon in gaseous dielectrics/mediums is known as corona.

Streamer:

- a long, narrow strip of material used as a decoration or flag
- a Pennon, ribbon attached at one end and floating or waving at the other

Figure 1.1 Diamond ring with long extension of the solar corona seen at total eclipse taken by the author in 1995.

- column of light shooting up in aurora
- any long narrow wavy strip resembling or suggesting a banner floating in the wind
- a long extension of the solar corona visible only during a total solar eclipse
- Aurora Borealis
- anything which streams
- stream of light shooting upward from the horizon, as in some forms of the aurora borealis

The partial breakdown (PB) phenomenon in gaseous dielectrics at hemispherical rods, spherical or similar electrodes appear like a streamer or a shower of discharge, are known as streamer corona.

Aurora:

- luminous atmospheric (prob. electrical) phenomenon radiating from earths northern or southern magnetic pole; down; colour of sky at sunrise
- Roman Goddess of dawn (morning)
- a luminous phenomenon that consists of streamers or arches of light appearing in the upper atmosphere of a planet's polar regions and is caused by the emission of light from atoms excited by electrons accelerated along the planets magnetic field lines.
- the sporadic radiant emission of light from the upper atmosphere over middle and high latitudes

Auroras are spectacular displays of luminous radiation in the sky near polar regions, their symmetry defined by the earth's magnetic field. Aurora lights are emitted when atoms in the ionosphere are struck by high energy electrons coming from the sun [1.7].

The well known "Faraday Glow" is nothing but emission of light from atoms excited by electrons accelerated along a tube having atmospheric pressures, as in high latitudes at an altitude of hundreds of kilometers above the ground (earth), on application of voltage.

Aurora australis:

- an aurora that occurs in earth's southern hemisphere
- the southern lights
- streamers of coloured light seen in the sky near the South Pole origin: Latin

Aurora borealis:

- an aurora that occurs in earth's northern hemisphere
- the northern lights
- streamers of coloured light seen in the sky near the North Pole origin: Latin
- the northern down in Latin, meaning the light generated by electrons and ions bombarding the upper atmosphere at high latitudes.

1.6 CAPACITANCE AND CAPACITOR

Conductors have resistance; coils have inductance; and dielectrics have capacitance. A dielectric between two electrodes gives rise to a capacitor having a capacitance. The exact value of capacitance (in Farads) of a capacitor is difficult to determine analytically. It depends upon the shape and size of the electrodes, the volume of the dielectric between them, and the condition of the dielectric.

Figure 1.2 shows a typical parallel plate capacitor. The capacitance "C" of this capacitor is analytically calculated as,

$$C = \varepsilon_o \varepsilon_r \frac{A}{d} \qquad F$$

where

ε_o: is the absolute permittivity or dielectric constant equal to 8.854 10^{-12} or $1/36\pi \times 10^{-9}$ Farads/m.

ε_r: the relative permittivity number, a dimension less quantity which is a function of the temperature of the dielectric and also the magnitude and frequency of the voltage applied to it.

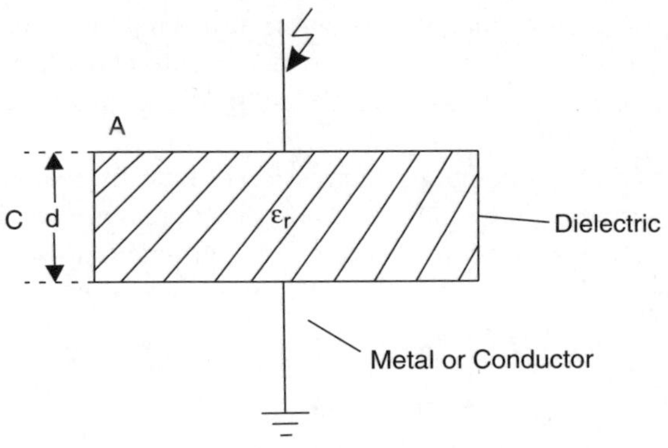

Figure 1.2 A Capacitor.

A: area of the plates (considered to be identical) in sq.m

d: gap distance between the plates in m

This analytical formula for the calculation of capacitance has been derived with a very important assumption that the electric field between the plates is a "uniform" field. However, if the two plates are of limited size, the fringing effect of the plate ends would not render unifor the field inbetween. Hence, many authors have described it to be valid for two "infinite" size plates in the literature. In that case, the field in the "center" of the plates may be uniform but when the area "A" tends to infinity, this formula is not valid for determining capacitance of this capacitor. Even if one considers two very large area plates, the field may be uniform only in the middle of the plates, not throughout the area "A". Uniform field between two electrodes is only an ideal condition, one which is very difficult to achieve in practice.

Another lacunae in this formula is that "ε", the permittivity of the dielectric is often considered to be a constant. As mentioned, the relative permittivity varies with temperature and applied magnitude of voltage and its frequency. Since, $\varepsilon = \varepsilon_o \varepsilon_r$, it would be wrong to describe ε to be a constant.

The formula for the calculation of capacitance of the parallel plate capacitor should therefore be applied for a rough estimation of the capacitance. It is always advisable for the actual value of capacitance formed by a dielectric between two electrodes to be determined by measurement.

1.6.1 Stray Capacitance

A capacitor, depending upon its physical location, forms capacitance with other wholly or partially conducting bodies.

As shown in Figure 1.3, the stray capacitances could be constituted by one or more dielectrics. The stray capacitances may vary in magnitude with respect to the location of the main capacitor. Air is the dielectric which constitutes most often the stray capacitances. To minimize the effect of stray capacitance, often screens (grounded concentric electrodes) are used in practice.

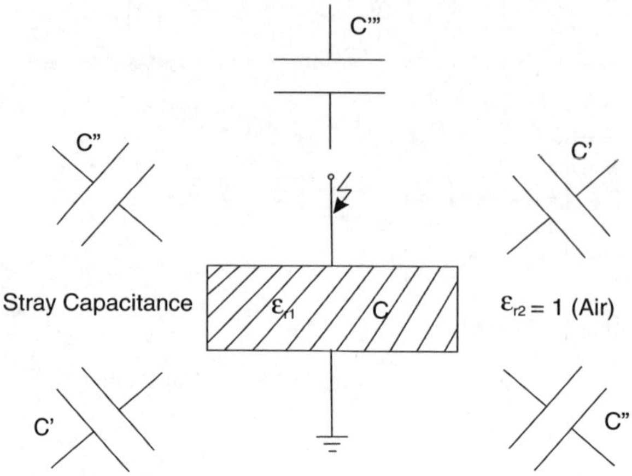

Figure 1.3 A Capacitor with its stray capacitance.

REFERENCES

[1.1] "Webster's Ninth New Collegiate Dictionary", Merriam-Webster Inc. (1991).
[1.2] "The Concise Oxford Dictionary of Current English", Fifth Edition, Oxford University Press (1972).
[1.3] "New Webster's Dictionary", College Edition, Surjeet Publications (1985).
[1.4] "Compact Oxford Dictionary Thesaurus and Wordpower Guide", Indian Edition, Oxford University Press (2006).
[1.5] Neff, Herbert P., Jr., "Basic Electromagnetic Fields", Herper International Edition (1987).
[1.6] Cheng, David K., "Field and Wave Electromagnetics", Second Edition, Pearson Education, India (2002).
[1.7] Dooling, Dave, "Stormy Weather in Space", IEEE Spectrum, June 1995.

CHAPTER 2

ELECTRIC FIELDS, THEIR CONTROL AND ESTIMATION

To optimally design insulation that could provide long and satisfactory performance of electric equipment, it's important to understand electric field intensity in high voltage engineering. A systematic approach, with the help of electric field theory, develops a vivid understanding of the behavior of dielectrics under various field conditions.

The electric field, produced due to potential on a body, stresses the dielectric (electric insulation) with "electric stress". The parameter that determines the magnitude of electric stress on the dielectrics is known as the "electric field intensity". The performance of a dielectric strongly depends upon the field configuration and the magnitude of electric field intensity with which it is stressed.

The electric charge is considered static when there is no movement of charge. This is possible only when the dielectrics have no or negligible conduction of current. Unlike in metals, where the charge is turbulent, it can be considered to be relatively stationary in all dielectrics when static voltage is applied.

The fields produced by static charge or direct voltage is known as "electrostatic field", whereas the field produced by power frequency alternating voltage is described as "quasi-stationary electric field". Both these fields are, however, often assumed to be without any space charge and not influenced by the movement of charge carriers for analysis. Such fields, also described as streamlined, rotation free or curlfree fields, are interesting to analyze. This chapter classifies the field configurations, and then describes different analytical and numerical methods of field estimation. Methods of stress control and numerical optimization techniques of electric stress are also explained.

2.1 ELECTRIC FIELD INTENSITY, "E"

Faraday described the space around a magnet to be filled with "lines of magnetic force". Similarly, the region around an electrified object can be considered filled with "lines of electric force". To Faraday, these lines existed as mechanical structures in the surrounding medium (the dielectric) and could exert force on an object placed

High Voltage and Electrical Insulation Engineering, First Edition. Ravindra Arora, Wolfgang Mosch.
© 2011 Institute of Electrical and Electronics Engineers. Published 2011 by John Wiley & Sons, Inc.

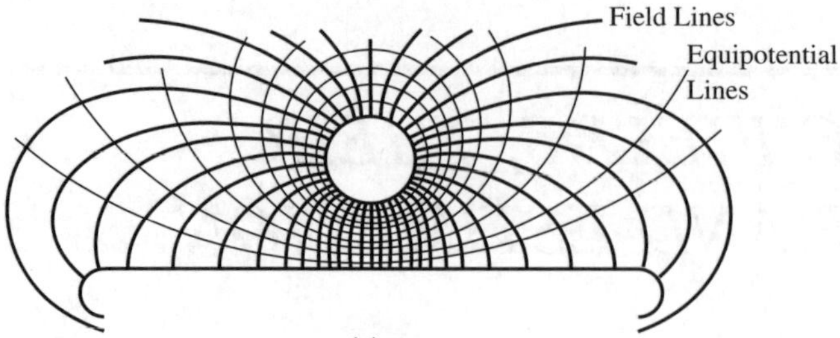

(a)

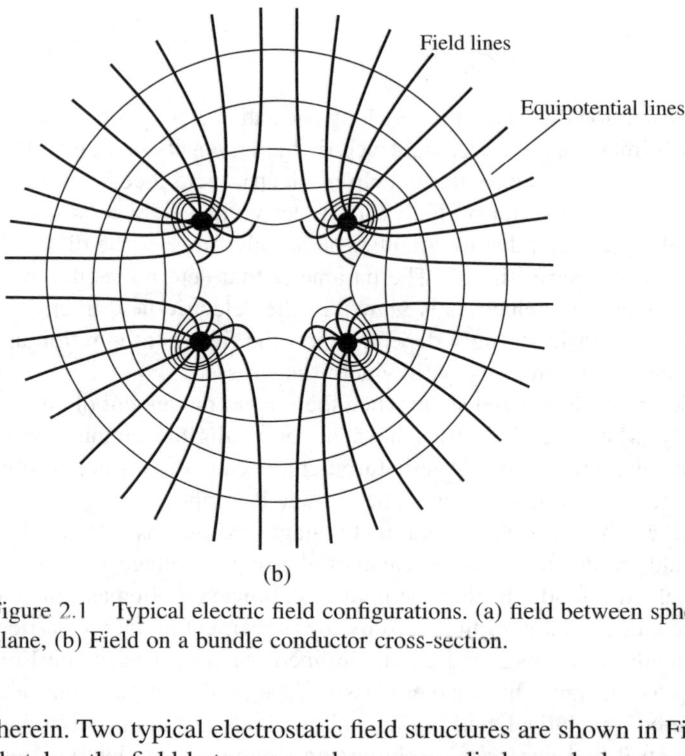

(b)

Figure 2.1 Typical electric field configurations. (a) field between sphere or cylinder and plane, (b) Field on a bundle conductor cross-section.

therein. Two typical electrostatic field structures are shown in Figure 2.1. Figure (a) sketches the field between a sphere or a cylinder and plane, and Figure (b) shows the field on a cross section of a bundle of four conductors. The sketches of these field configurations neglect the effect of ground.

The "electric field intensity", also known as the "electric field strength", is defined as the electrostatic force F per unit positive test charge q, placed at a particular point p in a dielectric. It is denoted by E, and expressed in the unit "Newtons per Coulomb", that is, the force per unit charge.

Since the potential is expressed in "Joules per Coulomb (J/C)", or "Newton-meter per Coulomb (Nm/C)", which is defined as "Volt", the electric field intensity is measured in its more common practical units of "Volt per meter" (V/m or kV/cm). It is often expressed in kV/mm also.

The electric field intensity is often more specifically mentioned as "electric stress" experienced by a dielectric or an electrical insulating material. The potential

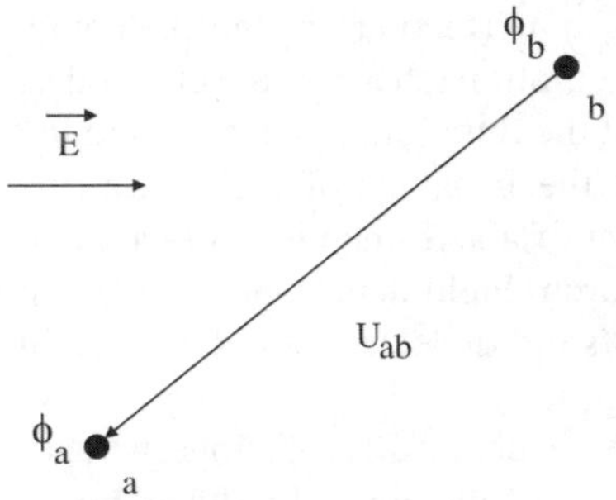

Figure 2.2 Field between two points.

difference U_{ab} between two points a and b (as shown in Figure 2.2, having scalar potentials ϕ_a and ϕ_b in an electric field $\vec{E}$, without any concentration of like polarity charge, that is, space charge) is defined as the work done by an external source in moving a unit positive charge from **b** (last named) to **a** (first named),

$$U_{ab} = -\int_b^a |\vec{E}|\, dx = (\phi_a - \phi_b) \tag{2.1}$$

U_{ab} is positive if the work is done in carrying the unit positive charge from **b** to **a**. The magnitude of electric field intensity is, therefore, given by the value of the rate of change of potential with distance. The maximum magnitude of the field intensity can therefore be obtained when the direction of the increment of distance is opposite to the direction of $\vec{E}$. In other words, the maximum value of the rate of change of potential is obtained when the direction of $\vec{E}$ is opposite to the direction in which the potential is increasing most rapidly,

$$\left.\frac{dU_{ab}}{dx}\right|_{max} = -|\vec{E}|_{max} \tag{2.2}$$

Equation (2.2) serves to provide a physical interpretation of the process of finding the electric field intensity from the scalar potential ϕ. The operator on ϕ by which $\vec{E}$ is obtained is thus known as the "gradient". The relationship between ϕ and $\vec{E}$ is written as,

$$\vec{E} = -grad\ \phi \tag{2.3}$$

The electric field intensity is, therefore, numerically equal to the "potential gradient", another name for $\vec{E}$ commonly mentioned in the literature.

2.2 BREAKDOWN AND ELECTRIC STRENGTH OF DIELECTRICS, "E_b"

The qualitative definition of "electric strength" of a dielectric is the "maximum electric stress a dielectric can withstand without rendering it to rupture completely". A quantitative definition is, however, complicated because a large number of factors affect the electric breakdown of a dielectric. These factors include the composition of dielectric material, presence of impurities and imperfections in the dielectric, pressure, humidity, temperature, electric field configuration (shape of the electrodes, their size and gap distance), electrode material and the duration, magnitude and the waveform of the applied voltage. To understand the behavior of dielectrics, the mechanisms of insulation breakdown and the affect of above-mentioned factors on breakdown is an important field of study and research. Field dependent breakdown mechanisms in gaseous (also vacuum), liquid, and solid dielectrics are discussed in Chapters Three to Seven.

In a time varying **ac** power frequency field (the so called quasi stationary field), the maximum electric stress occurs at the peak value of the applied voltage, not at its **rms** value. This must always be kept in mind while dealing with other types of voltages, namely lightning, switching and the fast transient impulse voltages, since their magnitudes are given only in their peak values.

2.2.1 Partial Breakdown in Dielectrics

On applying sufficient voltage, extreme electric field enhancement at a particular or at several locations within a dielectric may develop depending upon the electrode configurations. At such location/locations the dielectric could get stressed excessively to the extent that it may lose its insulating capability restricted only to this region. This gives rise to breakdown of the insulation locally known conventionally as "partial discharge" (PD). In a real sense, it is "partial breakdown" (PB) in the dielectric. Although the term "partial discharge" (PD) has been very widely conceptualized, it does not express the actual phenomenon taking place within the dielectric. Hence, Partial Breakdown (PB), the more appropriate terminology expressing the phenomenon, is followed in this book.

The voltage applied across the dielectric, at which such an activity begins locally somewhere in the dielectric, is known as "partial breakdown inception voltage", U_i. PB takes place in all the dielectrics but only under extremely nonuniform field conditions. When stable PB occurs at a free electrode in gaseous dielectrics the phenomenon is known as "Corona". Inside a solid or a liquid dielectric, it is described as "Internal Breakdown". When PB takes place on the surface of a solid or a liquid, but in gas, the phenomenon is known as "Surface Breakdown", also commonly known as "Tracking".

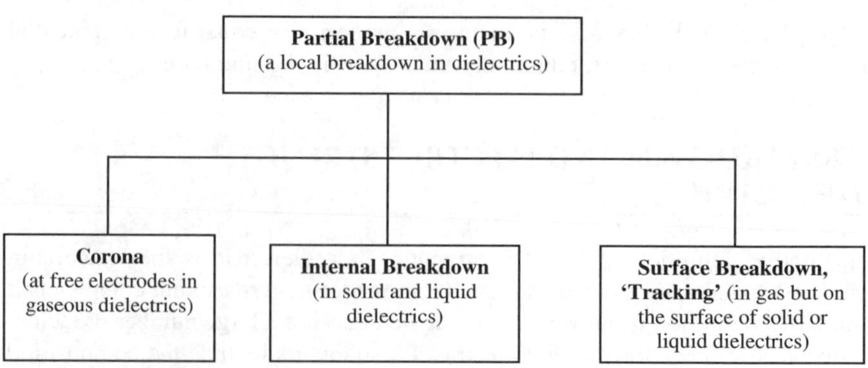

Figure 2.3 Classification of Partial Breakdown (PB).

2.3 CLASSIFICATION OF ELECTRIC FIELDS

The electric field configurations can be classified in broad sense into two forms, that is; uniform and non uniform fields, as shown in Figure 2.4.

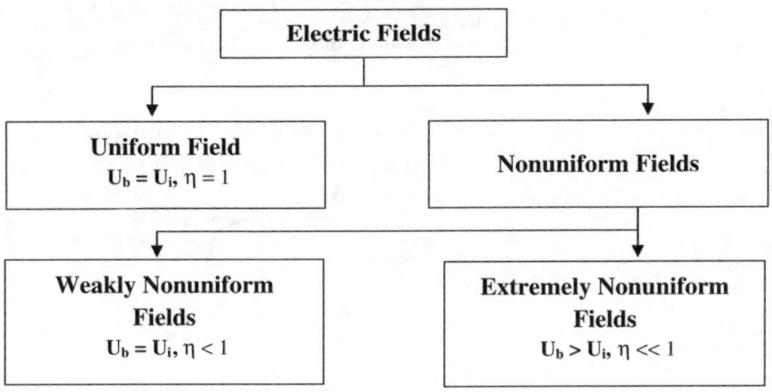

Figure 2.4 Classification of electric fields.

In a "uniform field", the potential is linearly distributed. The equi-potentials and the field lines make perfect square in the main field region. Hence the electric field intensity is constant throughout the space between the two electrodes, Figure 2.5 (a). An important characteristic of the uniform field is that the insulation breakdown in such a field always takes place without any partial breakdown proceeding within the dielectric. In other words, it can be said that in uniform field configuration, the breakdown voltage, U_b is equal to the partial breakdown inception voltage, U_i. The degree of uniformity "η" (defined in section 2.3.1) of uniform field is numerically equal to one, Figure 2.5 (a). A uniform field can be achieved between two electrodes especially designed for the given gap distance with the help of Rogowsky or Borda profiles [2.1, 2.2].

On the basis of the particular behavior of dielectrics in nonuniform fields, the electric field configuration can be further distinguished between weakly and extremely nonuniform fields. As in uniform fields, in weakly nonuniform fields also, no stable partial breakdown takes place before the complete breakdown in the dielectrics. Hence, in such fields too, the breakdown voltage, U_b is equal to the partial breakdown inception voltage, U_i. The degree of uniformity, η of such fields is less than one and its lower value limit depends upon the physical conditions of the dielectric between the electrodes. Typical examples of such fields are the radial fields between two coaxial cylinders and concentric spheres, and the field between two adjacent spheres shown in Figure 2.5 (b).

The electric field configurations having an extreme nonlinear distribution of potential in the dielectric between the two electrodes results, in extreme nonuniformity in electric stress in the dielectric. Unlike in uniform and in weakly nonuniform fields, the breakdown of dielectrics in "extremely nonuniform fields" always takes place after stable partial breakdown phenomenon is able to set in. These PBs are rendered unstable only just before the global breakdown in the dielectric. Hence in such electric field configurations, the complete breakdown voltage U_b is much

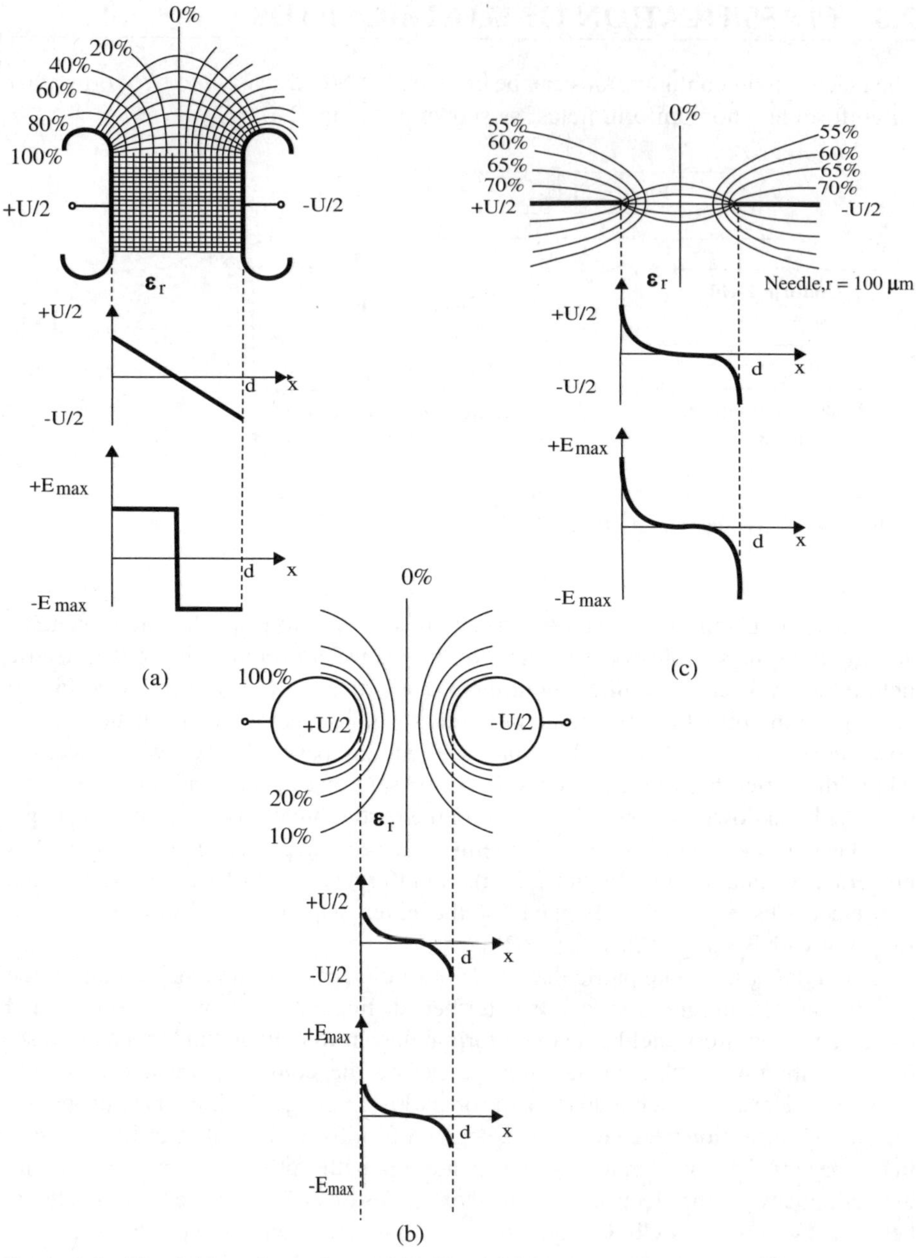

Figure 2.5 The field configurations (a) Uniform field between two parallel plates (b) Weakly nonuniform field between two adjacent spheres (c) Extremely nonuniform field between needle-needle electrodes.

greater than the partial breakdown inception voltage U_i. This field configuration is very important as it is the most unfavorable, still widespread, type of electric field in the power system. For example at the tip of a sharp electrode, the dielectric is subjected to very high electric stress, but elsewhere in the same insulation system between the electrodes, it may be stressed moderately. Figure 2.5 (c) shows a typical "extremely nonuniform field" in the dielectric between two symmetrical needle-needle electrode system.

A dielectric has the highest electric strength when it is subjected to uniform electric field. However, it is very difficult to realize a uniform field in practice. It is accomplished only for experimental purposes in research laboratories with tremendous efforts and utmost care. The size of the electrodes may have to be increased extraordinarily large depending upon the gap distance. Even a minor irregularity in the electrode surface may change the field characteristic.

Although, the extremely nonuniform field configuration is the most commonly prevalent, thc weakly nonuniform field is achieved often with moderate efforts in coaxial insulation systems, for example, in cables, gas insulated systems (GIS), etc.

2.3.1 Degree of Uniformity of Electric Fields

The degree of uniformity η (eta), introduced by Schwaiger in 1922 [2.3] as measure of the uniformity of a field, is defined as follows:

$$\eta = \frac{\hat{E}_{mean}}{\hat{E}_{\max}} = \frac{\hat{U}}{d} \cdot \frac{1}{\hat{E}_{\max}} \tag{2.4}$$

or

$$\hat{U} = \hat{E}_{\max} \cdot \eta \cdot d$$

where $\hat{E}_{mean}$ and $\hat{E}_{\max}$ are the peak values of the mean and the maximum field intensities in a dielectric respectively. $\hat{U}$ is the peak value of potential difference applied between the two electrodes at a distance d apart. η is also known as "Schwaiger factor".

The value of η also represents the degree of use of the dielectric between the two electrodes. A higher value of η represents better use of the insulating properties of the dielectric. It compares the ideal condition of electric field intensity (uniform field between electrodes at the same gap distance d apart) with the existing actual maximum field intensity. Thus η, a dimensionless quantity enables a comparison of the uniformity of field configurations formed between different electrodes. The value of η lies between,

$$0 \leq \eta \leq 1$$

With the knowledge of the value of η for a particular field configuration, the maximum electric field intensity or the maximum electric stress on a dielectric can easily be estimated. η serves as a ready reference that is important information for insulation design in equipments. However, for determining the exact magnitude of maximum electric stress, numerical estimation techniques, as discussed later, have to be applied for different shapes of electrodes used at different locations in the equipment.

The reciprocal of η is denoted by "f", which represents the "degree of non-uniformity" of an electric field, a concept described by some authors in the literature [2.4, 2.5]. Schwaiger also introduced "p", the geometrical characteristic factor for electrode configurations. He established that it is possible to represent η as a function of "p" described as,

$$p = \frac{r+d}{r} \tag{2.5}$$

where

$$(1 \leq p < \infty$$

and

$$\eta = f(p) \tag{2.6}$$

where r is the smallest radius of curvature of electrodes, and d the shortest gap distance between the two electrodes under consideration.

For some common and practical electrode configurations, the equation (2.6) is represented graphically in Figure 2.6. These are known as "Schwaiger Curves".

In Figure 2.6, the curves 1 and 2 represent curved surface electrodes; 3, 4 and 5 are for cylindrical electrodes; and 6, 7, and 8 for spherical electrode systems.

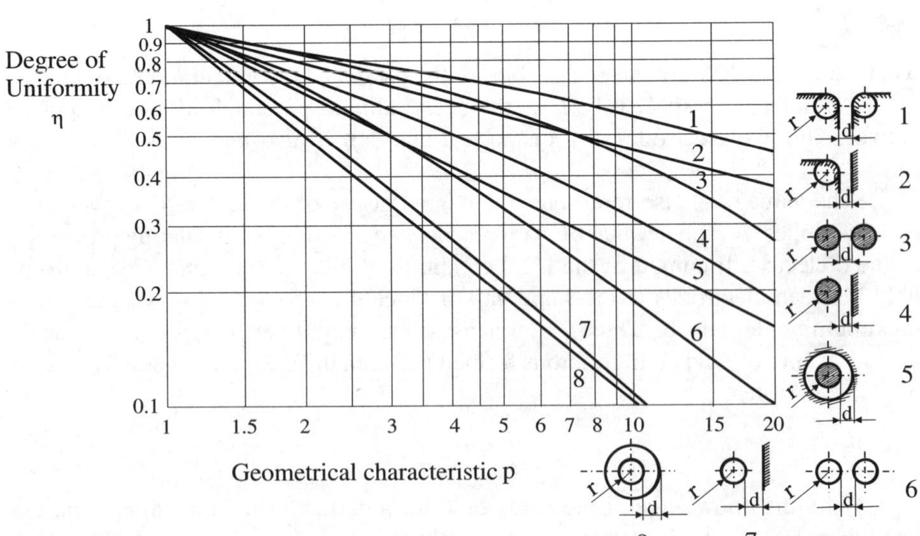

Figure 2.6 Schwaiger curves for fields with spherical, cylindrical and curved electrode configurations.

For a fixed value of "p", following important basic relationship between different electrode systems depending upon the value of η are observed from these curves,

(a) fields between cylindrical electrode systems; cylinder-cylinder (3) cylinder-plane (4), concentric cylinders (5) etc., have a higher value of η, that is, they are more uniform than the fields in spherical electrode systems; sphere-sphere (6), sphere-plane (7), concentric spheres (8), etc.

(b) a symmetrical electrode system, for example, sphere-sphere or cylinder-cylinder, has a higher value of η than the corresponding unsymmetrical system, that is sphere-plane or cylinder-plane systems.

(c) the field between two similar electrodes, cylinders or spheres, placed adjacent to each other is more uniform or has a higher value of η than when the electrodes are placed in coaxial or concentric formation.

2.3.1.1 Effect of Grounding on Field Configuration In Figure 2.5 all the field configurations shown are formed between identical pair of electrodes having symmetrical voltage applied across the dielectric. In such cases, the electric field configuration has symmetry on either side of the zero potential line, a straight line that falls exactly in the middle of the dielectric between the two electrodes. The equipotential lines are evenly distributed in the dielectric on either side of this center line. However, in practice such symmetrical systems are rarely present.

Asymmetry in electric field configurations is introduced in two ways: firstly, when the two electrodes are not identical, and secondly when the voltage applied across them is not symmetric. An example of the second case is when one of the two electrodes is grounded. Unsymmetrical electrode combinations, such as sphere-plane, rod-plane etc., applied with unipolar voltage on one electrode and the other electrode grounded, are typical examples of asymmetry.

The analytical methods of field intensity estimation have limitations. These are applicable only for the electrode shapes, which can be represented by geometrical equations. For analysis, it is also necessary that the electrode system has a symmetry around the axis of rotation and that the electrodes are applied in symmetrical voltages. The development of numerical methods has enabled the estimation of electric field patterns for un-symmetrical configurations of electrodes and voltages in a simple way. The equipotential lines in such cases are not symmetrically distributed in the main field region. This gives rise to distortion in the field, hence a lower degree of uniformity, that is, a lower value of η.

Figure 2.7(a) shows the equipotential lines between identical pairs of electrodes, sphere-sphere, applied with symmetrical voltage $+U/2$ and $-U/2$ respectively. In this case the equipotential lines can be seen distributed symmetrically on either side of the horizontal plane representing the 0% voltage line in the middle of the two electrodes. Figure 2.7(b) shows the same electrode pair applied an equal magnitude of voltage U on the top sphere and the lower sphere being grounded. It can be observed from this figure that the distribution of equipotential lines in this case is unsymmetrical. These lines have a tendency to shift closer to the high voltage electrode. It results in higher maximum field intensity $E_{\max}$ in the dielectric for the

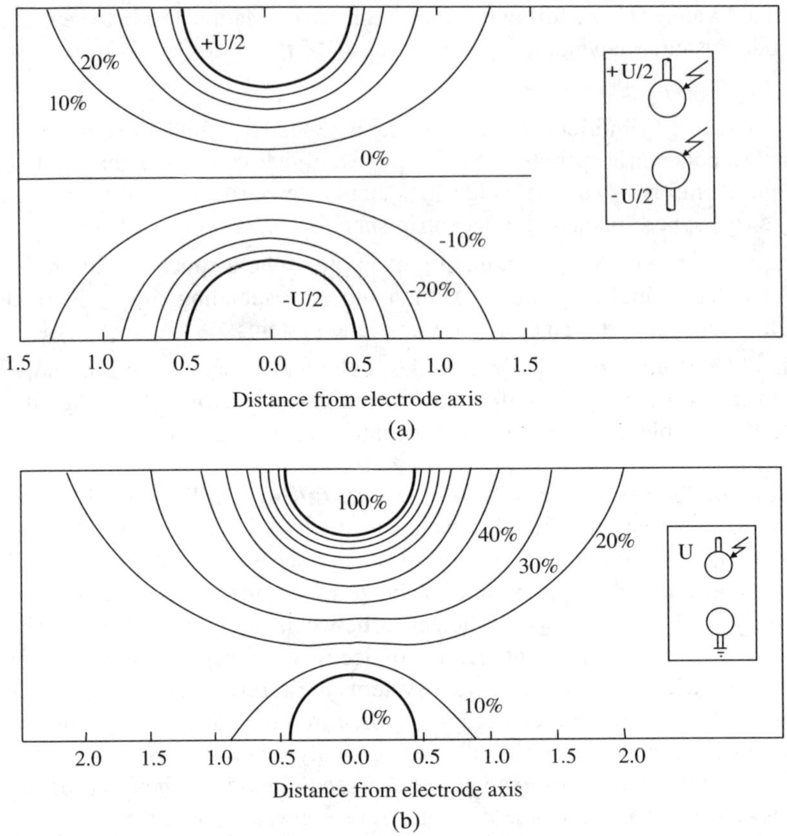

Figure 2.7 Effect of grounding on symmetrical electrode system showing equipotential lines on (a) symmetrical voltage application (b) grounding the lower electrode.

same magnitude of voltage applied across the gap as in the first case. One can conclude that the grounding of one of the electrodes introduces more nonuniformity in the electric field, hence a lower value of η.

The same effect described above can also be observed in the case of unsymmetrical electrode pair system of rod-plane shown in Figure 2.8(a) and (b) [2.6].

2.4 CONTROL OF ELECTRIC FIELD INTENSITY (STRESS CONTROL)

The more uniformity in a field, the better the use of the dielectric. An ideal use can be accomplished only in case of a totally uniform field, that is, when η is equal to one. However, it is not possible in practice. Nonuniformity in field leads to uneconomical use of the insulating materials. More is the nonuniformity, higher the electric stress in the dielectric. Insulation design in an equipment should be done

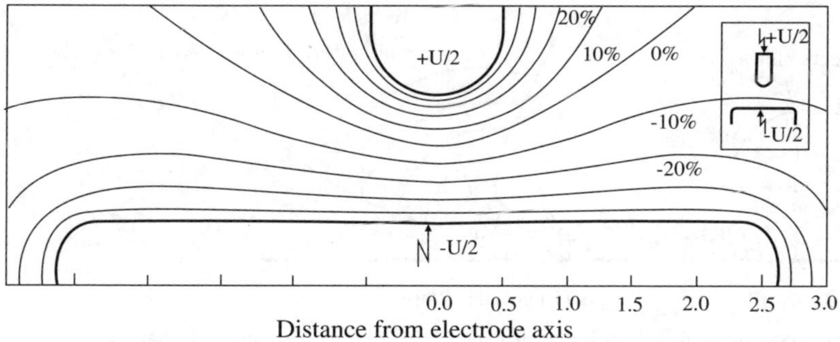

Equipotential lines for Rod-Plane electrode ; Symmetrical voltage.

(a)

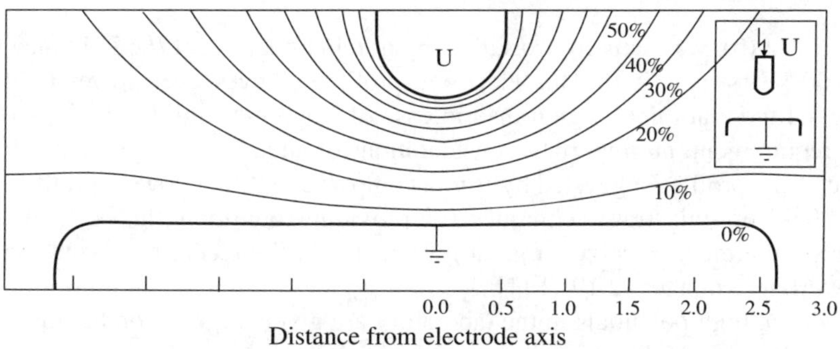

Equipotential lines for Rod-Plane electrode ; Asymmetrical voltage.

(b)

Figure 2.8 Effect of grounding on unsymmetrical electrode system equipotential lines on (a) symmetrical voltage application (b) grounding the lower electrode.

with due consideration to the estimated maximum electric field intensity. It is possible to achieve a higher degree of uniformity of fields by giving suitable shapes and sizes to various electrodes in the equipment.

An abrupt interruption of electrodes at either anode or cathode in high voltage equipment leads to concentration of electric field at the brims. It results in a tremendous enhancement of electric stress on the dielectric. The volume of dielectric in the vicinity thus becomes highly vulnerable to complete breakdown of insulation. In order to avoid this and to obtain an altogether higher degree of uniformity for such fields, the electrodes must be shaped suitably. For stress control, in principle the electrodes are extended and/or given forms so that higher field intensity than in the main field region does not appear anywhere in the dielectric. To achieve this, Rogowski suggested in 1923 [2.7, 2.8] a shape by which the electrodes could be extended, known as "Rogowski Profile", Figure 2.9(a). One can see in this figure that the field intensity continuously reduces beyond the main field region. The analytical method to design such electrode contours for given gap distance is given in Appendix I. Another shape of the electrode credited to Borda known as "Borda

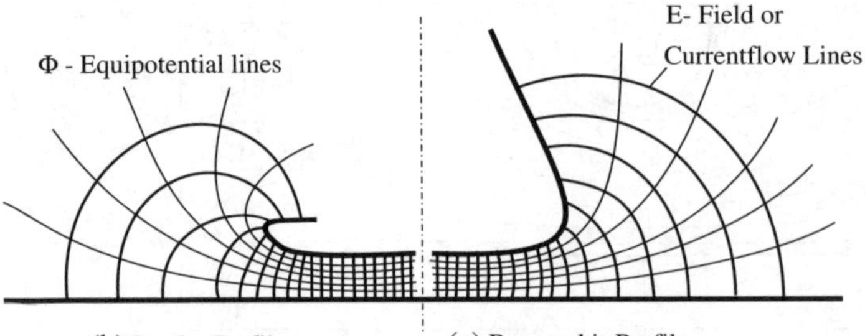

Figure 2.9 Equipotential and field (current flow) lines between plane and shaped electrodes.

Profile", Figure 2.9(b), was actually workd out by him in as early as 1766 in France, more than 200 years ago [2.9]. This electrode profile achieves even lower field intensity beyond the main field region than in case of Rogowski profile.

Space requirement, an important consideration for apparatus design, is much lower in case of the profiles suggested by Borda compared to the Rogowski profiles. The electric field optimization techniques for providing optimum shapes to the electrodes have, therefore, received a great importance in the recent time with the increase in working voltages [2.10, 2.11].

Electrodes at high potentials in the laboratory are given large, smooth, dome-like bodies or other shapes, like toroid rings, to bring down the electric stress on the atmospheric air (dielectric). The modern trend in such electrode design includes "segmented electrodes". These constitute a number of small, identical, smooth discs brought together to give rise to a large desired continuous shape as per requirement. The curvatures of the individual segment discs are worked out by optimization of the suggested profiles. Figure 2.10 shows both: the single metallic body and the segmented electrodes used in High Voltage test apparatus for stress control. These measures are necessary not only to prevent any partial breakdown occurring in the laboratory but also to check Electro Magnetic Interference (EMI) during measurements.

Extended shapes of electrodes, also known as "shields", are suitably provided on high voltage apparatus for electric stress control as shown in Figure 2.11. Sharp contacts are often enveloped by a large diameter hemispherical electrode having an aperture, or provided with concentric toroidal rings (doughnut shaped rings). Instead of wires, tubular electrodes of large diameters are used for the connections at sub-stations and high voltage laboratories, which bring down the field intensity at higher voltages considerably. Large spheres used at bends are provided with smooth holes for the connection of these circular and tubular electrodes.

Use of screen (also known as shield or concentric conductor) over the insulation in coaxial high voltage cables is made to control the electric stress. The field achieved in these screened cables is radial and generally a weakly nonuniform field. For making cable joints and terminations, the screen is extended in the form of a cone, known as "stress cone". This helps in achieving a more uniform distribution

(a)

(b)

Figure 2.10 Photographs of stress control arrangements in HV dc test equipment, courtesy Highvolt, Dresden, Germany (a) Dome and toroid shaped electrodes (b) Large shaped bodies with segmented electrodes.

of electric stress in the dielectrics at the end terminations and the joints, as shown in Figure 2.11.

It is a common practice to use bundles of two or more conductors at the same potential instead of a single conductor in transmission lines and bus bars above 220 kV. It brings down the electric stress, i.e. a method of stress control, (Figure 2.1 b). As the transmission voltages are increasing, bundles with eight or even more number of conductors are being used in practice.

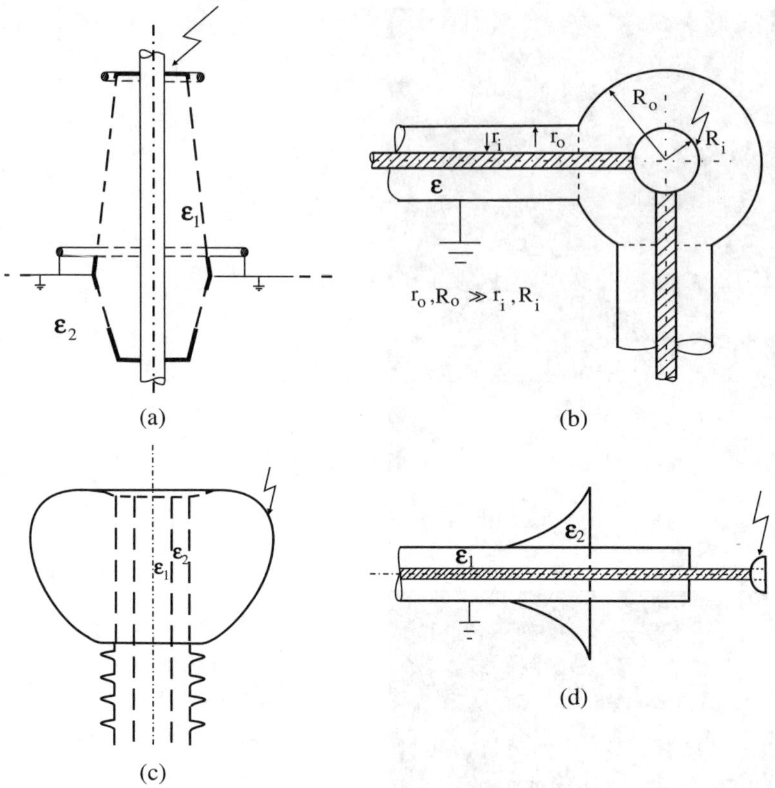

Figure 2.11 Extended shapes of electrodes for stress control (a) A bushing with toroids (b) Right angle turn of a bus bar in gas insulated switchgear (GIS), (c) High Voltage electrode on a condenser (d) Stress cone on a screened (shielded) cable end termination.

In high voltage bushings, potential transformers, and cable terminations, the insulation is provided with capacitive grading in order to achieve a well-distributed potential gradient leading to a more uniform field distribution in the dielectric. It is attained by inserting a few concentric conductive layers of calculated dimensions at appropriate positions in the dielectric, known as “floating screens” to control the electric stress as shown in Figure 2.12. These floating electrodes work as equipotentials. An economic use of the insulating material is thus achieved by evenly distributing the electric stress in the complete dielectric.

A modest rule of thumb to control electric stress in high voltage apparatus is to avoid sharp points and edges. Symmetrical, smoothly shaped and large electrodes are preferable. It must be borne in mind that even the roughness on metallic surfaces can lead to distortions in the field at higher voltages especially when the gap distance is small. Furthermore, microprotrusions in solid dielectrics may grow with PB and penetrate deeper in the dielectric when the electric field intensity exceeds a certain minimum value and cause local field enhancement. These must be

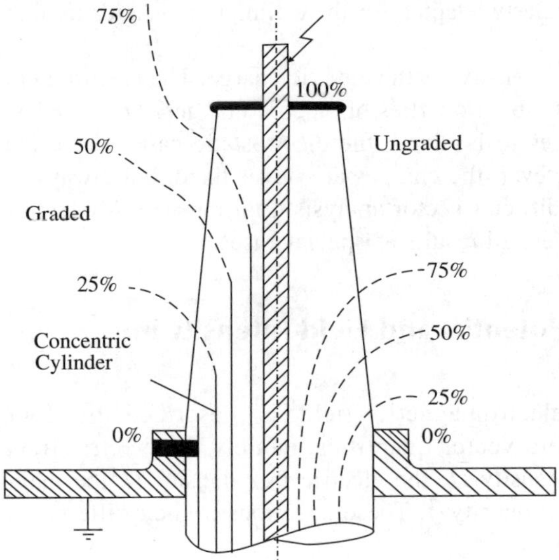

Figure 2.12 Potential distribution in a bushing with and without capacitive grading [2.4, 2.12].

prevented from developing during manufacturing and subsequently in service and also maintenance.

2.5 ESTIMATION OF ELECTRIC FIELD INTENSITY

The increase in electric stress on insulation beyond a certain magnitude may lead to partial or total breakdown of the insulating material. Such critical magnitudes of electric stress are briefly explained in section 2.2. These are discussed in detail for different types of dielectrics in Chapters 3 to 6. However, one of the most important practical aspects of high voltage engineering is to prevent any kind of electrical breakdown of the insulation and achieve a long serviceable life of an apparatus. For a good design, the knowledge of distribution of electric stress in insulating materials between any two electrodes must be acquired. It also helps in achieving better use of the insulating materials and therefore reduces the overall size and cost of the equipment. The permissible electric stress in equipment is interlinked with total electric field distribution in the space between the electrodes and the electric strength of the dielectric in use.

In case of fields with rotational symmetry and in which the electrode configurations can be represented by simple mathematical equations, the estimation of electric stress is accomplished analytically. On the other hand, it is complicated to obtain complete physical description of asymmetric three-dimensional electric fields by spatial geometry of surfaces. Techniques using electrolytic tanks and conductive graphite paper were employed for such field configurations. However, these methods are not sufficiently accurate but are time-consuming and clumsy, and hence have been slowly discarded. With the rapid development of computational techniques, numerical methods have been popularly adapted for the estimation of fields in high voltage equipment.

The source of "electric field intensity" is the electric charge. The electric field theory is intimately associated with the properties of electric charges. The simplest approach is considering the charges to be fixed, the electrostatic case, where the Coulomb's law is applicable. However, the charges are never fixed, but always in motion. Hence they pose a more difficult vector analysis. Starting with Maxwell's equations, one can obtain all the desired results as special cases.

2.5.1 Basic Equations for Potential and Field Intensity in Electrostatic Fields

According to Maxwell, the electromagnetic field is described by four equations co-relating the five field vector quantities, namely the electric field intensity $\vec{E}$, the magnetic field intensity $\vec{H}$, the electric flux density $\vec{D}$, the magnetic flux density $\vec{B}$ and the current density $\vec{J}$. These equations in their differential forms are given as following:

1. $$Curl\ \vec{H} = \vec{J} + \frac{\partial \vec{D}}{\partial t} \quad or \quad Curl\ \vec{H} = \kappa \vec{E} + \varepsilon \frac{\partial \vec{E}}{\partial t} \tag{2.7}$$

2. $$Curl\ \vec{E} = -\frac{\partial \vec{B}}{\partial t} \tag{2.8}$$

3. $$div\ \vec{B} = 0 \tag{2.9}$$

4. $$div\ \vec{D} = \rho \tag{2.10}$$

where κ (kappa) is the specific conductivity, ε the permittivity and ρ the volume charge density of the dielectric medium in which the field is considered.

Fields produced by power frequency alternating voltage are not electrostatic but are regarded to be quasi-stationary electrostatic, basically a turbulence-free field without any space charge. Therefore, the time dependent quantities as well as the specific conductivity in the above equations are redundant and can be totally ignored for such fields. Only equations (2.8) and (2.10) are needed to be considered. For the electrostatic and for quasi-electrostatic fields mentioned above, these two equations acquire even more simple forms as following:

$$Curl\ \vec{E} = 0 \tag{2.11}$$

$$div\ \vec{D} = 0 \tag{2.12}$$

It may be mentioned here that the equation (2.3), $\vec{E} = -grad\ \phi$ is derived from equation (2.11).

Considering Maxwells' 4th equation (2.10), a physical fact is revealed that the source of an electric flux must be represented by charge "q". If a closed surface area $\vec{A}$ covers a volume V in which no source of charge is present ($\rho = 0$), then the electric flux entering this volume is equal to the flux leaving. On the other hand, if a source of charge is present ($\rho \neq 0$), then there is a change in the flux leaving, affected by the magnitude of the source.

Therefore, the surface integral over a closed surface area $\vec{A}$ having flux density $\vec{D}$ gives the total charge enclosed in this area; hence:

$$\int_A \vec{D} \cdot d\vec{A} = q \tag{2.13}$$

This is known as the first fundamental rule of an electrical field. The relation given by Gauss between surface and volume integrals of flux density $\vec{D}$ is as follows:

$$\int_A \vec{D} \cdot d\vec{A} = \int_V div\, \vec{D}\, dV \tag{2.14}$$

Substituting equation (2.10) in (2.14), we obtain:

$$\int_A \vec{D} \cdot d\vec{A} = \int_V \rho\, dV = q \tag{2.15}$$

The relationship between the vector field $\vec{E}$ and the electric flux density $\vec{D}$ is given including ε of the dielectric medium as:

$$\vec{D} = \varepsilon \vec{E} \tag{2.16}$$

where $\varepsilon = \varepsilon_o \varepsilon_r$, ε_r is the relative permittivity number, a dimensionless quantity depending upon the dielectric material and its physical conditions, and ε_o is known as the absolute permittivity, a constant referred to vacuum, $\varepsilon_o = 8.855 \cdot 10^{-12}$ As/Vm or F/m. Substituting equation (2.16), in (2.10), we obtain:

$$div\,(\varepsilon \vec{E}) = \rho \tag{2.17}$$

but since $\vec{E} = -grad\ \phi$, equation 1.17 can, therefore, be written as:

$$div\,\{\varepsilon(-grad\ \phi)\} = \rho \tag{2.18}$$

or

$$div\ grad\ \phi \equiv \nabla^2 \phi = \Delta\phi = -\frac{\rho}{\varepsilon} \tag{2.19}$$

This is known as Poisson's equation for potential in a field, which may have distortion due to space charge. The space charge is defined as "the concentration of like polarity charge". For a field undistorted by any space charge concentration, i.e., having uniformly distributed charge ($\rho = 0$), this relation reduces to:

$$div\ grad\ \phi \equiv \Delta\phi = 0 \tag{2.20}$$

The Laplace operator "$\Delta = div\ grad$" was introduced by Murphy as early as 1833. The differential equation (2.20) is known as the "Laplace's potential equation" of the field undistorted by any space charge concentration. The Laplac equations in cartesian, cylindrical and spherical coordinates are given in Table 2.1. The final forms of these equations depend upon the direction/directions in which potential ϕ varies [2.13].

TABLE 2.1 Laplace's Differential Equation in different coordinate systems [2.14]

Coordinate System	Laplace Equation
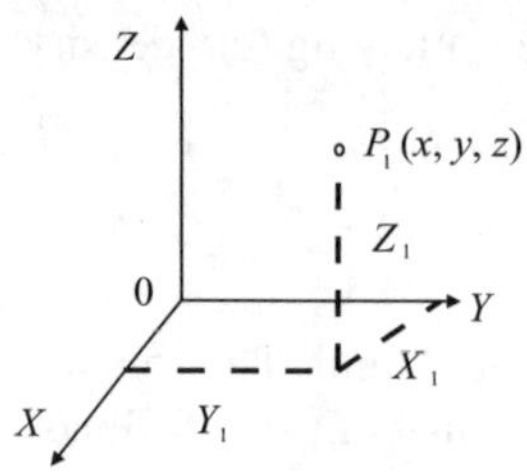 a) Cartesian coordinate	
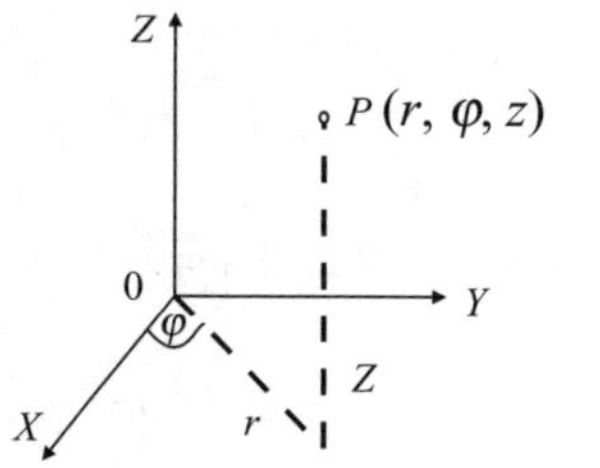 b) Circular cylinder coordinates	
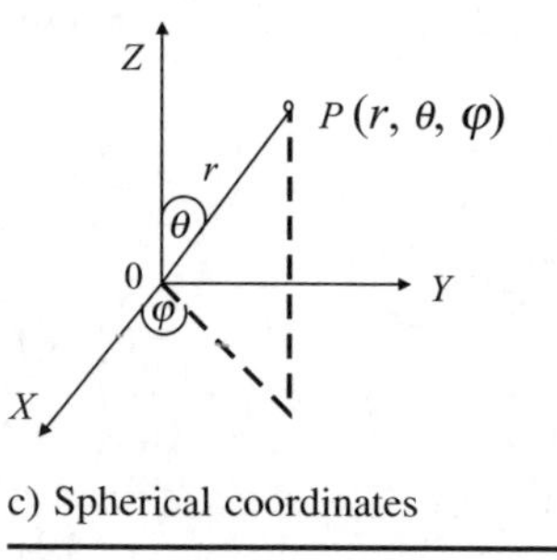 c) Spherical coordinates	

Solution of Poisson's equation (2.19) is possible only in certain simple cases of electrode geometry describable by mathematical relations, and for which the relationship between ρ and $\vec{E}$ is known. However, in most of the cases in practice, it is not possible to establish the dependence of ρ with $\vec{E}$. Approximated mathematical models simplified with assumptions may resolve such cases. The solution of Poisson's equation, that is, for fields with space charge, still remains an insoluble problem. Measurement of the field intensity on the actual electrode system is the only possible method of estimation of fields in such cases.

$$\Delta\phi(x, y, z) = \frac{\partial^2\phi}{\partial x^2} + \frac{\partial^2\phi}{\partial y^2} + \frac{\partial^2\phi}{\partial z^2} = 0 \qquad (2.21)$$

$$\Delta\phi(r,\varphi,z) = \frac{1}{r}\cdot\frac{\partial}{\partial r}\left(r\frac{\partial\phi}{\partial r}\right) + \frac{1}{r^2}\cdot\frac{\partial^2\phi}{\partial\varphi^2} + \frac{\partial^2\phi}{\partial z^2} = 0 \tag{2.22}$$

$$\Delta\phi(r,\theta,\varphi) = \frac{1}{r^2}\cdot\frac{\partial}{\partial r}\left(r^2\frac{\partial\phi}{\partial r}\right) + \frac{1}{r^2\sin\theta}\cdot\frac{\partial}{\partial\theta}\left(\sin\theta\cdot\frac{\partial\phi}{\partial\theta}\right) + \frac{1}{r^2\sin^2\theta}\cdot\frac{\partial^2\phi}{\partial\varphi^2} = 0 \tag{2.23}$$

The Laplace equation (2.20) may have a direct solution in some special cases as discussed in sub-section 2.5.2.1; otherwise it needs the numerical solution described in section 2.5.4. Computer programs are developed for the numberical estimation of fields between complicated electrode geometries and multidielectric insulation systems. Simulation of three-dimensional fields is achieved by a number of mathematical functions in different coordinate systems and solved numerically [2.9].

Superposition methods and mirror image techniques are also applied for the estimation of simple fields in practice. These are used to calculate fields analytically by the direct application of Gauss's law, eliminating the need for a formal solution of Laplace equation [2.9].

While estimating the electric field intensity, it must be understood that the electric charge rests only in dielectrics, whereas it is restless in the electrodes as well as on the surface of the dielectrics. Hence the electric field prevails only in the dielectrics. The electrodes form the known equipotentials.

2.5.2 Analytical Methods for the Estimation of Electric Field Intensity in Homogeneous Isotropic Single Dielectric

For the estimation of electric field intensity in complicated electrode geometries, a number of methods have been developed involving transformation of different coordinate systems. Suitability of a particular coordinate system for the given electrode system must therefore be established. Analytical methods involving direct solution of Laplace's equations and the techniques using distribution of charge are explained with examples for simple electrode geometries used in practice in the following paragraphs.

2.5.2.1 Direct Solution of Laplace Equation

2.5.2.1.1 Parallel Plate Condenser Consider a parallel plate condenser with infinitely large plates. This assumption is made in order to nullify the effect of distortion of field at the brims of the plates. The field in the main field region between the two plates can then be assumed to be uniform, as shown in Figure 2.5 (a). The variation of potential between two such plates, therefore, depends upon only one variable/direction. Considering the cross section of the field in cartesian coordinates, we have:

$$\phi(x,y,z) = \phi(x)$$

since

$$\frac{\partial \phi}{\partial y} = \frac{\partial \phi}{\partial z} = 0$$

and the Laplace equation (2.21) in cartesian coordinates for this case reduces to:

$$\Delta \phi = \frac{d\phi^2}{dx^2} = 0 \tag{2.24}$$

The solution of this equation is as follows:

$$\frac{d\phi}{dx} = K_1$$

and

$$\phi(x) = K_1 x + K_2$$

where K_1 and K_2 are the constants of integration, determined by the boundary conditions of potential.

From the Figure 2.5 (a), at $x = 0$:

$$\phi = +\frac{U}{2}, \quad \therefore K_2 = +\frac{U}{2}$$

and at

$$x = d, \phi = -\frac{U}{2}, \quad \therefore K_1 = -\frac{(U/2 + K_2)}{d} = -\frac{U}{d}$$

Hence, the potential distribution is given by:

$$\phi(x) = -\frac{U}{d} x + \frac{U}{2} \tag{2.25}$$

The field intensity is determined by equation (2.3) varying only in one direction:

$$\left|\vec{E}\right| = E(x) = -grad\ \phi(x) = -\frac{d\phi(x)}{dx}$$

or

$$E_x = -\frac{U}{d}, \quad \text{hence} \quad \left|\vec{E}\right| = \frac{U}{d} \tag{2.26}$$

Therefore, the magnitude of field intensity or electric stress in the dielectric is constant throughout the space in between the two plates and the potential varies linearly in the gap, as shown in Figure 2.5(a).

2.5.2.1.2 Concentric Sphere Condenser The electric field distribution in this case is symmetrical in all directions with reference to the center of the spheres. The field lines or the current flow lines are radial, therefore the field intensity in the dielectric is a function of only one parameter, the distance r from the center,

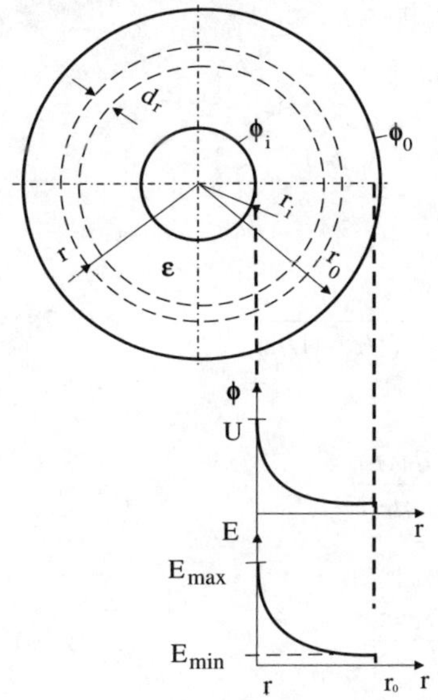

Figure 2.13 Field in concentric sphere condenser.

as shown in Figure 2.13. Considering the spherical coordinates described in Table 2.1:

$$\phi(r, \theta, \varphi) = \phi(r)$$

since

$$\frac{\partial \phi}{\partial \theta} = \frac{\partial \phi}{\partial \varphi} = 0$$

the Laplace equation (2.23) in spherical coordinates for this case reduces to:

$$\Delta \phi = \frac{1}{r^2} \cdot \frac{d}{dr}\left(r^2 \frac{d\phi}{dr}\right) = 0 \tag{2.27}$$

Solution of equation (2.27) is possible by substitution:

Let

$$r^2 \frac{d\phi}{dr} = K_1$$

then

$$\phi(r) = -\frac{K_1}{r} + K_2$$

where K_1 and K_2 are the constants of integration determined by the boundary conditions of the electrode potentials.

If the potentials at inner and outer spherical electrodes are ϕ_i and ϕ_o, respectively, and the potential difference between them is U, then,

$$\phi(r_i) = \phi_i;\ \phi(r_o) = \phi_o;\quad \text{and} \quad \phi_i - \phi_o = U$$

hence,

$$K_1 = \frac{U}{(1/r_o - 1/r_i)};\quad K_2 = \frac{U}{(r_i/r_o - 1)} + \phi_i$$

therefore

$$\phi(r) = -\frac{K_1}{r} + K_2 = \phi_i + \frac{U}{(r_i/r_o - 1)} - \frac{1}{r} \cdot \frac{U}{(1/r_o - 1/r_i)} \tag{2.28}$$

The field intensity is determined by equation (2.3),

$$\left|\vec{E}\right| = E(r) = -grad\ \phi(r) = -\frac{d\phi(r)}{dr}$$

or

$$E_r = \frac{1}{r^2} \cdot \frac{U}{(1/r_o - 1/r_i)} = \frac{U}{r^2} \cdot \frac{r_o \cdot r_i}{(r_o - r_i)} \tag{2.29}$$

E_r has its maximum magnitude for smallest value of r, that is, in the dielectric at inner electrode where $r = r_i$:

$$E_r\big|_{\max} = E_{r_i} = \frac{U \cdot r_o \cdot r_i}{r_i^2 (r_o - r_i)}$$

or

$$E_{\max} = \frac{U \cdot r_o}{r_i (r_o - r_i)} \tag{2.30}$$

The minimum magnitude of field intensity will be for the largest value of r, that is, in the dielectric at outer electrode surface where $r = r_o$.

The partial expression $U/(r_o - r_i)$ in equation (2.30) gives the magnitude of field intensity as if the field between the two electrodes was a uniform field. This expression is, therefore, equal to the magnitude of mean field intensity in concentric sphere condenser:

$$E_{mean} = \frac{U}{(r_o - r_i)} \tag{2.31}$$

Hence, the Schwaiger factor for concentric sphere electrode system works out to be:

$$\eta = \frac{E_{mean}}{E_{\max}} = \frac{r_i}{r_o}$$

2.5.2.1.3 Coaxial Cylindrical Condenser For a coaxial cylindrical electrode system shown in Figure 2.14, cylindrical coordinates are considered. With the same arguments as in the case of concentric spheres:

$$\phi(r, \varphi, z) = \phi(r)$$

since

$$\frac{\partial \phi}{\partial \varphi} = 0; \quad \frac{\partial \phi}{\partial z} = 0$$

The Laplace equation (2.22) in cylindrical coordinates for the given field reduces to:

$$\Delta \phi = \frac{1}{r}\frac{d}{dr}\left(r\frac{\partial \phi}{dr}\right) = 0 \tag{2.32}$$

Solution for this equation is given by substituting:

$$\phi(r) = K_1 \ln r + K_2$$

and if

$$\phi(r_i) = \phi_i;\ \phi(r_o) = \phi_o;\ \text{and } \phi_i - \phi_o = U$$

then,

$$K_1 = \frac{U}{\ln(r_i/r_o)};\ K_2 = \phi_i - \frac{U}{\ln(r_i/r_o)}\ln r_i$$

and

$$\phi(r) = \phi_i + \frac{U}{\ln(r_i/r_o)}(\ln r - \ln r_i) \tag{2.33}$$

Further, since:

$$\left|\vec{E}\right| = E(r) = -grad\ \phi(r) = -\frac{d\phi(r)}{dr}$$
$$\therefore\quad E_r = \frac{1}{r}\frac{U}{\ln(r_o/r_i)} \tag{2.34}$$

and

$$E_r\big|_{\max} = E_{r_i} = \frac{U}{r_i \ln(r_o/r_i)}$$

or

$$E_{\max} = \frac{U}{r_i \ln(r_o/r_i)} \tag{2.35}$$

As in the case of concentric spheres, for the coaxial cylindrical condenser, the magnitude of mean field intensity is also given by the expression:

$$E_{mean} = \frac{U}{(r_o - r_i)}$$

The Schwaiger factor or the degree of uniformity of the field in this case is given by the expression:

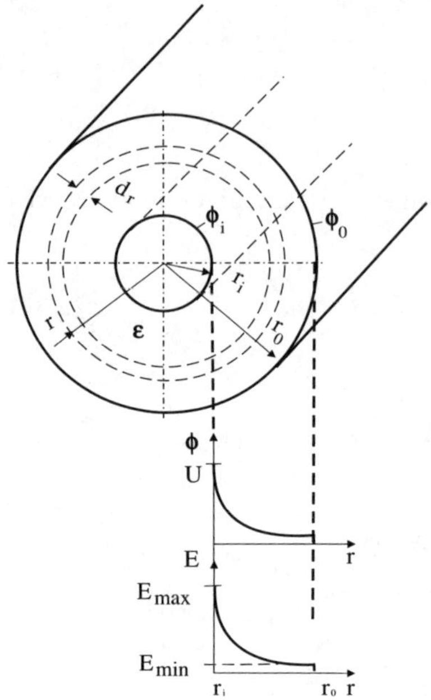

Figure 2.14 Field between coaxial cylindrical electrodes system.

$$\eta = \frac{E_{mean}}{E_{\max}} = \frac{r_i \ln r_o / r_i}{(r_o - r_i)}$$

The minimum magnitude of field intensity will be in the dielectric at the outer concentric electrode and it is obtained by putting $r = r_o$ in the equation (2.34).

2.5.2.2 "Gaussian Surface" Enclosed Charge Techniques for the Estimation and Optimization of Field Gauss's law states that, "the net electric flux passing out of any closed surface is equal to the net positive charge enclosed by that surface". Mathematically it relates the net flux out of a closed surface with the uniformly distributed net charge within that closed surface. In other words, the integral of the normal component of the flux density over the closed surface surrounding the region under consideration is equal to the volume within the closed surface containing the net charge as described by the equations (2.14 & 2.15).

This technique is applied by considering an unknown charge at a very thin surface element. The potential difference between the two electrodes across the dielectric is obtained by the line integral of electric field intensity. On eliminating the unknown charge, the desired relation for electric stress is obtained.

The two examples of concentric spheres and concentric cylinders are solved again to show the difference.

2.5.2.2.1 Concentric Sphere Condenser In Figure 2.13, let q_r be the charge on the surface of a thin concentric spherical dielectric element of wall thickness dr, at a distance r from the center. Then, from equation (2.15), we have:

$$q_r = D(r)A(r)$$

The area covered by this thin spherical electrode is equal to $4\pi r^2$. From equation (2.16), considering the magnitudes:

$$E(r) = \frac{D(r)}{\varepsilon} = \frac{q_r}{\varepsilon A(r)}$$

or

$$E(r) = \frac{q_r}{\varepsilon 4\pi r^2} \tag{2.36}$$

applying equation (2.1), and the boundary conditions,

$$\phi_i - \phi_o = U = -\int_{r_i}^{r_o} E(r)\,dr \tag{2.37}$$

or

$$U = \frac{q_r}{4\pi\varepsilon}\left(\frac{1}{r_i} - \frac{1}{r_o}\right) \tag{2.38}$$

or

$$U = E(r)r^2\left(\frac{r_o - r_i}{r_o r_i}\right)$$

$$\therefore E_r = \frac{U}{r^2}\cdot\left(\frac{r_o r_i}{r_o - r_i}\right) \tag{2.39}$$

Equations (2.29) and (2.39) are identical. Magnitudes of maximum and mean field intensities for this case can be found out by equations (2.30) and (2.31).

It would be interesting to learn how to obtain the lowest magnitude of $E_{\max}$ in the dielectric at the inner electrode surface for a given voltage and fixed outer electrode dimension. It is possible by applying direct optimization technique on equation (2.30). If the values of r_o and U are kept constant the minimum ratio of this expression can be obtained by differentiating it with respect to the inner radius r_i, and equating it to zero.

$$\left.\frac{dE_{ri}}{dr_i}\right|_{\substack{r_o=\text{constant}\\ U=\text{constant}}} = 0$$

Thus, the minimum field intensity at the surface of inner electrode r_i is obtained when:

$$\frac{r_o}{r_i} = 2 \tag{2.40}$$

This is known as optimum radii ratio for a concentric sphere electrode system. The optimum maximum field intensity for this electrode configuration at r_i is obtained by putting $r_o = 2\, r_i$ in equation (2.39) and it is given as:

$$E_{\max}(optimum) = \frac{2U}{r_i} \tag{2.41}$$

In a concentric sphere condenser, the radius of the inner sphere should be half of the radius of the outer sphere in order to achieve optimum maximum field intensity at the inner sphere surface, (Figure 2.13 and 2.11 (b)).

The relation for the capacitance of an electrode system separated by a dielectric is:

$$C = \frac{q}{U} \tag{2.42}$$

Hence, the capacitance for a concentric sphere condenser can be determined by equation (2.38) as follows:

$$C = 4\pi\varepsilon \frac{r_o r_i}{(r_o - r_i)} \tag{2.43}$$

2.5.2.2.2 Coaxial Cylindrical Condenser For a coaxial cylindrical system of length l, the surface area $A(r)$ of a thin element of cylinder of thickness drat a distance r from the center is equal to $2\pi rl$ (Figure 2.14). Therefore, in this case:

$$E(r) = \frac{D(r)}{\varepsilon} = \frac{q_r}{\varepsilon A(r)} = \frac{q_r}{\varepsilon 2\pi rl} \tag{2.44}$$

where q_r is the total charge enclosed on the cylindrical element under consideration. On applying equation (2.1) and the boundary conditions,

$$\phi_i - \phi_o = U = \int_{r_i}^{r_o} E(r)\,dr \tag{2.45}$$

or

$$U = \frac{q_r}{2\pi\varepsilon l}\int_{r_i}^{r_o} \frac{1}{r}\,dr$$

or

$$U = \frac{q_r}{2\pi\varepsilon l} \ln \frac{r_o}{r_i} \tag{2.46}$$

hence,

$$E_r = \frac{U}{r \ln(r_o/r_i)} \tag{2.47}$$

Electric stress on the dielectric has its maximum magnitude where r is smallest, that is at $r = r_i$, equation (2.35) and it is minimum where r is largest, that is at $r = r_o$. The expression for the mean stress is the same as in case of concentric spheres, equation (2.31).

If U is given in kV and r in mm, then E, is given in kV/mm. This is a common unit of field intensity for power cables and other equipments.

The capacitance per unit length l of a coaxial cylindrical condenser is obtained from equations (2.42) and (2.46), and it is given as:

$$C/l = \frac{2\pi\varepsilon}{\ln(r_o/r_i)} \tag{2.48}$$

It has the units of the permittivity ε, which is normally given in F/m.

The optimum field intensity at the inner electrode r_i, for a given outer electrode r_o, can be determined from equation (2.35) by differentiating this equation with respect to r_i, considering r_o and U to be constant and equating it to zero:

$$\left.\frac{dE_{r_i}}{dr_i}\right|_{\substack{r_o=\text{constant}\\ U=\text{constant}}} = \frac{d}{dr_i}\left[\frac{U}{r_i \ln(r_o/r_i)}\right] = 0$$

The optimum ratio of r_o to r_i for the lowest field intensity in the dielectric at the inner electrode surface in between the coaxial cylinders is thus given by:

$$\frac{r_o}{r_i} = e = 2.718\ldots\ldots\ldots\ldots \tag{2.49}$$

The optimum maximum field intensity for such an electrode configuration under this condition is given by:

$$E_{\max}(optimum) = \frac{U}{r_i} \tag{2.50}$$

The comparison of equation (2.41) and (2.50) for equal dimension of r_i and the conditions for optimization of electrodes proves that the field between coaxial cylinders is more uniform than between concentric spheres.

An important observation must be brought forward here for discussion. Although the potential acquired by the dielectric at high voltage electrode is that of the inner electrode, the actual potential in the dielectric at the outer electrode is not zero even when it is grounded. It should have some value depending upon the geometrical gradient between the electrodes. Hence there is an abrupt fall in potential at the outer surface of the dielectric due to grounding. It gives rise to a finite magnitude of minimum field intensity, $E_{\min}$, in the dielectric depending upon the gradient of potential. The distortion in electric field introduced due to grounding, as discussed in section 2.3.1.1, can not be taken into consideration in such analysis. It is a limitation of this method of field estimation.

Figure 2.15 shows the variation of "η", the degree of uniformity of fields between coaxial cylindrical and spherical electrode systems with respect to the variation in their inner radius r_i, that is, on varying the thickness of insulation for a fixed dimension of outer electrode.

The above two electrode configurations are typical examples of radial fields, which may represent weakly nonuniform fields under particular physical conditions. All single core power cables at and above 11 kV have this kind of radial field pattern. The circular solid or stranded conductor, which is the inner electrode, carries the

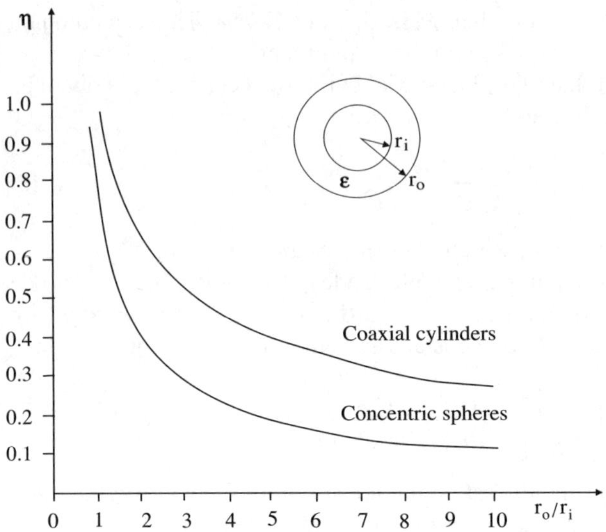

Figure 2.15 Degree of uniformity "η", in concentric electrode systems.

high voltage. The outer cylindrical electrode over the insulation is normally grounded and serves as a screen, also known as "shield" or "concentric conductor".

In Table 2.2, the expressions for maximum and mean field intensities, degree of uniformity η given by Schwaiger and for optimum value of E_{max} are brought together for some common electrode configurations in practice as well as in high voltage laboratories. In the first three cases of parallel plates, concentric spheres and coaxial cylinders, the enclosed charge technique is applied in determining the electric stress. However, for the last three cases of sphere opposite a sphere, parallel conductors and needle-needle electrode combinations, the mirror image technique involving reflection of point charge is applied. These derived results are taken from [2.9, 2.15 and 2.16]. The optimum field intensities are worked out in [2.17, 2.18].

In most of the practical and experimental cases, potential ϕ_o at the outer electrode is kept at zero, described in common parlance as earthed or grounded. The potential ϕ_i at the inner electrode is, therefore, equal to the applied voltage U. In all such estimations the effect of grounding of one of the electrodes can not be taken into consideration.

2.5.3 Analysis of Electric Field Intensity in Isotropic Multidielectric System

In most of the high voltage apparatus, the insulation system is comprised of more than one isotropic dielectric material. For example, transformers have oil (liquid), paper, wood and ceramic (solid) dielectrics. Gas Insulated Systems (GIS) have ceramic, Teflon, epoxy resin and polyethylene (solid) and sulphurhexafluoride (SF_6) or its mixture with other gases as gaseous dielectrics. Behavior of individual dielectric materials in an electric field is distinguished by their permittivities "ε" ($\varepsilon = \varepsilon_o \varepsilon_r$). At any such interface, an abrupt change in ε_r (the relative permittivity number) takes

TABLE 2.2 Electric stress in a dielectric between two electrodes of different configurations

Electrode System	$E(r)$ or $E(x)$	$E_{\max}$	E_{mean}	$\eta = \dfrac{E_{mean}}{E_{\max}}$	$E_{\max/\text{Optimum}}$	Observation
Paralled Plates 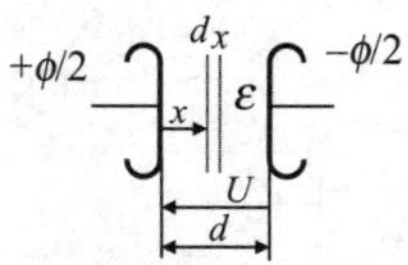	$\dfrac{U}{d}$	$\dfrac{U}{d}$	$\dfrac{U}{d}$	1	$\dfrac{U}{d}$	Uniform field
Concentric Spheres 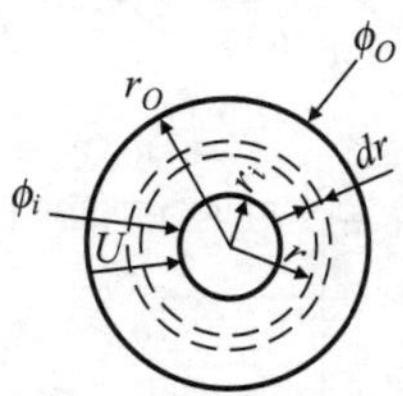	$\dfrac{Ur_o r_i}{r^2(r_o - r_i)}$	at $r = r_i$ $\dfrac{Ur_o}{r_i(r_o - r_i)}$	$\dfrac{U}{(r_o - r_i)}$	$\dfrac{r_i}{r_o}$	at $r = r_i$ $\dfrac{2U}{r_i}$ when $r_o/r_i = 2$	Radial field
Coaxial cylinders 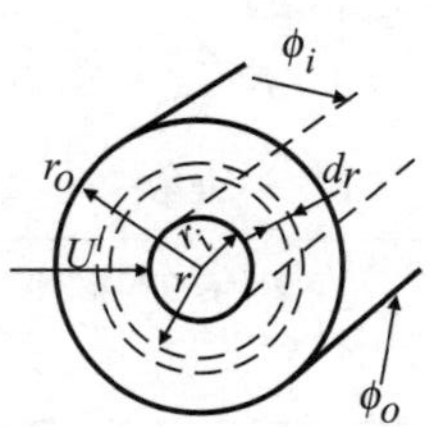	$\dfrac{U}{r\ln(r_o/r_i)}$	at $r = r_i$ $\dfrac{U}{r_i \ln(r_o/r_i)}$	$\dfrac{U}{(r_o - r_i)}$	$\dfrac{\ln(r_o/r_i)}{\left(\dfrac{r_i}{r_o} - 1\right)}$	at $r = r_i$ $\dfrac{U}{r_i}$ when $r_o/r_i = e$	Radial field

(Continued)

TABLE 2.2 Continued

Electrode System	$E(r)$ or $E(x)$	$E_{\max}$	E_{mean}	$\eta = \frac{E_{mean}}{E_{\max}}$	$E_{\max/\text{Optimum}}$	Observation
Sphere-Sphere 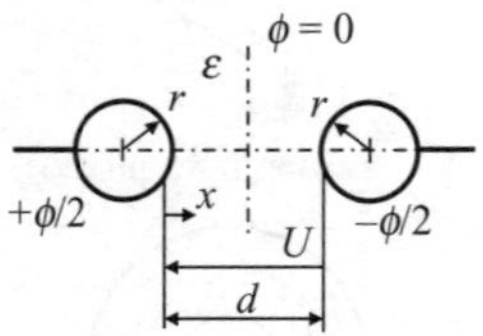	(is given in a different form)	at $x = 0,d$ if $d > r$ $\approx 0.9\frac{U}{2d}\left(2+\frac{d}{r}\right)$	$\frac{U}{d}$	$\approx \frac{2}{0.9\left(2+\frac{d}{r}\right)}$	at $x = 0,d$ $\approx 3.67\frac{U}{d}$ when $\frac{d}{r} = 2.08$	Nonuniform field, but for $d < 2r$, a weakly non uniform field is achieved.
Parallel Conductors (Cylinders) 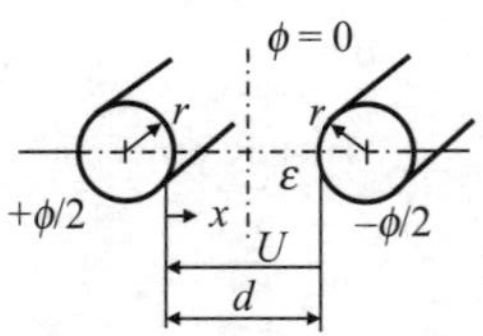	$\frac{U\sqrt{\left(\frac{d}{2r}\right)^2+\frac{d}{r}}}{d\left[\left(\frac{x}{r}-\frac{x^2}{dr}+1\right)\ln\left(1+\frac{d}{2r}+\sqrt{\left(\frac{d}{2r}\right)^2+\frac{d}{r}}\right)\right]}$	at $x = 0,d$ $\approx \frac{U}{2r\ln\frac{d}{r}}$ if $\frac{d}{r} >> 1$	$\frac{U}{d}$	$\approx \frac{2r\ln\frac{d}{r}}{d}$ if $\frac{d}{r} >> 1$	at $x = 0,d$ $\approx 1.59\frac{U}{d}$ when $\frac{d}{r} = 3.88$	Nonuniform field, having $E_{\max}$ at the electrodes and $E_{\min}$ in the middle
Needle-Needle (hyperbolic) 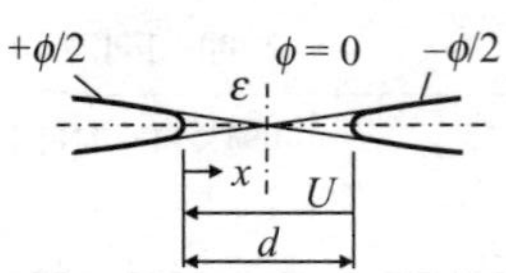	$\frac{U\cos\alpha}{d\left[1-\left(\frac{2x}{d}-1\right)^2\cos^2\alpha\ \ln\left(\cot\frac{\alpha}{2}\right)\right]}$	at $x = 0,d$ $\frac{U\cos\alpha}{d\left[1-\cos^2\alpha\ \ln\left(\cot\frac{\alpha}{2}\right)\right]}$	$\frac{U}{d}$	$\frac{(1-\cos^2\alpha)\ln\left(\cot\frac{\alpha}{2}\right)}{\cos\alpha}$		Extremely nonuniform field having $E_{\max}$ at the electrode tips and $E_{\min}$ in the middle.

α-asymptotic angle

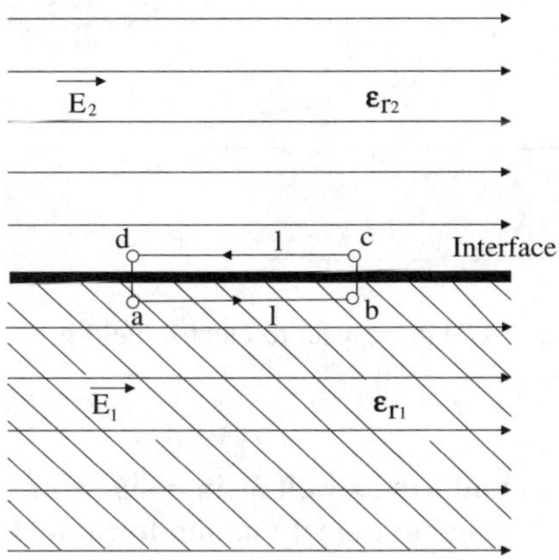

Figure 2.16 Path of Integration for field intensity with longitudinal interface.

place. The variation in parametric values of permittivity of dielectric materials in multidielectric system leads to different potential as well as field distribution in individual dielectrics. The actual field intensities and the insulating properties of individual dielectrics affect the behavior of the composite dielectric system.

The behavior of the composite insulation system of two or more dielectrics depends upon the direction of the dielectric-dielectric interface/interfaces with respect to the direction of the electric field. The direction of an interface could be longitudinal, perpendicular or diagonal. Accordingly, the respective interfaces between two dielectrics are described and their fundamental phenomenon affects explained. In the following analysis, it is taken for granted that the interface under consideration is ideally formed, i.e. a clean, exact and elaborate contact is achieved between the two dielectric surfaces.

2.5.3.1 Field with Longitudinal Interface

As seen in Figure 2.16, the interface between two dielectrics having relative permittivities ε_{r1} and ε_{r2} fall parallel to the direction of field lines $\vec{E_1}$ and $\vec{E_2}$ in the two dielectrics. Let us consider a rectangular closed path of integration abcda, falling parallel to the field at the interface.

According to equation 2.1, the integration of the field over a closed path abcda, Figure 2.16, is given as,

$$\oint \vec{E} \cdot \vec{dl} = 0 \tag{2.51}$$

If the sides bc and da of the rectangular path of integration are reduced to zero (that means the path falls on the interface itself) and sides ab = cd = l, then we obtain

$$E_1 l + 0 - E_2 l + 0 = 0$$

or

$$E_1 = E_2 \tag{2.52}$$

This proves that the magnitudes as well as the directions of the fields in the two dielectrics, having longitudinal interface with respect to the direction of the field, remain unchanged.

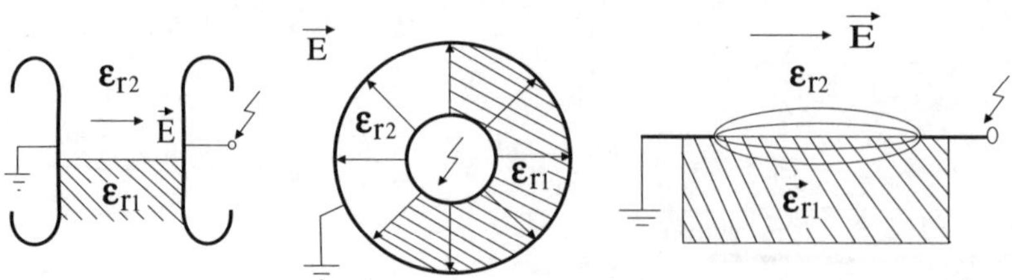

Figure 2.17 Dielectric interface in longitudinal direction with respect to electric field in different electrode configurations of classified fields.

Such longitudinal interfaces may appear in uniform as well as in nonuniform fields. Examples of such cases in uniform, weakly and extremely nonuniform field configurations in practice are shown in Figure 2.17. The fundamental condition of the field remains unchanged on either side of the interface. This leads to an undisturbed distribution of potential as well as field intensity in the two dielectrics.

Let us consider the breakdown/flashover in such a multi-dielectric system of two dielectrics. Ideally the breakdown should always occur in the dielectric having lower breakdown strength. For an undisturbed field distribution, the breakdown voltage should be measured the same as for the weaker single dielectric system. However, in practice it is not always true. Generally, it is observed that the breakdown voltage U_b of a multidielectric system is even lower than the breakdown voltage of single dielectric system with weaker dielectric, that is,

U_b **(multidielectric)** < U_b **(single dielectric with lower electric strength)**

An explanation of this phenomenon is rendered by the local field enhancement due to field distortions. It is extremely difficult in practice to achieve an ideal interfacial contact between any two surfaces. If care is not taken, microscopic voids/slits are formed at these interfacial contacts. The roughness of electrode materials and unevenness of dielectric surfaces may lead to imperfect contact between the interfacial surfaces of two dielectrics, as well as between electrodes and the dielectrics. Such defects, along with contaminants on the surfaces, if any, result in local field intensity enhancement. This kind of local field distortion at interfaces affects the breakdown voltage in uniform and weakly nonuniform fields more adversely than the partial breakdown inception voltage U_i in extremely nonuniform fields.

Hence, in case of extremely nonuniform fields with longitudinal interface of dielectric, the following general approximation regarding the partial breakdown inception voltage U_i can be taken for granted,

U_i **(multidielectric)** ≈ U_i **(single dielectric with lower electric strength)**

2.5.3.2 Field with Perpendicular Interface This is the case when a dielectric-dielectric interface falls directly along an equipotential surface in the field. Figure 2.18 shows an interface between two dielectrics ε_{r1} and ε_{r2} falling perpendicular (normal) to the direction of the field lines, that is, along the equipotential surface. Let $\vec{E}_1$ and $\vec{E}_2$ be the field intensities, and $\vec{D}_1$ and $\vec{D}_2$ the electric flux densities in the two dielectrics, respectively.

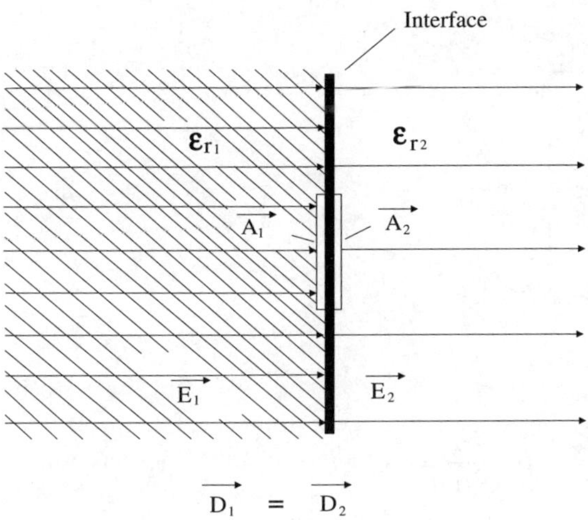

Figure 2.18 An interface between the two dielectrics perpendicular to the field.

Consider two very thin closed surface area elements $\vec{A_1}$ and $\vec{A_2}$ covering equal volumes in the dielectrics ε_{r1} and ε_{r2} respectively, falling on either side of the interface. If these volume elements contain no source of charge ($\rho = 0$), which is true for electrostatic fields, then from the first fundamental rule of electrical fields, equations 2.10 and 2.15 reduce to:

$$\oiint_A \vec{D} \cdot \vec{dA} = 0 \tag{2.53}$$

Therefore, the electric flux entering this area is equal to the flux leaving. Hence we obtain by integration:

$$\vec{D_1} \cdot \vec{A_1} + \vec{D_2} \cdot \vec{A_2} = -D_1 A_1 + D_2 A_2 = 0$$

Since the areas A_1 and A_2 are equal and the components of flux densities D_1 and D_2 are falling perpendicular to the interface:

$$D_1 = D_2 \tag{2.54}$$

and from the relation $D = \varepsilon E$, equation (2.16), we get:

$$\frac{E_1}{E_2} = \frac{\varepsilon_{r2}}{\varepsilon_{r1}} \tag{2.55}$$

This shows that in multidielectric insulating system with a dielectric-dielectric interface falling perpendicular to the field, the field intensities in the two dielectrics are inversely proportional to their relative permittivity numbers. The field intensity or the electric stress on the dielectric with smaller ε_r is proportionately more enhanced. However, the basic field distribution characteristic in the two dielectrics remains unaffected even by the introduction of different dielectrics.

Figure 2.19 shows a parallel plate condenser with two dielectrics ε_{r1} and ε_{r2}, forming a perpendicular interface in uniform field. The potential difference U, between the two plates, is given by:

$$U = E_1 d_1 + E_2 d_2 = U_1 + U_2 \tag{2.56}$$

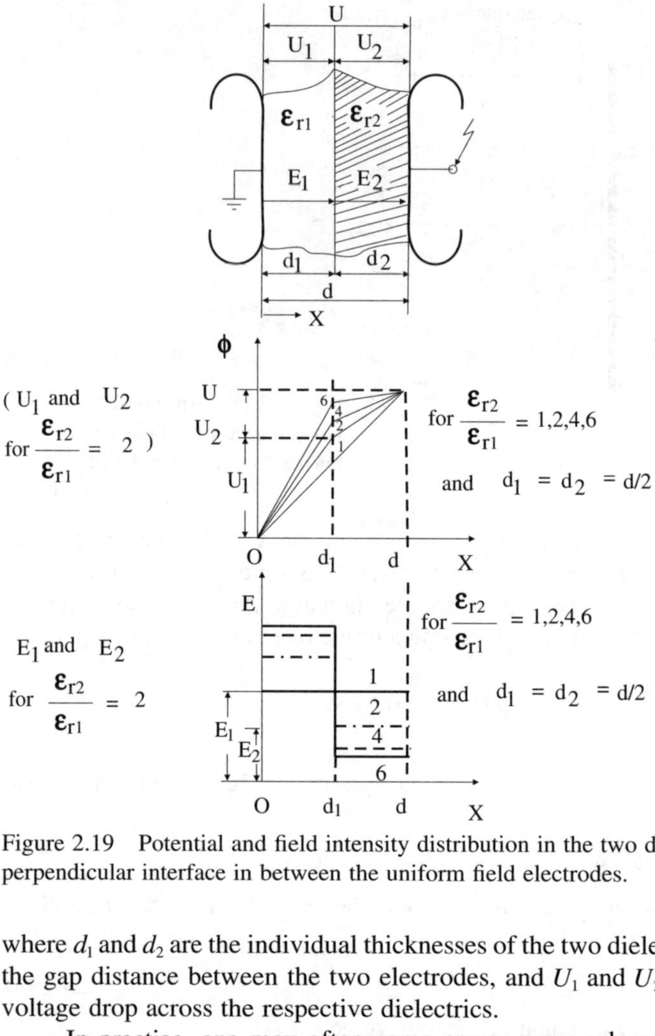

Figure 2.19 Potential and field intensity distribution in the two dielectrics forming perpendicular interface in between the uniform field electrodes.

where d_1 and d_2 are the individual thicknesses of the two dielectrics, and $(d_1 + d_2) = d$ the gap distance between the two electrodes, and U_1 and U_2 are the magnitudes of voltage drop across the respective dielectrics.

In practice, one may often come across cases where a solid dielectric with higher relative permittivity number ε_{r_2} comes in contact with air having lower $\varepsilon_{r_1}, (\varepsilon_{r_1} = 1)$. As seen in Figure 2.19, the field intensity E_1 in the dielectric having lower permittivity increases with the increasing ratio of $\varepsilon_{r_2}/\varepsilon_{r_1}$.

By introducing equation (2.55) in (2.56), the magnitudes of E_2 for an applied voltage U between the plates can be obtained as follows:

$$E_1 = \frac{U}{\varepsilon_{r_1}(d_1/\varepsilon_{r_1} + d_2/\varepsilon_{r_2})} = \frac{U}{d}\frac{\varepsilon_{r_2}/\varepsilon_{r_1}}{[d_1/d(\varepsilon_{r_2}/\varepsilon_{r_1} - 1) + 1]} \tag{2.57}$$

and

$$E_2 = \frac{U}{\varepsilon_{r_2}(d_1/\varepsilon_{r_1} + d_2/\varepsilon_{r_2})} = \frac{U}{d}\frac{1}{[d_1/d(\varepsilon_{r_2}/\varepsilon_{r_1} - 1) + 1]} \tag{2.58}$$

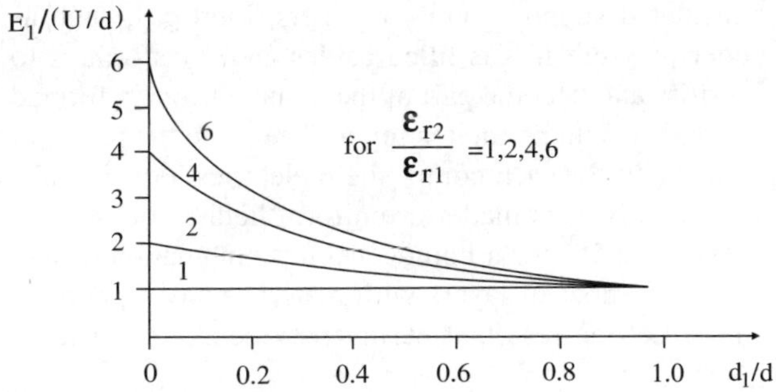

Figure 2.20 Field intensity with increasing thickness of one dielectric in a twin dielectric parallel plate condenser.

For a similar electrode arrangement as in Figure 2.19, equation (2.57) is plotted for increasing thickness of the dielectric d_1 with respect to the total gap distance d and for different values of the ratio ($\varepsilon_{r2}/\varepsilon_{r1}$) in Figure 2.20.

Figure 2.20 shows the rise in field intensity E_1 in the dielectric having lower ε_{r1}(air or any other gas) in a twin dielectric system compared with the field intensity in a single dielectric system U/d. It reveals that the rise in field intensity is much steeper for smaller thickness of the dielectric $d_1$and also for higher permittivity ε_{r2} of the other dielectric.

This phenomenon poses a very common problem in high voltage apparatus having solid dielectrics. Air pockets, of arbitrary shape and size known as voids, are very commonly present within solid dielectrics in high voltage apparatus like generators, transformers, power cables, bushings, insulators, capacitors, current and voltage measuring transformers, etc. Similarly, air bubbles may also be present in liquid dielectrics. It is essential to avoid such voids or bubbles in solid or liquid insulation systems because a sharp rise in field intensity may lead to the problems of partial breakdown. In an apparatus with imperfect contact between the electrode and the solid dielectric, air/gaseous pockets (microvoids) or even slits may be formed acquiring very high field intensity. This can prove quite dangerous in GIS and power cables as partial breakdown may take place in the region with enhanced field intensity causing damage to the solid insulation. On the other hand, this property can be suitably utilized for reducing the electric stress in the solid dielectrics by introducing an appropriately designed air corridor in equipments, for example in bushings.

2.5.3.2.1 Effective Permittivity of Composite Dielectrics In practice, homogeneously formed suitable combinations of different dielectrics, solid-liquid, solid-solid, and solid-gaseous are frequently used in high voltage apparatus. Such a dielectric combination is known as "composite dielectric". For example, resin or wax based impregnating compounds or different hydrocarbon oils alone are widely used to impregnate paper insulation in power cables, bushings, capacitors and transformers. Mica and fiberglass reinforced plastic materials are commonly used in electric machines and for insulated supports and containers. Inert gas, such as nitrogen, and SF_6 are used under pressure in gas filled cables and transformers to fill the micro-voids of paper porosity and become part of the homogeneously formed composite dielectric. At this stage, it would be interesting to determine the resultant or effective mean value of permittivity for such composite dielectrics. For the sake of convenience and simplicity the analysis is made for uniform fields.

Consider equations (2.57) and (2.58) for a homogeneous combination of two dielectrics, subdivided into infinite number of layers with materials having permittivities ε_1 and ε_2 in uniform field. Let the resultant permittivity be ε_{res}. For such a material,

$$D = \varepsilon_{res} E$$

where D and E are the macroscopic mean values. Since the field intensities E_1 and E_2 in microscopic multiple layers in uniform field remain unchanged, then,

$$D = \varepsilon_{res} E = \varepsilon_1 E_1 = \varepsilon_2 E_2$$

by introducing this relation in equation (2.57) or (2.58) and rearranging, we obtain:

$$\varepsilon_{res} E = \frac{U}{d} \cdot \frac{1}{\left(\frac{d_1/d}{\varepsilon_1} + \frac{d_2/d}{\varepsilon_2} \right)}$$

In this expression U/d represents the mean value of field intensity E within the composite dielectric. The relation d_1/d and d_2/d represent the proportionate volumes of the respective dielectrics, hence these can be replaced by v_1 and v_2. Therefore,

$$\varepsilon_{res} = \frac{1}{(v_1/\varepsilon_1) + (v_2/\varepsilon_2)} \tag{2.59}$$

for a combination of n dielectrics, it can be written as,

$$\varepsilon_{res} = \frac{1}{(v_1/\varepsilon_1) + (v_2/\varepsilon_2) + \ldots\ldots (v_n/\varepsilon_n)} \tag{2.60}$$

where

$$\sum_{i=1}^{n} v_i = 1 \quad \text{or } 100\% \text{ volume.}$$

Thus the resultant permittivity of a homogeneously formed composite dielectric would depend upon the volume proportion and permittivity of the individual dielectrics. Although the above example has been explained for uniform field configuration, the phenomenon is also true for weakly nonuniform fields.

2.5.3.3 *Field with Diagonal Interface* Often the dielectric-dielectric interface in high voltage apparatus is neither parallel nor normal to the direction of electric field. In this case, the dielectric interface falls diagonally at an angle, let us say "α" to the field direction as shown in Figure 2.21.

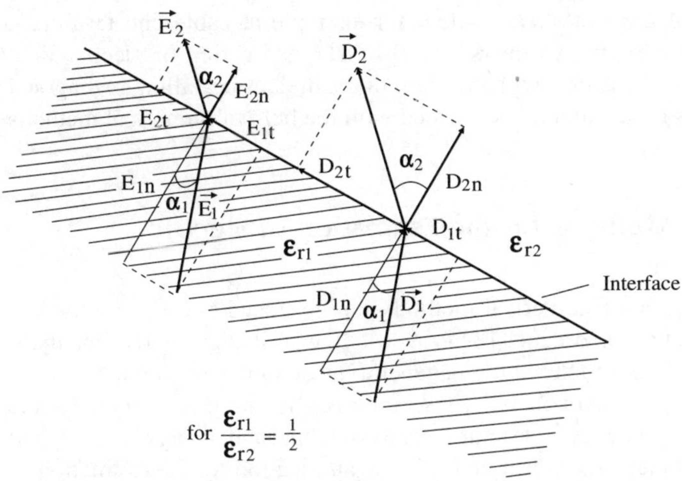

Figure 2.21 Field, flux density and their components for the interface falling diagonally in their direction of the field.

This is the most common and general position of a dielectric interface in any electrode system of high voltage apparatus. The field analysis for diagonal interface is a combination of the two cases of parallel and normal field directions with respect to the interface discussed above.

The electric field intensity $\vec{E}_1$ and flux density $\vec{D}_1$, both in the same direction in a dielectric having relative permittivity ε_{r1}, meet an interface with another dielectric at an angle of incidence α_1. The directions of the field and flux density vectors in the second dielectric, having relative permittivity ε_{r2}, change by refraction. The two vectors $\vec{E}_1$and $\vec{D}_1$ can be split into their normal E_{1n}, D_{1n} and tangential E_{1t}, D_{1t} components falling in perpendicular and longitudinal direction on the interface respectively, as shown in Figure 2.21. For the normal components, the diagonal interface behaves like an interface perpendicular to the field, and for the tangential components, like a longitudinal interface. According to the theory developed for longitudinal and perpendicular interfaces, equations (2.52) and (2.55) can be written as,

$$E_{1t} = E_{2t} \quad \text{and} \quad \frac{E_{1n}}{E_{2n}} = \frac{\varepsilon_{r2}}{\varepsilon_{r1}}$$

If it is assumed that no free sources of charge are present at the interface and only polarization charges define the boundary conditions, then the angles or incidence α_1 and refraction α_2 can be related as follows,

$$\frac{\tan\alpha_1}{\tan\alpha_2} = \frac{E_{1t}/E_{1n}}{E_{2t}/E_{2n}} = \frac{E_{2n}}{E_{1n}} = \frac{D_{2n}/\varepsilon_2}{D_{1n}/\varepsilon_1} = \frac{D_{1t}}{D_{2t}} = \frac{\varepsilon_{r1}}{\varepsilon_{r2}} \tag{2.61}$$

Thus, we can say that at diagonal interface a refraction of field lines takes place, which is analogous to the phenomenon in optical physics. A direct solution of the Laplace's equation for such a case is practically impossible. In practice, one comes across diagonal interfaces most often, for example at cable end terminals, machine slots, switchgears, transformers and GIS. The estimation of electric field intensity in such electrode systems with interface of the dielectrics falling in diagonal direction to the field is more suitably performed with the help of numerical methods described in the next section.

2.5.4 Numerical Methods for the Estimation of Electric Field Intensity

The history of application of numerical methods can be traced back to the time of Gauss (1777–1855) and Boltzmann (1844–1906). It was in the 1940s that these methods received more attention for solving general linear and even nonlinear partial differential equations [2.19 and 2.20]. These methods have since received wide application in civil engineering (structural analysis), thermal analysis, and fluid mechanics. Numerical methods later found suitable application for the estimation of electric fields. The development in their computational techniques has become a continuous process.

Application of numerical methods for the estimation of electric field intensity in a dielectric in between given electrode system involves solving partial differential equations, including the Laplace equation, by systematic approximation. A number of methods have been developed since the rise of computers. The approach has endeavoured to evolve more generalized methods, which could cater to the need for two or three dimensional fields with unsymmetrical boundary conditions (electrode configurations) and with insulating materials having different permittivity. The numerical methods have inherent advantages as well as disadvantages with respect to the particular problem to be solved. The suitability of the computational techniques should, therefore, be taken into consideration. Efforts are made to reduce complexities in computational algorithm as well as requirement of memory and time by the computer. The programs are designed to be user friendly.

Unlike the case of electrodes in free air, for compressed gas insulation systems (GIS), that is, the metal clad switchgears with SF_6 and other gas insulation systems, it is extremely important to achieve uniformity in field in different parts of the apparatus to the greatest extent possible. With the development of complete substations with compressed gas up to the highest system voltage of 1100 kV, the exact knowledge of electric field distribution in three dimensional irregular (unsymmetrical) electrode systems has become a necessity for providing appropriate shapes to the electrodes and thus optimizing the size of the apparatus. It is extremely important that the GIS are designed to prevent any stable PB at the rated working conditions. Numerical methods for estimating electric fields have, therefore, achieved considerable importance.

For estimating electric fields, there are basically two numerical methods being pursued widely over the world. These are Finite Element Method (FEM) and Charge Simulation Method (CSM). They are based upon entirely different techniques. These have advantages and disadvantages over each other. With the development of computational techniques, very large number of versions of these methods have been developed by the researchers to suit different applications [2.21, 2.22].

2.5.4.1 Finite Element Method (FEM) Finite Element Method (FEM) is the driving force and dominating tool behind today's computational science and engineering [2.21]. This method was first developed in the United States in the 1940s for estimating the stresses in aircraft structures [2.9]. Later it was developed for the estimation of electric field distribution in dielectrics. With the availability of software package programs, it is the method used most widely in the world for the electric field estimation.

Finite Element Method (FEM) in general is based on transforming the differential equations in integral form and then using an approximation. One way to transform these equations is to find a function that minimizes an energy integral. The basic approach of FEM for electric field estimation involves the factual characteristic of an electrostatic field that the total energy enclosed in the whole field region acquires a minimum value. In other words, the potential ϕ, under given conditions of electrode surface, should make the enclosed energy function to be at a minimum for a given dielectric volume "V"; therefore:

$$W = \int_V \frac{1}{2}\varepsilon(grad\ \phi)^2\, dV \rightarrow \text{minimum} \tag{2.62}$$

which is obtained by solving the basic potential equation (2.3). W is the electrical energy stored in the volume of dielectric under consideration.

Hence, in this method the field between the electrodes under consideration is subdivided into finite number of discrete sized elements. The behavior of these elements is specified by a number of parameters, for example, the potential. The number of nodal points and elements established within the mesh are assigned identifying integer numbers. The shape of these discrete elements is suitably chosen to be triangular for two dimensional representations. For three dimensional field configurations, "tetrahedron", a pyramid-like solid structure with four plane triangular faces, shown in Figure 2.22, is found appropriate.

The size (dimensions) and the orientation of triangles, as well as tetrahedrons, irregularly distribute over the generated mesh in the field region depending upon the magnitude of the potential gradient. At the locations where higher field intensity or electric-stress exist, discrete elements of smaller size cover the region. Figure 2.23

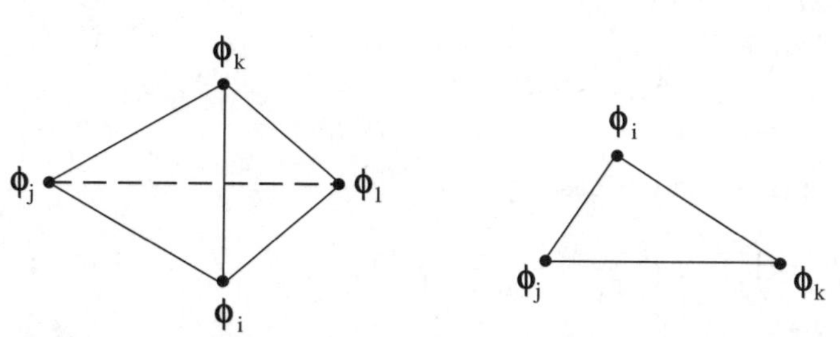

A Tetrahedron and a triangular finite element

Figure 2.22 Triangular and Tetrahedron shaped finite discrete elements.

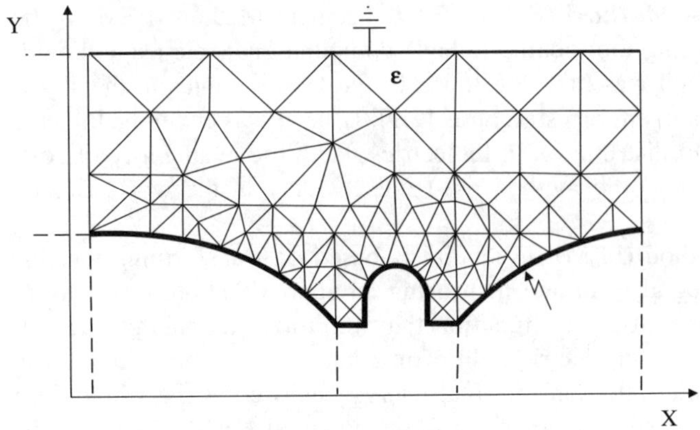

Triangular discrete elements (FEM)

Figure 2.23 Field computation by FEM in two dimension using triangular discrete elements.

shows the field between two electrodes covered with irregular triangular discrete elements.

Let us consider an electrostatic field, undistorted by any space charge concentration, in a single isotropic dielectric between two electrodes. Then the potential ϕ would be determined by the boundaries, that is, the metal electrode surfaces. The above equation (2.62) for electrical energy W, stored within the whole region of such a Laplacian field is given in cartesian coordinates as follows,

$$W = \iiint_V \left[\frac{1}{2} \varepsilon \left\{ \left(\frac{\partial \phi}{\partial x} \right)^2 + \left(\frac{\partial \phi}{\partial y} \right)^2 + \left(\frac{\partial \phi}{\partial z} \right)^2 \right\} \right] dx\, dy\, dz \tag{2.63}$$

It should be noticed that for a small volume element $dV = (dx\, dy\, dz)$, the expressions $(1/2\varepsilon\Delta^2\phi)$ within equation (2.63) represent the energy densities per unit volume in a particular direction.

If it is assumed that the potential distribution does not change in the z direction, that is, reducing the problem into a two dimensional case, then the total energy W_A stored within the area A per unit length located between the two electrode boundaries can be given by the equation (2.62),

$$W_A = \int_A \frac{1}{2} \varepsilon (grad\ \phi)^2\, dA$$

and according to equation (2.63) in this case by,

$$W = z \iint_A \left[\frac{1}{2} \varepsilon \left\{ \left(\frac{\partial \phi}{\partial x} \right)^2 + \left(\frac{\partial \phi}{\partial y} \right)^2 \right\} \right] dx\, dy \tag{2.64}$$

where z is a constant. W/z gives the energy density per elementary area dA.

Figure 2.24 shows part of the field under consideration. For the triangular element (e) having nodes i, j and k, basic functions for the potential distribution

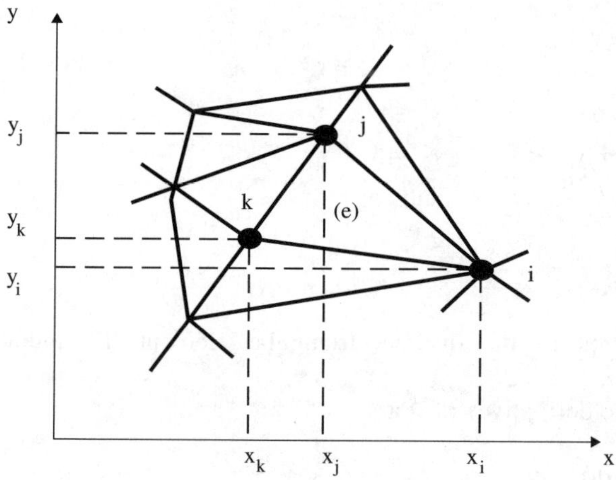

Figure 2.24 Part of a two dimensional field subdivided into irregular triangular elements on x-y plane.

$\phi(x, y)$ within this element are desired to be calculated. Normally a linear dependency of ϕ on x and y is assumed, which gives rise to the first order approximation,

$$\phi(x, y) = \phi = a_1 + a_2 x + a_3 y \tag{2.65}$$

It is worth mentioning that for a more exact estimation of field, higher order polynomials can be applied, for example, a square or a cubic equation, Speck [2.23], then

$$\begin{aligned}\phi(x, y) = \phi = a_1 + a_2 x + a_3 y + a_4 x^2 + a_5 xy + a_6 y^2 \\ + a_7 x^3 + a_8 x^2 y + a_9 xy^2 + a_{10} y^3\end{aligned} \tag{2.66}$$

This however, leads to an increase in computation time and analysis becomes more complex without adding much to the accuracy of the calculation.

Equation (2.65) implies that within the element the potentials are linearly distributed and the field intensity is constant. For the element (e) under consideration, the unknown potentials at respective nodes i, $j$and k are given by the equations,

$$\begin{aligned}\phi_i &= a_1 + a_2 x_i + a_3 y_i \\ \phi_j &= a_1 + a_2 x_j + a_3 y_j \\ \phi_k &= a_1 + a_2 x_k + a_3 y_k\end{aligned} \tag{2.67}$$

and the coefficients a_1, a_2 and a_3 may be determined by applying Cramer's rule:

$$\begin{aligned}a_1 &= \frac{1}{2\Delta_e}(\alpha_i \phi_i + \alpha_j \phi_j + \alpha_k \phi_k) \\ a_2 &= \frac{1}{2\Delta_e}(\beta_i \phi_i + \beta_j \phi_j + \beta_k \phi_k) \\ a_3 &= \frac{1}{2\Delta_e}(\gamma_i \phi_i + \gamma_j \phi_j + \gamma_k \phi_k)\end{aligned} \tag{2.68}$$

where

$$
\begin{aligned}
\alpha_i &= x_j y_k - x_k y_j & \beta_i &= y_j - y_k \\
\alpha_j &= x_k y_i - x_i y_k & \beta_j &= y_k - y_i \\
\alpha_k &= x_i y_j - x_j y_i & \beta_k &= y_i - y_j \\
\gamma_i &= x_k - x_j & & \text{and} \\
\gamma_j &= x_i - x_k & 2\Delta_e &= \alpha_i + \alpha_j + \alpha_k \\
\gamma_k &= x_j - x_i & &= \beta_i \gamma_j - \beta_j \gamma_i
\end{aligned}
$$

The symbol Δ_e represents area of the discrete triangular element (e) under consideration.

From equation (2.65), the derivatives of ϕ are,

$$\frac{\partial \phi}{\partial x} = a_2 = f(\phi_i, \phi_j, \phi_k)$$

and

$$\frac{\partial \phi}{\partial y} = a_3 = f(\phi_i, \phi_j, \phi_k) \tag{2.69}$$

In order to minimize the energy within the field region under consideration, only derivatives of the energies with respect to the potential distribution in each element are of particular interest here. For the element (e) under consideration, W_e is the energy enclosed within the element, then the energy per unit length W_e/z in the direction z, denoted by $W_{\Delta e}$can be given by equation (2.64) as follows,

$$W_{\Delta e} = \frac{W_e}{z} = \frac{1}{2}\Delta_e \varepsilon \left[\left(\frac{\partial \phi}{\partial x} \right)^2 + \left(\frac{\partial \phi}{\partial y} \right)^2 \right]_e \tag{2.70}$$

since $\iint_A dx\,dy$ provides the area Δ_e of the element under consideration.

In order to minimize the energy in this specific element (e) the desired expression can be derived by differentiating equation (2.70) partially with respect to ϕ_i, ϕ_j and ϕ_k. Thus, taking into account equations (2.68) and (2.69), the differentiation with respect to ϕ_i gives the following expression,

$$
\begin{aligned}
\frac{\partial W_{\Delta e}}{\partial \phi_i} &= \frac{1}{2}\Delta_e \varepsilon \left(2a_2 \frac{\partial a_2}{\partial \phi_i} + 2a_3 \frac{\partial a_3}{\partial \phi_i} \right) \\
&= \frac{1}{2}\varepsilon (a_2 \beta_i + a_3 \gamma_i) \\
&= \frac{\varepsilon}{4\Delta_e} [(\beta_i^2 + \gamma_i^2)\phi_i + (\beta_i \beta_j + \gamma_i \gamma_j)\phi_j + (\beta_i \beta_k + \gamma_i \gamma_k)\phi_k]
\end{aligned} \tag{2.71}
$$

The set of all the three equations with respect to ϕ_i, ϕ_j and ϕ_k can be expressed in matrix form,

$$
\begin{aligned}
\frac{\partial W_{\Delta e}}{\partial \phi_e} &= \frac{\varepsilon}{4\Delta_e}\begin{bmatrix} (\beta_i^2+\gamma_i^2) & (\beta_i\beta_j+\gamma_i\gamma_j) & (\beta_i\beta_k+\gamma_i\gamma_k) \\ (\beta_j\beta_i+\gamma_j\gamma_i) & (\beta_j^2+\gamma_j^2) & (\beta_j\beta_k+\gamma_j\gamma_k) \\ (\beta_k\beta_i+\gamma_k\gamma_i) & (\beta_k\beta_j+\gamma_k\gamma_j) & (\beta_k^2+\gamma_k^2) \end{bmatrix}\begin{bmatrix}\phi_i\\ \phi_j\\ \phi_k\end{bmatrix} \\
&= \frac{\varepsilon}{4\Delta_e}\begin{bmatrix} (h_{ii})_e & (h_{ij})_e & (h_{ik})_e \\ (h_{ji})_e & (h_{jj})_e & (h_{jk})_e \\ (h_{ki})_e & (h_{kj})_e & (h_{kk})_e \end{bmatrix}\begin{bmatrix}\phi_i\\ \phi_j\\ \phi_k\end{bmatrix} \\
&= [h]_e\{\phi\}_e
\end{aligned} \tag{2.72}
$$

The matrix $[h]_e$ is known as the "stiffness matrix" for the individual element (e). It geometrically locates the element and contains the functional sensitivity with respect to the potentials. It also contains the permittivity of the dielectric material.

The energy enclosed in the individual element $W_{\Delta e}$ is dependent upon the node potentials of this element (ϕ_i, ϕ_j, ϕ_k). If we denote the total energy in the whole field of given elements by W_Δ, the relation for minimizing the energy within the complete system can be given as,

$$\frac{\partial W_\Delta}{\partial \{\phi\}} = 0 \tag{2.73}$$

where $\{\phi\}$ is the total potential vector for all the nodes within a given system.

Considerations to take into account the contributions to the potential of triangular element (e) made by the neighbouring elements still remain to be applied. Figure 2.24 shows the triangular element (e) being surrounded by other triangular elements. Any node potential within such a system depends upon the potentials of the surrounding nodes. The number of surrounding nodes to a node is dependent upon the network and may vary from node to node within a network. However, it is always a small number of nodes.

Let us consider node k of the discrete element (e) given number 5 as shown in Figure 2.24. Four triangular elements, (1) to (4), meet at this node. The nodes adjacent to node 5 are numbered 1 to 4. For this set of elements, following relation is obtained on applying equation (2.73),

$$\frac{\partial W_\Delta}{\partial \phi_5} = 0 \tag{2.74}$$

where W_Δ is the total energy of the system with four elements and $\phi_1 \ldots \phi_5$ are the potentials of the nodes 1 to 5 respectively.

Applying equation (2.72) to each of these four elements, that is, replacing the index e by individual numbers as given in Figure 2.25, the above equation works out by adding the contribution from each element as follows:

$$
\begin{aligned}
\frac{\partial W_\Delta}{\partial \phi_5} = [&(h_{ik})_1\phi_2 + (h_{jk})_1\phi_1 + (h_{kk})_1\phi_5 \\
&+ (h_{ik})_2\phi_3 + (h_{jk})_2\phi_2 + (h_{kk})_2\phi_5 \\
&+ (h_{ik})_3\phi_4 + (h_{jk})_3\phi_3 + (h_{kk})_3\phi_5 \\
&+ (h_{ik})_4\phi_1 + (h_{jk})_4\phi_4 + (h_{kk})_4\phi_5] \\
= 0&
\end{aligned} \tag{2.75}
$$

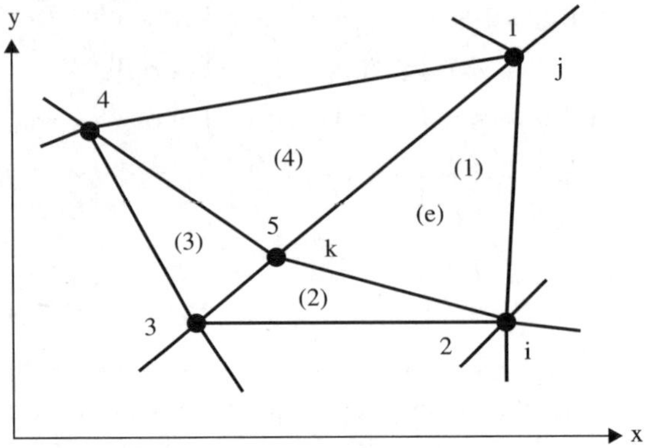

Figure 2.25 Node k of element (e), shown as node No. 5 connected to four triangular elements (1) to (4).

this equation can also be written in the form,

$$H_{15}\phi_1 + H_{25}\phi_2 + H_{35}\phi_3 + H_{45}\phi_4 + H_{55}\phi_5 = 0 \tag{2.76}$$

The potential ϕ_5 can be calculated if the potentials ϕ_1 to ϕ_4 are known, which depend upon the larger network. Therefore, for every unknown potential, a corresponding equation has to be evolved. For a system in which Laplace equation is applicable, that is, the space charge free electrostatic fields, the solution may be written as,

$$\frac{\partial W}{\partial\{\phi\}} = 0 = [H]\{\phi\} \tag{2.77}$$

which represents the assembly of the whole set of minimizing equations.

Many FEM based software programs have been developed. For example; FEMLAB, QuickField5.4, Mafia4, Finite Integration Technique (FIT) etc. These require the details of the electrode geometry/configuration and properties of the dielectric/dielectrics (ε_r). For two-dimensional problems, the choice exercised for the number of discrete elements determines the size of the triangles in different regions of the field. A plot of equi-potential lines can be worked out for the entire field region between the electrodes for a desired percentage of potential difference between the lines.

The FEM leads to comparatively simpler techniques for estimating fields at highly curved and thin electrode surfaces with different dielectric materials. However, its suitability remains mainly for weakly nonuniform and symmetrical fields, which can be easily represented by two dimensional geometries. For three-dimensional complicated field configurations, there are reservations in its application.

2.5.4.2 Charge Simulation Method (CSM) Derived from the "method of image charge" introduced by Kelvin in 1872, Steinbigler [2.24] developed this method for digital computation of electric fields in his dissertation in 1969. This method is accepted as one of the most superior and suitable for two and three dimensional field configurations with electrode systems of any desired shape. It is also found suitable for insulation systems comprising more than one dielectric. The CSM has proved to be a highly accurate method as it even permits to take into consideration the effect of the surroundings, including the grounded electrodes, on the main field in high voltage insulation system.

Coulomb's law states that the force between two small charged objects, having charges q_1 and q_2, in vacuum, separated by a distance r which is large compared to the largest dimension of either one, is directly proportional to q_1 and q_2 and inversely proportional to the square of r, or

$$F = K\frac{q_1 q_2}{r^2}$$

The law asserts that the force depends linearly on the charge on either object. Thus, the force on one charged object due to the presence of several other charged particles can be found as the superposition (sum) of the forces on the one charged particle due to the others acting separately. Hence, it can be said that the Coulomb forces are additive, the base for the principle of "superposition" [2.14].

The basic principle of Charge Simulation Method (CSM) lies in the concept that the electric field under consideration is formed by superposition of many individual fields. For the calculation of electrostatic fields by this method, the distributed charge on the surface of a dielectric with the conductor is replaced by n discrete fictitious individual charges. Also known as "simulation charge", these are arranged suitably inside the conductor, or in other words, outside the region in which the field is to be computed, that is, the dielectric. The method implies that the total potential function of the field at any point due to a number of individual charges can be determined by the summation of scalar potentials produced by individual charges. The potential contributions made by individual fictitious charges are obtained by the solution of Laplace's and Poisson's equations. The shape of the individual fictitious charges or the "simulation charges", can be chosen as point, line (finite) or ring (circular) form, Figure 2.26. These are placed suitably within an electrode system, outside the space in which the field is to be computed. The potential at any point on the electrode surface is calculated by the potential function of these individual charges. For two-dimensional fields, point and line charges are found suitable. However, for three dimensional electric field calculations, ring charges with partially

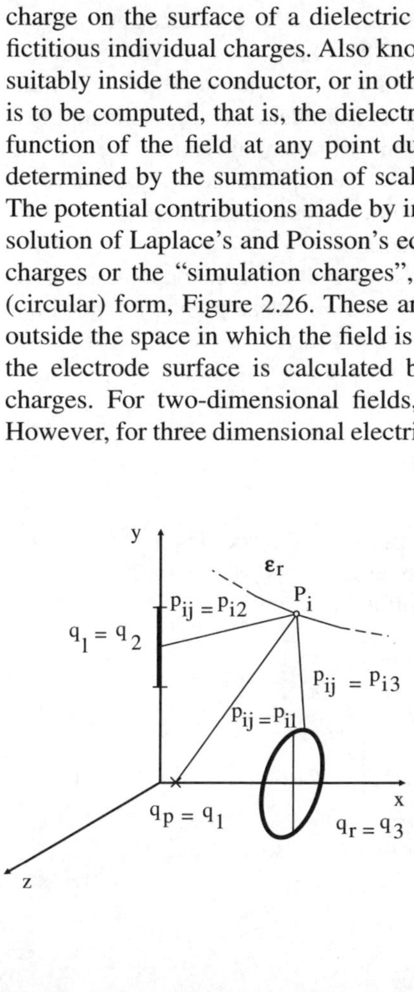

Figure 2.26 Three types of individuals fictitious charges with respect to a point on the conductor surface.

constant charge density or continuously variable charge density are found more suitable [2.25, 2.27].

In order to determine the magnitude of these fictitious individual charges, n numbers of points are chosen on the surface of the conductor (contour check points). The potential, determined by the superposition of field contributions made by these individual charges at any contour point on the conductor surface, must be equal to the known conductor potential ϕ_c since a conductor is equipotential.

Let q_j be one of the n individual fictitious charge, and ϕ_i the potential of an arbitrary point P_i within the dielectric system, independent of the coordinate system. From the principle of superposition, the potential ϕ_i at the point P_i is determined by the individual potential contributions made by different charges as follows:

$$\phi_i = \sum_{j=1}^{n} p_{ij} q_j \tag{2.78}$$

where p_{ij} are known as "potential coefficients", which need to be calculated for different types of individual charges by a particular solution of Laplace's equations.

The procedure for estimating the electric field in the dielectric between any particular electrode system involves reproducing the metallic electrode contour as a known equipotential surface. As the shape of the electrodes reproduces accurately; the field estimation is achieved more efficiently. The known electrode boundaries are ascribed by a number of discrete check points suitably located on its contour. The number, their location and the type of individual simulation charges are modified in such a way that the checkpoints acquire the known potential. Thus, the electrode contour, a known equipotential surface, needs to be reproduced as accurately as possible. In the process, the magnitudes of these fictitious charges need to be calculated depending upon the known electrode potential.

Equation (2.78) is determined for each discretely located checkpoint assigned on the contour of the electrode configuration. At any of these points, the potential ϕ_i results by the superposition of potentials due to individual charges. Under the given conditions when ϕ_i is equal to the known conductor potential ϕ_c, the equation (2.78) may be rewritten as,

$$\sum_{j=1}^{n} p_{ij} q_j = \phi_c \tag{2.79}$$

where q_j is the discrete charge and p_{ij} the associated potential coefficient. Application of this equation for n number of contour points leads to a system of n linear equations for n unknown charges obtained in the following form,

$$\begin{bmatrix} p_{11} & p_{12} & \cdots & p_{1n} \\ p_{21} & p_{22} & \cdots & p_{2n} \\ \vdots & \vdots & \vdots & \vdots \\ p_{n1} & p_{n2} & \cdots & p_{nn} \end{bmatrix} \begin{bmatrix} q_1 \\ q_2 \\ \vdots \\ q_n \end{bmatrix} = \begin{bmatrix} \phi_1 \\ \phi_2 \\ \vdots \\ \phi_n \end{bmatrix} \tag{2.80}$$

or

$$[p][q] = [\phi_c]$$

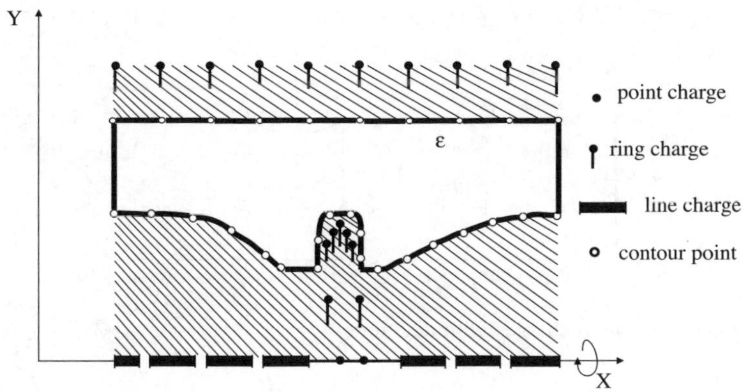

Figure 2.27 Arrangement of point, line and ring forms of simulation charges within an electrode system.

By solving this set of equations for $[q]$, the magnitudes of the individual fictitious charges are obtained that determine the electrode potentials at the chosen discrete contour points. Only then can the electrode potential in the vicinity of these points be approximated. It may still differ from ϕ_c. It is, therefore, necessary to check whether the calculated set of charges fit the boundary conditions and the desired potential is acquired all over the electrode contour. This is done by again using equation (2.78) to determine potentials at a number of check points on the electrode surface of known potential. The difference between the computed and the assigned electrode potentials gives a measure of the accuracy of the simulation and hence the appropriateness of the number and location of fictitious charges. The chosen system of individual fictitious charges should be accordingly modified in number and/or in position to be able to obtain the exact electrode configuration. This can be achieved by application of optimization techniques [2.10, 2.25].

Having adopted a suitable and dependable charge system placed in the given electrode system as shown in Figure 2.27, the potentials and field intensities at any desired point in the dielectric can be determined. The field intensity is obtained by the superposition of potential gradients, contributed by each individual charge. It is more appropriate to calculate the individual directional components of field intensity separately. In certain coordinate systems, the component of field in x direction $\vec{E}_x$ for n number of charges is given by,

$$\vec{E}_x = \sum_{j=1}^{n} \frac{\partial p_{ij}}{\partial x} q_j = \sum_{j=1}^{n} (f_{ij})_x q_j \tag{2.81}$$

where f_{ij} are known as “field intensity coefficients” in x-direction.

According to the general electrostatic field theory, the potential ϕ at a point P, distant d from a single point charge q is given by,

$$\phi = \frac{q}{4\pi\varepsilon d} = pq \tag{2.82}$$

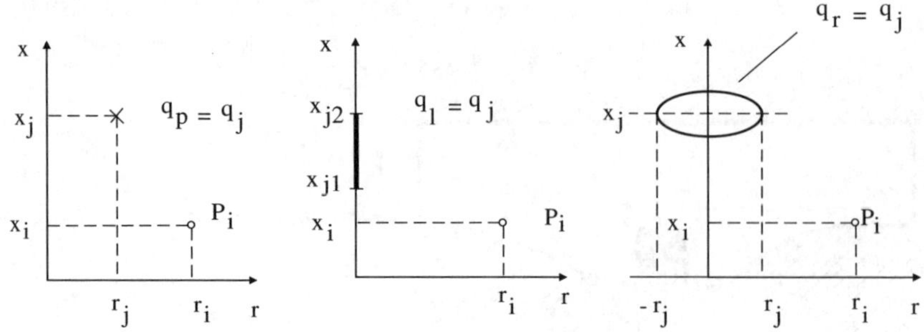

Figure 2.28 Individual simulation charges with notations.

From the analogy of equations (2.79) and (2.82) the general expression for the "potential coefficient" of a point charge at distance d can be written as,

$$p = \frac{1}{4\pi\varepsilon d} \tag{2.83}$$

It is assumed that the three types of commonly used discrete charges—that is, point, finite straight line, and toroidal or circular (ring)—have constant charge densities and a total charge "q_j" centered on the axis of symmetry. "Potential coefficients" are calculated by analytical solution. These three types of charges are shown in Figure 2.28 and their "potential coefficients" for point P_i are given by the following equations.

Point Charge:

$$p_{ij} = \frac{1}{4\pi\varepsilon\sqrt{(r_i - r_j)^2 + (x_j - x_i)^2}} \tag{2.84}$$

Finite Straight Line Charge:

$$p_{ij} = \frac{1}{4\pi\varepsilon(x_{j2} - x_{j1})} \ln \frac{x_{j2} - x_i + \sqrt{r_i^2 + (x_{j2} - x_i)^2}}{x_{j1} - x_i + \sqrt{r_i^2 + (x_{j1} - x_i)^2}} \tag{2.85}$$

Ring Charge:

$$p_{ij} = \frac{1}{2\pi^2\varepsilon} \frac{k}{4\sqrt{r_j r_i}} K(k)$$

where

$$K(k) = \int_0^{\pi/2} \frac{d\psi}{\sqrt{i - k^2 \sin^2\psi}} \quad \text{and} \quad k = \sqrt{\frac{4r_j r_i}{(r_j + r_i)^2 + (x_i - x_j)^2}}$$

Proper selection of the suitable types of simulation charges and their appropriate arrangement is a very important aspect of this method of field estimation. A particular type of simulation charge is found to more suitably reproduce a given shape of electrode. For example, for cylindrical electrodes, finite line charges are suitable, whereas for fields with axial symmetry having projected circular structures, ring charges are better. Spherical or similar electrodes are best evolved by point and ring charges. One could use other forms of simulation charges, but the complexity of computation in most cases increases with the complexity of simulation charges. This is because the potential coefficients become more difficult to compute.

The algorithm of computer programming for CSM involves the following main steps:

1. Selection of the type and number of individual simulation charges and their appropriate location within the electrode system.
2. Location of "contour points" on the surface of the electrodes. These points should be chosen in such a way that they are adequately allocated to the critical sections of the contour such as, curves, corners, etc. Larger numbers of contour points are needed to be assigned near such critical locations.
3. Calculation of "potential coefficients", $p_{ij}$of different types of individual charges for the contour points.
4. Inversion of potential coefficient matrix and multiplication with the potential vector of the contour point in order to calculate the charge vector.
5. Selection of "check points" on the electrode surface. The potential at these points is calculated to determine any deviation from the actual potential. The checkpoints should be located to accurately reflect the errors, bearing in mind that there is a likelihood of maximum error somewhere near the middle of two contour points.
6. On the basis of the magnitude of error calculated at the checkpoints, collocation of some individual simulation charges should be done within the electrode system. One may apply any other objective criteria for error minimization to achieve higher accuracy.
7. Choice of a suitable method for the calculation of equipotential lines is made based on the electrode contours having known equipotential surfaces.
8. The electric field at any desired point is then calculated analytically by superposition.

CSM can be applied suitably for the estimation of not only the two-dimensional fields with or without axial symmetry but also for three-dimensional fields without axial symmetry. Electric field estimation of fields with space charge and having more than one dielectric is also possible by this method with reasonable amount of computation [2.26, 2.27].

The inherent feature of CSM to simulate curved and rounded surfaces of electrodes or interfaces of different dielectric materials in a rather simple fashion makes it a very appropriate method for field estimation, especially for complicated electrode geometries in practice. This method has proved to be highly accurate too. Because of the involvement of lesser number of linear equations, in general the field estimation by CSM requires a relatively lower computation time of digital computer compared with FEM. Suitability of CSM for three-dimensional fields without any symmetry has also been established with a reasonable amount of computational expense. The estimation of field by CSM is most accurate as it is in a position to take into consideration the surroundings/electrodes at ground potential.

Some shortcomings of CSM were reported in the literature by Reister, Sato, et al. [2.28, 2.29]. While applying a check, there is no way to predetermine the optimum position of simulation charges for the best results. The accuracy in CSM essentially depends upon the proper choice of the shapes and location of the simulation charges. Once the charges are located, values of charges are determined by the method itself. However, locating the charges requires the personal experience and judgment of the operator.

Hence, it was desirable to deploy some optimization technique to determine the appropriate locations of simulation charges to make this method independent of personalized experience and judgment. It could also reduce the member of required charges for accurate reproduction of the electrode surface and thus considerably reduce the time required for computation.

Use of "Genetic Algorithms", (GAs), a particular class of evolutionary algorithms used in computer science to find solutions of optimization and search problems, has been found to be an appropriate tool to find optimum location and also number of simulation charges to achieve very high accuracy of CSM [2.30, 2.31, 2.32, 2.33]. With a lesser number of simulation charges, faster convergence is achieved by using CSM coupled with GA.

Figure 2.29 shows equipotential lines and field intensity arrows calculated by CSM coupled with GA for an identical rod-rod electrode when the voltage is applied on the upper rod and the lower one is grounded [2.35]. This figure shows the effect

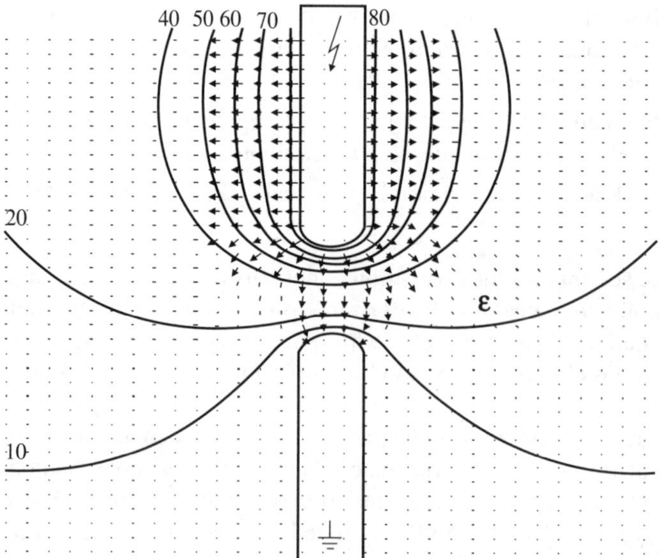

Figure 2.29 Equipotential lines and field intensity by CSM for identical rod-rod electrode with lower rod grounded [2.34].

of grounding one of the identical electrodes on the field distribution in the dielectric. Field in this case is not identical on either side of the center line between the two electrodes.

Further, due to the requirement of a finite gap distance between the location of a charge and electrode contours (equipotential line), it is difficult to apply CSM as such for thin electrodes like foils, plates or coatings. Highly irregular and complicated electrode boundaries with sharp bends have also been found difficult to be reproduced by this method. In order to overcome the above mentioned problems, a later version of CSM known as "Surface Charge Simulation Method" (SCSM) was developed by Singer [2.35].

The Charge Simulation Method (CSM) has undergone considerable development in the last three decades. It has emerged as a very efficient and accurate numerical method for electric field estimation. The method has been developed for field analysis of a wide variety of high voltage insulation systems including three-dimensional and multiple dielectric arrangements. It is also suitable for the optimization of electrodes as well as insulator contours, Malik [2.27].

2.5.5 Numerical Optimization of Electric Fields

In high voltage engineering, it is desirable to minimize the maximum electric stress in the apparatus in order to achieve better performance of the dielectrics under given working conditions. Hence, for a good design of insulation system in power apparatus, it is advisable to develop suitable optimization technique taking into consideration not only the electric field distribution but also the criteria of performance of the particular insulation; gaseous, solid or liquid used in the apparatus [2.43].

The procedure for optimization of electric fields in high voltage apparatus involves nothing but optimization of electrode configurations. Numerical optimization techniques in high voltage electrode and insulation design were first introduced in the early 1970s [2.10]. New techniques have constantly been evolved to improve the design and facilitate the conceptualization of optimal electrode configurations [2.31–2.46].

Although attempts have been made to develop optimization techniques applicable to electrodes and dielectric profiles by FEM [2.31], most of the work is dedicated to CSM because of its inherent suitability for optimization. Methods involving modification of contour elements [2.32] and development of an experimental electrode system model [2.33] have also been reported in the literature.

Besides achieving an economical design, a considerable improvement in dielectric behavior, and therefore a higher life expectancy of high voltage equipment can be anticipated by the optimization of field intensity. The object of optimization of electrode configurations, in general, is to determine the forms or dimensions of electrodes that may possibly lead to a more uniform distribution of the field in the complete dielectric/electrode surface while attempting to keep the magnitude of the field intensity as low as possible.

The criteria for optimization techniques can be based upon either the partial breakdown (PB) inception field intensity "E_i" or upon the field intensity required for local breakdown prevailing at the location of interest. The PB inception field intensity mainly depends upon the dielectric material/medium, its pressure and the electrode configuration. Therefore, different optimum electrode configurations may workout when the optimization exercise is performed for different dielectrics, Figure 2.31. For example, at atmospheric pressure "E_i" for gaseous dielectrics can be considered to be their "intrinsic strength", that is, the breakdown field intensity in uniform field for 1 cm gap. It is well known that in uniform and weakly nonuniform fields, the field intensity required for PB inception is the same as for breakdown. Therefore, it is the electrode configuration that can be optimized to achieve the lowest possible value of applied electric field E. At the rated voltage, the field intensity E varies over the apparatus depending upon the shape of the electrodes at different locations. If E_i is higher than the applied field intensity E all over an equipment, no PB would occur. In other words, an electrode having many projected shapes can be said to be optimized if at a given voltage the maximum value of E/E_i all over the dielectric is achieved as small as possible.

Since the maximum value of E/E_i depends upon three parameters—the electrode shape, position, and size—three different kinds of optimization possibilities exist. The optimum shape of an electrode contour can be characterized by,

$$E/E_i = \text{constant} = (E/E_i)\,\text{max} < 1 \tag{2.86}$$

Most of the optimization methods are based upon an iterative process of shaping the contour until equation (2.86) is satisfied. The shape at the individual location thus evolved is the optimum one. While applying CSM this may be achieved by displacement of contour points perpendicular to the surface [2.10, 2.39]; by changing the position of the so called "optimization charges" within the electrode [2.35, 2.36, 2.37, 2.43]; by modification of contour elements [2.32] or by the construction of experimental electrode system model [2.33, 2.38, 2.39]. Optimization by electrode contour modification, taking into consideration of the electrode area/volume effects on the breakdown, is introduced in [2.44]. Possible applications of neural network in optimization techniques is described in [2.45, 2.46]. A brief introduction to some of these optimization techniques is made in the following sections.

2.5.5.1 Optimization by Displacement of Contour Points In this procedure, a constant magnitude of field intensity is achieved in the electrode system by an iterative displacement of contour points. On the particular electrode configuration surface to be optimized, the field intensity distribution should first be determined on the proposed electrode contour, which may appear to be most nonuniform. In order to optimize the distribution of field intensity, displacing the contour points in a perpendicular direction to the surface then modifies the curvature of this contour. It is achieved step by step through an iterative process depending upon the difference in the calculated and the desired magnitudes of the field intensity. Figure 2.30 shows the block diagram of this procedure.

CSM is applied for the estimation of field intensity. Having determined the field intensity distribution for the complete electrode system, the field intensity at the location of optimization is compared with the desired field intensity. Necessary charges in the contour curvature are then introduced. The new field intensity

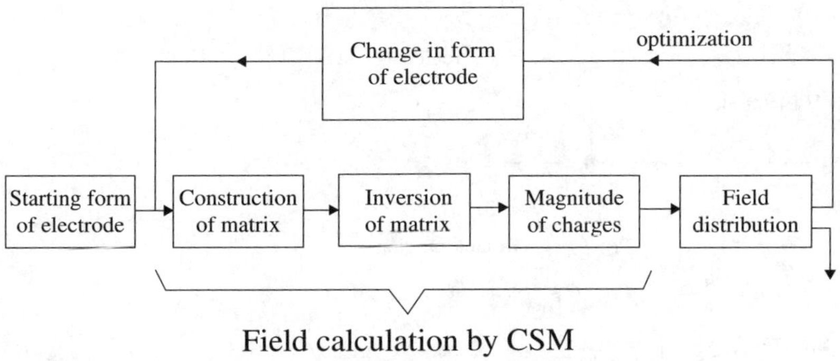

Figure 2.30 Optimization by displacement of contour points.

distribution is worked out again. After a number of iterative processes, an optimized shape of the electrodes can be obtained [2.10, 2.39]. A disadvantage of this method of optimization is that, the whole computation process of the field calculation is to be repeated for every iteration, resulting in a longer time required by the computer.

2.5.5.2 Optimization by Changing the Positions of Optimization Charges and Contour Points This procedure of optimization of electrode contour involves a direct intrusion into the simulation charge system of CSM. The region of electrode configuration requiring optimization should be identified in the whole electrode system after having determined the field intensity distribution all over the system. Normally in the charge simulation method of field estimation, selecting a number of contour points and placing a set of simulation charges of unknown magnitudes at known positions in the electrode system reproduce an identified region of electrode configuration. However, the optimization method has a slightly modified approach. Here, the optimization region (Figure 2.31) is reproduced by a proposed "*n*" number of contour "optimization points" in the same way as in CSM. This region is then provided with a set of "*n*" optimization charges, not only at fixed positions but also of known magnitudes obtained corresponding to a partial solution of the mathematical formulation. The process of optimization proceeds further with iterative steps explained as follows:

1. The optimization charges, which affect the field intensity in the optimization region more than other simulation charges, are shifted in such a way that the field intensities at the optimization points agree with the magnitudes chosen initially, not very low compared to the existing value.
2. By shifting the optimization charges, the potentials at the contour points in the region do not coincide any more with the given magnitude. The magnitudes of the other (simulation) charges must, therefore, be changed in order to fulfill the new potential requirements at the contour points.
3. The new equi-potential line having the given potential is calculated with the help of a complete fresh set of additional optimization charges. This is taken to be the new optimized contour of the electrode. However, the field intensity in the region of optimization may still not exactly coincide with the desired

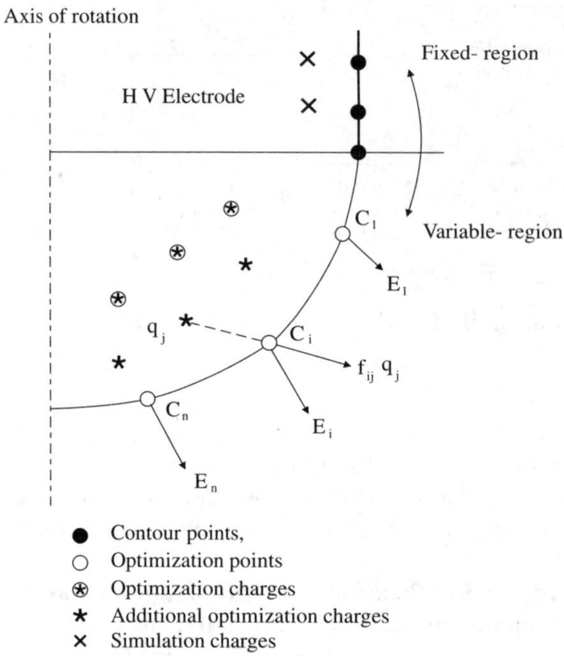

Figure 2.31 Identifying the location of optimization charges and optimizationregion in an electrode configuration.

value; that is, equation (2.86) may not be satisfied even after changing the shape of the electrode and the charge magnitudes in the other region.

4. The optimization points (Figure 2.31) should now be collocated at the newly obtained contour of the electrode. The position of the respective optimization charges must, therefore, be suitably corrected to begin anew the next iteration process.

The iterative processes are to be repeated with lower magnitudes of desired field intensities, until it converges or stops because of a too low and, therefore, an unattainable value of field intensity [2.42]. An optimum electrode contour can be obtained with the knowledge of E_i for the dielectric in use and by maximizing or keeping the E/E_i ratio to be constant in the apparatus. Figure 2.32 shows the flow-chart of the optimization process in principle.

2.5.5.3 *Optimization by Modification of "Contour Elements"* This method is based upon the qualitative correlation between the curvature of an electrode surface "C" and its electric field intensity. This correlation states that E/E_i increases with the growing electrode surface "C". Accordingly, the radii of the contour elements have to be enlarged in those regions where E/E_i is large, and is reduced where E/E_i is small [2.37].

This procedure is described by Welly [2.37] for a cylindrical rod and plane electrode configuration. The contour of a rod-end is to be optimized. Let r be the radius the rod and d its distance from the plane, Figure 2.33a. For an initial approximation, a hemispherical shape of the rod-end is chosen. Because of the symmetry

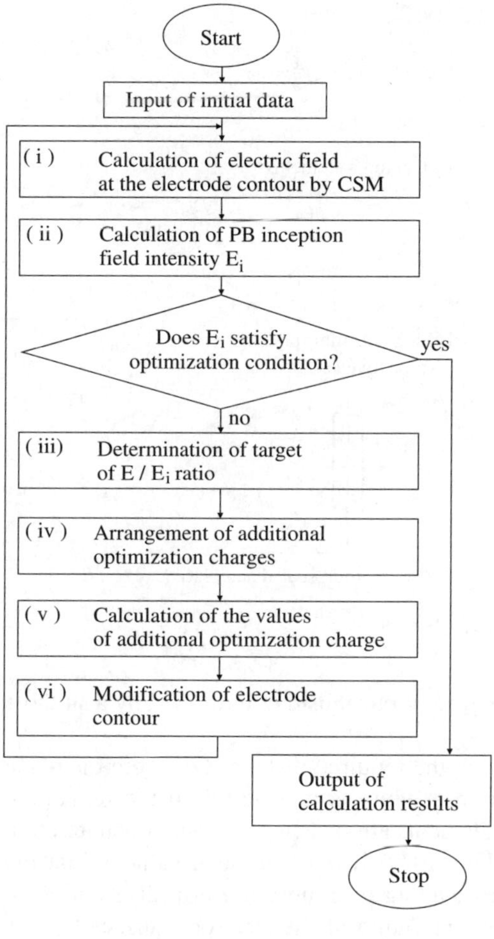

Figure 2.32 Flowchart of optimization technique taking into consideration of PB characteristic of the dielectric.

of rotation, the corresponding semicircle is divided into several sectors, Figure 2.33(b). Depending upon desired accuracy, the number of these sectors can be chosen. These arcs (parts of the circumference of the circle) on the periphery are called the "contour elements".

The first step is to calculate the mean field intensity E at each contour element. This may be done by any one of the numerical methods of field estimation. In Figure 2.33(c), contour elements are shown separately along with their corresponding magnitudes of E/E_i ratio.

In the next step, the radius of each element (which is same for all the elements in this particular case) is multiplied by their respective $(E/E_i)^n$ ratio, while the angles of the elements and therefore the shapes, remain unchanged. The exponent "n", which determines the number of iterations required for convergence, is recommended to be four by Welly [2.37]. This multiplication results in a new set of elements of different sizes, Figure 2.33(d). The new set of elements thus obtained are recomposed as shown in Figure 2.33(e), yielding a new contour. In order to attain

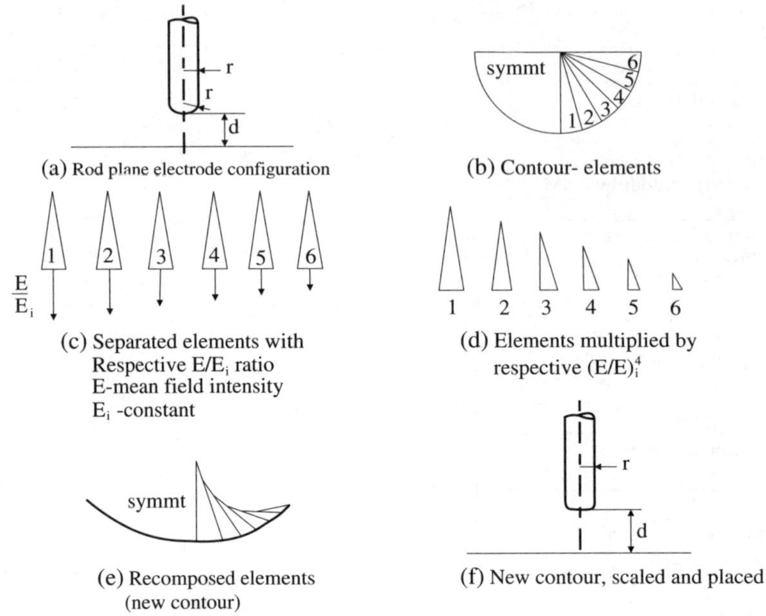

Figure 2.33 Optimization by contour elements.

the given rod radius *r* once again, this new contour must be multiplied by a suitable scale factor.

Finally, the new contour is placed at the required distance "*d*" from the plane as shown in Figure 2.33(f). The whole procedure is repeated until the remaining differences in E/E_i ratios at the contour elements are sufficiently small. Each iteration process flattens the surface regions where E/E_i exceeds its mean value, while the remaining regions become more curved leading to a more uniform distribution of E/E_i. In this process of optimization, the curvature of the electrode surface "*C* " is required only when E_i depends on it. In that case, instead of the element radii, the reciprocal of the curvature "*C*" is multiplied by $(E/E_i)^n$.

In most of the complex electrode configurations, the boundary conditions must also be taken into consideration in order to restrict the optimization to certain parts of the contour. However, even in the absence of boundary conditions, a total optimization including the shape, the position and the size is claimed to be feasible by this method.

2.6 CONCLUSION

The classification of electric fields described in this chapter is the basis for understanding the field-dependent behavior of dielectrics. The concept of "weakly non-uniform" field is unique. It is a very important concept to be taken into consideration for the highest voltage Gas Insulated Systems (GIS) and power cables. The analytical factor given by Schwaiger is another important evaluation parameter of the degree of uniformity of the field and utilization of the dielectric properties of the insulating materials. Numerical methods described are being widely applied for electric field estimation and optimization of electric fields in the highest voltage systems all over the world.

REFERENCES

[2.1] Mosch, W., Pilling, J., Hauschild, W., Eberhardt, M. and Speck, J., "Introduction to high voltage technology", Manual I-Determination of electric fields (I). Manuals for correspondence studies, VEB Verlag Technik, Berlin (1989) (in German).

[2.2] Mosch, W., Eberhardt, M., Hauschild, W., "Introduction to high voltage technology", Manual II-Determination of electric fields. Manuals for correspondence studies, VEB Verlag Technik, Berlin (1989) (in German).

[2.3] Schwaiger, A., "Contribution to the electric strength theory", Archiv für Electotechnik, 11 (1922), pp. 41–50 (in German).

[2.4] Alston, L.L., "High Voltage Technology", Oxford University Press (1968).

[2.5] Razevig, D.V., "High Voltage Engineering", (Russian) translated into English by Chaurasia, M.P., Khanna Publishers, Delhi (1996).

[2.6] Md. Abdul Gaffar Khan, "Investigation of Insulating Properties of Vacuum under High Voltage", Ph.D Thesis, Department of Electrical Engineering, IIT Kanpur, December (1995).

[2.7] Rogowski, W., "The electrical strength at the border of a parallel plate condenser, a contribution to the theory of spark electrodes and connecting leads", Archiv für Elektrotechnik, 12 (1923), pp. 1–15 (in German).

[2.8] Rogowski, W., Rengler, H., "Parallel plate spark gap with corrected brim shape", Archiv für Elektrotechnik, 16 (1926), pp. 72–75 (in German).

[2.9] Prinz, H., "High voltage fields", R. Oldenburg Verlag, Munich (1969) (in German).

[2.10] Singer, H., Grafoner, P., "Optimization of electrode and insulator contour", 2nd Int. Symp. on High Voltage Engineering, Zürisch, (1975) Report 13.03.

[2.11] Okubo, H., Amemiya, T., Honda, M., "Borda's profile and electrical field optimization by using charge simulation method", 3rd Int. Symp. on High Voltage Engineering, Milan, (1979) Report 11.16.

[2.12] Mukherjee, P.K., Roy C.K., "Computation of electric field in a condenser bushing by fictious area charges". Int. Symp. on High Voltage Engineering, Athens, Greece, (1983) Report 12.09.

[2.13] Moon, P., Spencer, D.E., "Field theory hand book", Berlin, Springer Verlag, (1971).

[2.14] Herbert P., Neff, Jr., "Basic Electromagnetic Fields", Harper & Row, Publishers, New York, Second Edition, (1987).

[2.15] Waddicor, H., "The Principles of electric power transmission", Asia Publishing House, (1964).

[2.16] Kuffel, E., Zaengl, W.S., Kuffel, J., "High Voltage Engineering" (Fundamentals), Second Edition, Newnes (2000).

[2.17] Bayer, M., Boeck, W., Möller, K., Zaengl, W.S., "High Voltage Engineering (Theoretical and Practical basics for the application)", Springer Verlag, Berlin, (1986) (in German).

[2.18] Mnemenlis, C., Nguyen, D.H., Shirmohammadi, D., "Field optimization of nonuniform fields gaps", IEEE Transactions of PAS, Vol. 104, No. 3, March (1985), pp. 718–726.

[2.19] Shortley, G.H., Weller, R., "Numerical solutions of Laplaces' equation", Journal of Applied Physics, 9 (1938), pp. 334–348.

[2.20] Shortley, G.H., Weller, R., Darby, P., Gamble, E.H., "Numerical solutions of Axissymmetrical Problems with Applications to Electrostatics and Torsion", Journal of Applied Physics, 18 (1947), pp. 116–129.

[2.21] Akin, J.E., "Finite Element Analysis", With Error Estimators, Elsevier Butterworth-Heinemann, UK, (2005).

[2.22] Zienkiewicz, O.C., Taylor, R.L., Zhu, J.Z., "Finite Element Method", Its Basis and Fundamentals, Elsevier Butterworth-Heinemann, UK, (2005).

[2.23] Speck, J., "Computation of electrostatic fields with FEM using cubic equation", Technical Journal of TU Dresden, 28 (1979) 3, pp. 597–605 (in German).

[2.24] Steinbigler, H., "Inception field intensity and utilization factor of electrode configurations with rotation symmetry in air", Dr.-Ing. Thesis, Technical University Münich (1969) (in German).

[2.25] Singer, H., Steinbigler, H., Weiβ, P., "A charge simulation method for the calculation of high voltage fields", Trans. IEEE, PAS 93 (1974), pp. 1660–1668.

[2.26] Steinbigler, H., Haller, D., Wolf, A., "Comparative Analysis of Methods for Computing 2-D and 3-D Electric Fields", IEEE Transactions on Electrical Insulation, Vol. 26, No. 3, June 1991, pp. 529–536.

[2.27] Malik, Nazar H., "A Review of the Charge Simulation Method and its Applications", IEEE Transactions on Electrical Insulation, Vol. 24, No.1, February 1989.

[2.28] Reister, M., Weiβ, P., "Computation of electric fields by use of surface charge simulation method", Int. Symp. on HV Engg., Athens (1983) Report 11.06.

[2.29] Sato, S., Bachmann, B., "A three dimensional high speed surface charge simulation method (3D-HSSSM)", Int. Symp. on HV Engg., Athens (1983) Report 11.08.

[2.30] Abouelsaad, M.M., El Bahy, M.M., "Accurate Field Computation of Needle-Plane Gaps using an Optimized Charge Simulation Method", Conference on Electrical Insulation and Dielectric Phenomena, 2000 Annual Report Conference on IEEE Electrical Insulation and Dielectric Phenomena.

[2.31] Liu, Xiaoming, Cao, Yundong, Wang, Erzhi, "Numerical Simulation of Electric Field with Open Boundary using Intelligent Optimum Charge Simulation Method", IEEE Transactions on Magnetics, Vol. 42, No. 4, April 2006.

[2.32] Deb, Kalyanmoy, "An Introduction to Genetic Algorithms", Kanpur Genetic Algorithm Laboratory (KanGAL), www.iitk.ac.in/kangal.

[2.33] Obitko, Marek, "Introduction to Genetic Algorithms", www.cs.felk.cvut.cz/~xobitko/ga.

[2.34] Dahiya, Manjeet, "Field Estimation using Charge Simulation method (CSM) Coupled with Genetic Algorithms (GAs)", B.Tech Project Report, Department of Electrical Engineering, IIT Kanpur, April 2006.

[2.35] Singer, H., "Computation of high voltage fields with the help of surface charges", Habilitation thesis, Technical University Münich (1973) (in German).

[2.36] Girdinio, P., Molfino, P., Molinari, G., Viviani, A., "Electrode and dielectric profile optimization by Finite Element or Finite Difference Methods", Int. Symp. on HV Engg. Athens (1983) Report 11.14.

[2.37] Welly, J.D., "Optimization of electrode contours in high voltage equipment using circular contour elements", Int. Symp. on HV Engg., Braunschweig (1987).

[2.38] Digmayer, M., "Optimization of multi-electrode systems with weakly nonuniform fields", Elektrie, 36 (1982) 8, pp. 398–403 (in German).

[2.39] Singer, H., "Computation of optimized electrode geometries", International Symp. on HV Engg., Milano (1979) Report 11.06.

[2.40] Metz, D., "A procedure for optimization of fields in high voltage apparatus", Dissertaion (1979), RWTH, Aachen, Faculty of Electrical Engineering (in German).

[2.41] Metz, D., "Optimization high voltage fields", International Symp. on HV Engg., Milano (1979) Report 11.12.

[2.42] Speck, J., "A method for the optimization of high voltage electrodes on the basis of numerical field calculation", Prace Naukowe Politechnik, Warsaw, 81 (1986).

[2.43] Kato, K., Okubo, H., "Optimization of HV Electrode Contour with the highest Gaseous Insulation Performance", IEEE Transactions on Dielectrics and Electrical Insulation, Vol. 4, No. 6, Dec. 1997.

[2.44] Kato, K., Han, X., Okubo, H., "Insulation Optimization by Electrode Contour Modification based on Breakdown Area/Volume Effects", IEEE Transactions on Dielectrics and Electrical Insulation, Vol. 8, No. 2, April 2001.

[2.45] Mukherjee, P. K., Trinitis, C., Steinbigler, H., "Optimization of HV Electrode System by Neural Networks using a new Learning Method", IEEE Transactions on Dielectrics and Electrical Insulation, Vol. 3, No. 6, Dec. 1996.

[2.46] Bhattacharya, K., Chakravorti, S., Mukherjee, P. K., "Insulator Contour Optimization by Neural Network", IEEE Transactions on Dielectrics and Electrical Insulation, Vol. 8, No. 2, April 2001.

CHAPTER 3

FIELD DEPENDENT BEHAVIOR OF AIR AND OTHER GASEOUS DIELECTRICS

Behavior of dielectrics when subjected to electric stress is a very important component of insulation system design in high voltage engineering. It is not only essential to know the existing maximum field intensity in the electrode system of an equipment, but also the maximum permissible electric stress for the dielectric in use. A perfect integration of these two may provide a reliable, economical, and durable insulation system.

Uniform fields, or ideal electric fields, have been a subject of considerable theoretical and practical fundamental research. However, uniform fields are not possible in practice, and thus a satisfying if not ideal solution is sought with weakly nonuniform field conditions. Although the most prevalent field condition in the power system is extremely nonuniform, efforts are made to achieve weakly nonuniform fields in some high voltage apparatus like power cables and Gas Insulated Systems (GIS). In case of uniform and weakly nonuniform fields, it is well known that the stable Partial Breakdown (PB) does not take place since the inception of PB immediately leads to complete breakdown in the dielectric. If the voltage applied to such an electrode configuration leads to breakdown of the dielectric, and U_i and U_b are the PB inception and breakdown (global or complete) voltages respectively, then,

$$U_i = U_b \text{ (for uniform and weakly nonuniform fields)}$$

Under conditions of given electrode systems, if E would be the field intensity in uniform field configuration, and $E_{\max}$ the maximum field intensity in weakly nonuniform electrode configuration, then the mean field intensity in the dielectric at breakdown "E_b" for such electrode systems can be related as following:

$$E \approx E_{\max} \approx E_{b(mean)} \text{ (for uniform and weakly nonuniform fields)}$$

However, the phenomenon in extremely nonuniform fields is quite different. Consider equipment comprising of gaseous and solid dielectrics, having a sharp metallic electrode (by default). On applying sufficiently high voltage to the equipment, an enhanced or distorted field, say $E_{\max}$ (equipment), will result at this

High Voltage and Electrical Insulation Engineering, First Edition. Ravindra Arora, Wolfgang Mosch.
© 2011 Institute of Electrical and Electronics Engineers. Published 2011 by John Wiley & Sons, Inc.

particular location. If it is equal to the field intensity required for breakdown (E_b) of the dielectric in uniform or weakly nonuniform fields mentioned above, PB would incept at this particular location having highest field intensity in the equipment.

Stable PB, a repetitive process, occurs only under extremely nonuniform field conditions. On increasing the applied voltage, the intensity of PB activity is also enhanced. Complete breakdown of the dielectric takes place at much higher applied voltage. If U_i is the voltage at which the local PB begins (PB inception voltage) and U_b is the voltage required to cause a global breakdown in the dielectric under these conditions, then in this case,

$$U_i << U_b \text{ (for extremely nonuniform fields)}$$

To achieve better use of dielectrics and to reduce the size of insulation systems, often high voltage apparatus are designed so that the rated service voltage is higher than the PB inception voltage U_i of the apparatus. Under such conditions, PB would occur in the equipment during normal service conditions. However, a breakdown of insulation is not likely to take place inspite of the existing maximum field intensity E_{max} (equipment) being higher than the electric strength E_b of the dielectric in uniform fields. For an optimum design of insulation system, it is therefore necessary to know the behavior of dielectrics, the physical process of PB and its development in the dielectrics leading to breakdown under different conditions of the field and the electrode systems.

Experimental results have shown that an analogy in breakdown processes exists not only among various gaseous dielectrics but also among gaseous, liquid and solid dielectrics. This chapter describes all the mechanisms of breakdown in gaseous dielectrics.

For the investigation of the development of breakdown process, one must distinguish among the three types of fields; namely the uniform, the weakly nonuniform, and the extremely nonuniform. While uniform and weakly nonuniform fields can be considered together due to their similar behavioral characteristics, the process in extremely nonuniform fields must be distinguished between short and long gap distances. As explained later, the breakdown processes in extremely nonuniform fields distinguish themselves not only for different electrode configurations and the gap distances, but also for the polarity and time variation of the applied voltages, direct current (dc), alternative current (ac), lightning impulse (li), switching impulse (si), etc.

Suitability of application of gaseous dielectrics in high voltage apparatus depends upon their physical, chemical, and electrical properties. The gases used as dielectrics are classified into two main groups, the electropositive and the electronegative. The electropositive gas molecules try to give up electrons forming positive ions, whereas in electronegative gases the molecules absorb free electrons forming negative ions. Pure nitrogen (N_2) was the most widely used and the cheapest electropositive gaseous dielectric. Among electronegative gases, sulphurhexafluoride (SF_6) is now used most extensively as a dielectric. The use of SF_6 gas has been continuously increasing since 1960s.

There are many other gases which have found use but to a limited extent in high voltage insulation techniques. These are, for example, oxygen (O_2), hydrogen

TABLE 3.1 Composition of the earth's atmosphere

Constituent	Percent by volume or by number of molecules of dry air
Nitrogen (N_2)	78.084
Oxygen (O_2)	20.946
Argon (A)	0.934
Carbon dioxide (CO_2)	0.031
Neon (Ne)	1.82×10^{-3}
Helium (He)	5.24×10^{-4}
Methane (CH_4)	1.5×10^{-4}
Krypton (Kr)	1.14×10^{-4}
Hydrogen (H_2)	5×10^{-5}
Ntrous oxide (N_2O)	3×10^{-5}
Xenon (Xe)	8.7×10^{-6}
Carbon monoxide (CO)	1×10^{-5}
Ozone (O_3)	upto 10^{-5}
Water H_2O (average)	upto 1

(H_2), carbon dioxide (CO_2), helium (He), neon (Ne), carbon tetrachloride (CCl_4), sodium vapour (Na) difluorodichloro methane or commercially known as Freon (CCl_2F_2) etc. However, air is the most important, freely available, and cheapest gaseous dielectric. The atmospheric air is in fact a mixture of a number of gases. Goody and Walker gave a detailed composition of earth's atmosphere in Table 3.1 in their book "Atmospheres".

The largest percentage content of atmospheric air is nitrogen (about 78%), which is an electropositive gas. The second largest constituent is oxygen (about 20%), which is a very weak electronegative gas. The content of hydrogen, an electronegative gas too, is so low (5×10^{-5}%) that for all practical purposes the air can be considered as an electropositive gas. Much theoretical as well as experimental research work, available in literature, has been performed on air to study the complicated discharge processes in gaseous dielectrics. That is why scientists have chosen air to develop a "model of discharge" or more appropriately termed as "breakdown mechanisms" for gaseous dielectrics, described in the following sections.

Insulation properties of gaseous dielectrics depend considerably upon their pressure. Investigations at different pressure conditions have been extensively carried out. A large number of units of pressure are mentioned in the literature by the authors. For convenient correlating, conversion factors for different units of pressure are brought together in Table 3.2.

3.1 FUNDAMENTALS OF FIELD ASSISTED GENERATION OF CHARGE CARRIERS

In a gas, electrons and ions are the electric charge carriers. Ions are produced from neutral molecules or atoms by ejection or attachment of an electron. Ejection of an

TABLE 3.2 Conversion factors for different units of pressure

Nomenclature	Pascal (Newton per meter square)	Bar	Torr	Pound force per square inch absolute	Millimeter of water column	Technical atmospheric	Standard Atmospheric
Symbol Unit	Pa (N/m^2)	bar	Torr (mm Hg)	psia (lb/in^2)	(mmH_2O)	at (kgf/cm^2)	atm
1 Pa	1	10^{-5}	0.750×10^{-2}	1.450×10^{-4}	0.102	1.020×10^{-5}	0.987×10^{-5}
1 bar	10^5	1	750	14.504	0.102×10^5	1.020	0.987
1 Torr	1.333×10^2	1.333×10^{-3}	1	1.933×10^{-2}	13.600	1.360×10^{-3}	1.316×10^{-3}
1 psia	6.894×10^3	6.894×10^{-2}	51.711	1	703.269	70.327×10^{-3}	68.051×10^{-3}
1 mm H_2O	9.806	9.806×10^{-5}	7.353×10^{-2}	1.422×10^{-3}	1	10^{-4}	9.676×10^{-5}
1 at	0.981×10^5	0.981	736	14.220	10^4	1	0.968
1 atm	1.013×10^5	1.013	760	14.695	0.103×10^5	1.033	1

electron from a neutral molecule leaves behind a positive ion, whereas absorption of an electron by a molecule produces a negative ion. The mass of an electron compared to that of a molecule is much lower (approximately 1/1840); therefore the mass of an ion can be considered to be equal to the corresponding molecule. Because of the lower mass, the drift velocity of an electron in atmospheric air is ~10^7 cm/sec, whereas a heavy ion only moves with a drift velocity of ~10^5 cm/sec. For a small gap distance between electrodes, ions can therefore be assumed not to have moved from the place at which they are generated.

The production of charge carriers from the neutral gas molecules is known as ionization process. There are many processes by which charge carriers are generated in gases, for example: ultra-violet irradiation, cosmic-rays, friction and movement, rubbing, electric field, corona discharge; ion, photon and metastable effects, and so on. The field-assisted ionization processes in a gas are the deciding factors leading to electrical breakdown.

One of the most significant features of the ionization process is the electron energy in different shells of a molecule. The total energy of an electron, while still attached to the molecule, can be divided into two types of energies. First the kinetic energy W_{ke}, which depends upon its mass and velocity, and second the potential energy W_{pot}, depending upon its charge in the Coulomb field of the nucleus of a molecule. These energies are given as:

$$W_{ke} = \frac{1}{2} m_e v_e^2 = \frac{1}{8\pi\varepsilon} \cdot \frac{e^2 z}{r_e} \tag{3.1}$$

and

$$W_{pot} = -\frac{1}{4\pi\varepsilon} \cdot \frac{e^2 z}{r_e} = -2W_{ke} \tag{3.2}$$

where m_e is the electron mass, v_e its velocity and ε the permittivity of the dielectric. z is the atomic number representing z electrons with negative elementary charge $e = -1.6 \times 10^{-19}$ A.s or C, lying in the discrete circular orbits r_e of an atom.

When an electron gets ejected out of an atom shell, that is, $r_e \to \infty$, the potential energy of the electron tends to be zero. Then the only energy it has is externally-acquired kinetic energy. The total energy with which an electron is attached to the nucleus of the molecule is given as follows:

$$\begin{aligned} W_{total} &= W_{ke} + W_{pot} = \frac{1}{2} W_{pot} \\ &= -\frac{1}{8\pi\varepsilon} \cdot \frac{e^2 z}{r_e} \end{aligned} \tag{3.3}$$

The binding energy of an electron in the nth shell to its nucleus is given as,

$$W_{total} = -13.61\, eV \cdot \frac{z^2}{n^2} = -W_I \tag{3.4}$$

This is the amount of energy required for releasing an electron from its molecule. Hence it is known as ionization energy "W_I" of an electron. If a free electron is

TABLE 3.3 Ionization energies for the first electron in gases

Gas	First Ionization energy $W_I(eV)$
N_2	15.6
SF_6	15.6
H_2	15.9
O_2	12.1
H_2O (vapour)	12.7
CO_2	14.4
He	24.0

absorbed by a molecule, forming a negative ion, this energy is released, known as "energy of recombination".

Gas molecules with more than one electron can be ionized multiple times. For the release of the first electron from an electrically neutral condition of a molecule, the energy required by different gases are given in Table 3.3. The energies required for the release of subsequent electrons are obviously higher.

The collision between gas particles are of two types, (i) elastic or simple mechanical collision, in which the energy exchange is always kinetic, and (ii) inelastic, in which some of the kinetic energy of the colliding particles is transferred into potential energy also.

Field-assisted generation of charge particles in gases involve mainly the following ionization process.

3.1.1 Impact Ionization

Impact or collision of particles amongst each other, accelerated under an electrical field, may lead to ionization or the formation of charge carriers from neutral gas molecules. The multiplication of charge carriers in gas takes place mainly by impact of electrons with neutral molecules known as α-process (also primary process). The positive ions make a moderate contribution to ionization only at solid insulation surfaces, which is known as β-process (also secondary process) shown in Figure 3.1.

In Figure 3.1 e is the electron charge and U_I the ionization potential of the atom or molecule and λ_e the mean free path of an electron in the direction of field intensity E.

The β-process does not play any significant role in gaseous dielectrics leading to breakdown except perhaps in case of surface breakdown or tracking. It is the α-process alone that plays the major role. When an electron gains more kinetic energy than required to cause ionization, W_I, it is capable of ionizing by impact, that is, ejecting an electron from a neutral molecule and thus leaving behind a positive ion. Electrons are in fact good ionizers of gas, while ions are not. To cause ionization, the incoming electron must have a kinetic energy greater than or equal to the ionization energy of the molecule (eU_I). However, not all electrons having gained energy $\geq (eU_I)$ cause ionization on collision with neutral molecules. Ionization by impact is actually a probability process.

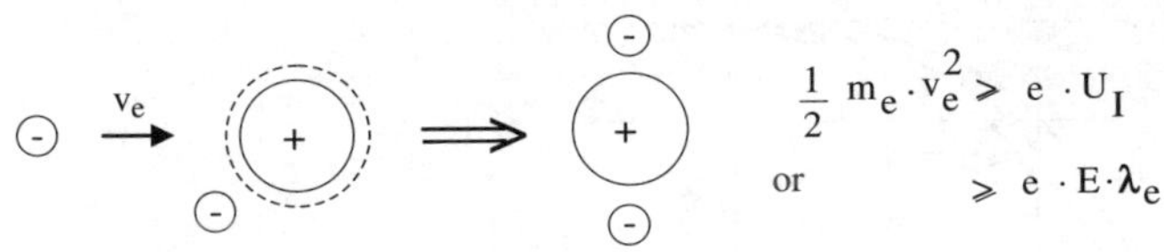

(a) Ionization by electron impact (α-process)

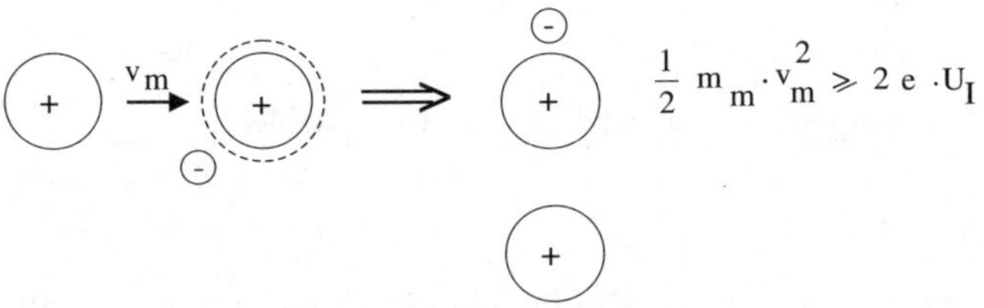

(b) Ionization by positive ion impact (β-process)

Figure 3.1 Impact ionization by electron and ion.

The mean number of ionizing collisions made by a single electron per centimeter drift across the gap (in electropositive gases) in a uniform field is denoted by α. Defined by Townsend, it is known as the Townsends's first or "primary ionization coefficient", which represents basically a probability process. For gas discharge, it is a very important coefficient strongly dependent upon the electric field intensity,

$$\alpha = f(E) \tag{3.5}$$

3.1.2 Thermal Ionization

If a gas is heated to a very high temperature, of the order of 10,000 K and above, the gas atoms or molecules acquire high velocity to cause ionization on collision with other atoms or molecules. The ionization may not take place on first impact, but mostly over a number of stages of excitation. The molecules excited by photon radiation also affect the ionization process. Thermal ionization is the principal source of ionization in flames arcs and to some extent in leader breakdown channels.

Saha [3.1] derived an expression for the degree of ionization θ in terms of gas pressure P and absolute temperature T, with the assumption that under thermodynamic equilibrium conditions, the rate of new ion formation must be equal to the rate of recombination. The expression is given as follows:

$$\frac{\theta^2}{1-\theta^2} = \frac{2.4\times10^{-4}}{p} T^{5/2}\, e^{-W_I/kT} \tag{3.6}$$

where P is the pressure in Torr, W_I the ionization energy of the gas, k Boltzmann's constant ($k = 1.38 \times 10^{-23}$ J/K), θ the ratio n_I/n, that is, the number of ionized to total particles, and T the absolute temperature in K. Equation 3.6 shows a strong dependence of θ on temperature T. θ will be negligible at room temperature. Thermal ionization becomes significant only at temperatures above 10,000 K, as shown in Figure 3.2, by Rieder [3.2].

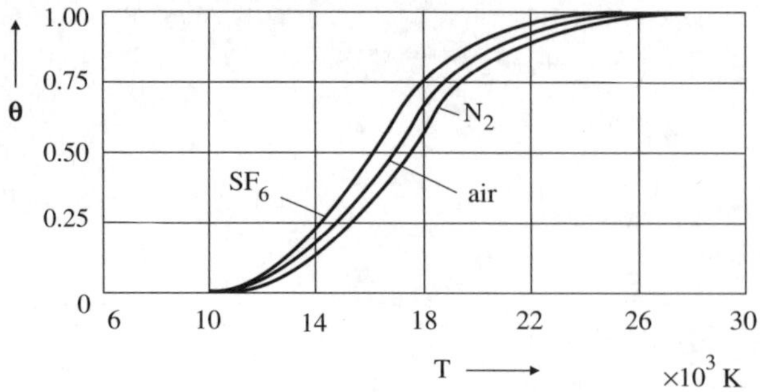

Figure 3.2 Degree of ionization of thermally ionized gases at 1 bar, Rieder [3.2].

During electrical breakdown in gases, thermal ionization has its significance only towards the final stage of breakdown. Because of the transformation of large amount of energy in the electrically conductive channel (known as "leader") towards the final stage of breakdown, an exceptionally high temperature rise in the core of the leader channel is possible.

3.1.3 Photoionization and Interaction of Metastables with Molecules

The continuation of the impact ionization process is possible only when the energy exchanged in collision is more than the energy required for ionization (eU_I). The electrons having energy lower than the ionization energy may excite the gas molecules to higher energy states on collision. Under this condition, an electron is raised from a lower energy level to a higher one. On recovering from the excited state in 10^{-7} to 10^{-10} sec., a molecule radiates a quantum of energy of photon $(h\nu)$. This energy in turn may ionize another molecule whose ionization potential energy is equal to or less than the photon energy. This process is known as photoionization and may be represented as $A + h\nu \rightarrow A^+ + e$, where A represents a neutral atom or molecule in the gas and $h\nu$ the photon energy. For photoionization to occur, the following condition must be met,

$$h\nu \geq e \cdot U_I$$

or the photon wavelength λ is,

$$\lambda \leq \frac{c_0 h}{e \cdot U_I}$$

as

$$\frac{c_0}{\lambda} = \nu$$

where c_0 is the velocity of light ($c_0 = 2.998 \times 10^8\ m/s$), h is Planck's constant $h = 6.63 \times 10^{-34}\ J \cdot s$) and ν is the frequency of light.

The photo excited hot electrons are luminescent. The energy level of the photo excited hot electrons is very high compared to their thermal energy level. These electrons relax via emission of longitudinal optical phonons.

Only a very strong radiation of light quantum (photon) having a short wavelength of less than 65 to 100 nm, can cause photoionization of a gas. In summary, the basic requirement for photoionization to occur is that the quantum energy of electromagnetic radiation must be greater than the ionization energy of the gas. If the photon energy is less than (eU_I), it may still be absorbed by the molecule and rise to a higher energy level. This process is known as photoexcitation.

For certain gas molecules, the lifetime in the excited state may extend to a few tens of milliseconds. This is known as "Metastable State" and the molecules under this state are referred as "Metastables". Metastables have a relatively higher potential energy and, therefore, ionize neutral particles on collision. However, the photons released by this reaction have too low an energy level to cause ionization in pure gases. They may release electrons on striking the cathode. This process is known as the cathode emission or secondary ionization.

Since the ionization caused by metastable interaction is accompanied by a time delay, it has been observed that these reactions are responsible for longer time-lags than usual in some gases.

The basic process of gas discharge is rather complicated. To consider the typical reactions with which the generation of charge carriers in gases take place, one refers mainly to the impact ionization or ionization by collision. Although the basic discharge process leading to breakdown has been explained by a number of experimental and theoretical works, it is still difficult to claim that a complete knowledge of the process is available. For a better understanding of the discharge process, certain simplifications and especially the qualitative considerations are necessary.

3.2 BREAKDOWN OF ATMOSPHERIC AIR IN UNIFORM AND WEAKLY NONUNIFORM FIELDS

The first and foremost process in a dielectric leading to breakdown is ionization. The development of an ionization process in gaseous dielectrics is, by nature, very systematic. It grows in the form of a swarm of charged particles known as "electron avalanche". The process is directly affected by the magnitude of electric field intensity in the dielectric. Depending upon the magnitude of the applied voltage, initially a conduction of current takes place. Breakdown or complete rupture of insulating properties of the dielectric is the advanced stage when an unlimited growth of conduction current is caused. This requires a vigorous ionization process. The basic discharge process leading to breakdown in weakly nonuniform fields is similar to the one in uniform fields.

Depending upon the local conditions, formation of "space charge" may take place in a dielectric irrespective of the field configuration. The space charge is nothing but the concentration of like polarity charge carriers. Since a significant distortion in a field could be caused by the space charge, a uniform field with a space charges may not remain uniform ideally. Under such conditions, the accurate knowledge of the distribution of potential and field in the dielectric is very difficult. However, to understand and analyse fields with space charges the simplest case that could be studied is a uniform field. There exists an analogy between the uniform field with space charge and extremely nonuniform fields.

3.2.1 Uniform Field with Space Charge

Field intensity and potential distribution have been analyzed in section 2.5 for electrostatic fields, that is, fields without space charge or source of charge, by solving the Laplace equation (2.20). If the volume charge density ρ_v is a function of the location and time, Poisson's equation (2.19) for the fields with space charge can be written as follows,

$$\Delta\phi = -\frac{\rho_v(x,t)}{\varepsilon} \tag{3.7}$$

The solution of equation (3.7) is extremely difficult even after making assumptions to simplify the $\rho_v(x)$ and $\rho_v(t)$. In order to understand the qualitative effect of space charge, an analysis is made, considering a simple electrode configuration of parallel plate condenser, having a constant space charge density ρ_v with respect to location and time (x, t), Figure 3.3.

Poisson's equation (2.19) for single dimension, that is for symmetrical uniform field with a gaseous dielectric having permittivity ε, is given as,

$$\frac{d^2\phi}{dx^2} = -\frac{\rho_v}{\varepsilon}$$

by integration we get,

$$\frac{d\phi}{dx} = -\frac{\rho_v}{\varepsilon}x + K_1$$

and

$$\phi = -\frac{\rho_v}{2\varepsilon}x^2 + K_1 x + K_2$$

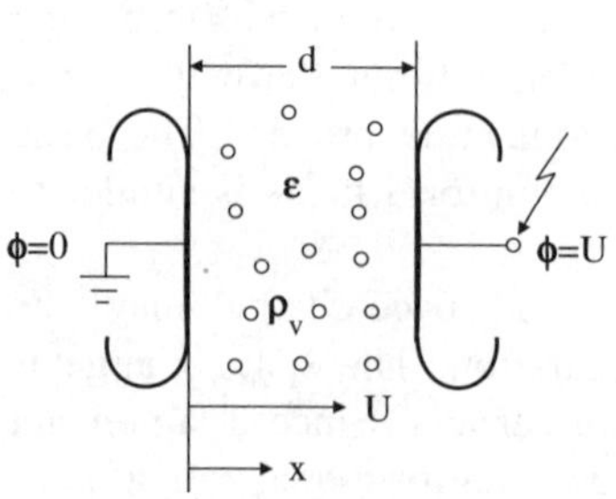

Figure 3.3 Parallel plate condenser with constant space charge density.

where K_1 and K_2 are constants of integration. On applying the boundary conditions,

a.) at $x = 0$, $\phi = 0$ and $K_2 = 0$

b.) at $x = d$, $\phi = U$ and $K_1 = \frac{U}{d} + \frac{\rho_v}{2\varepsilon} d$

the equation for potential distribution reduces to,

$$\phi = \frac{U}{d} x + \frac{\rho_v}{2\varepsilon} x(d - x) \tag{3.8}$$

The field intensity given by equation (2.3) for single direction, neglecting the –ve sign is,

$$E_x = \frac{d\phi}{dx}$$

$$E_x = \frac{U}{d} + \frac{\rho_v}{\varepsilon}\left(\frac{d}{2} - x\right) \tag{3.9}$$

The plots of equations (3.8) and (3.9) derived above for a parallel plate condenser with alternatively positive and negative space charges are shown in Figure 3.4. The following conclusions may be drawn regarding the effect of space charge on the field:

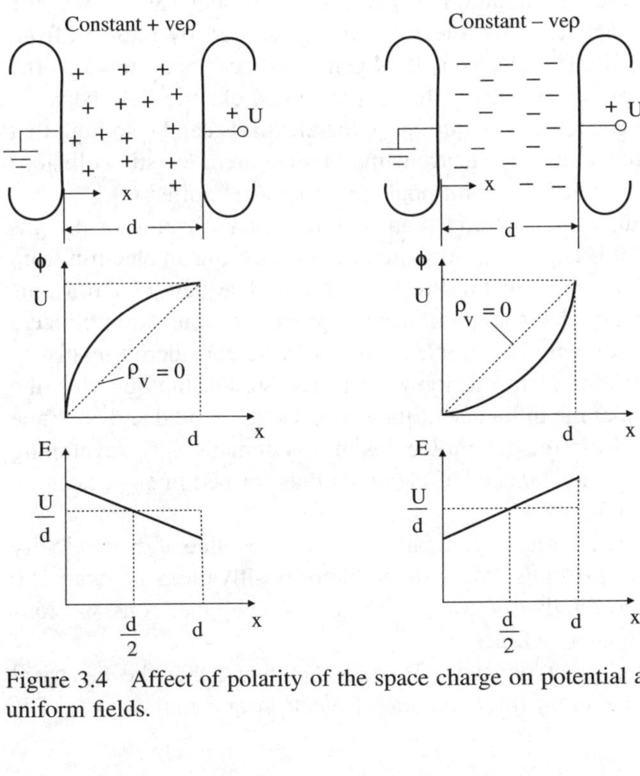

Figure 3.4 Affect of polarity of the space charge on potential and field distributions in uniform fields.

1. A field, original uniform, is affected by the space charge in such a way that the field intensity no longer remains constant.
2. A field intensity enhancement takes place towards the electrode having an opposite polarity potential compared to the polarity of the space charge.
3. A decrease in field intensity is caused towards the electrode having the same polarity potential as the space charge.

These conclusions are also valid in analogy for nonuniform fields, where strong space charges develop due to PB before the breakdown.

3.2.2 Development of Electron Avalanche

The electric field directly affects the charge carriers during the electrical breakdown of a gas. For a discharge process to begin, some charge carriers must, therefore, be present in the gas before applying the electric field. If not present, they must be generated externally, for example, by a very short wavelength light flash or by cosmic irradiation of the electrode surfaces. However, a very small number of charge carriers are produced this way and thus a considerably strong multiplication of charge carriers must be accomplished in order to achieve the required conductivity for an electrical breakdown in the gas.

Consider an electrode system with uniform field in atmospheric air. Let there be a *dc* voltage applied to this electrode system. Electrons are initially originated in the gap space either by ionization of neutral molecules, by photon irradiation from cosmic rays, or by ultraviolet illumination of cathode. At a later stage when the applied voltage gradually increases, the electrons are generated by photons from the avalanche discharge itself. The electrons thus generated accelerate towards the anode, gaining kinetic energy of movement from the applied electric field between the electrodes. The kinetic energy thus acquired by the electrons can be so high that on collision with neutral molecules the electrons may ionize them (elastic collision) or render them to a higher excited or vibrational state (inelastic collision).

When an electron gains more energy than required for ionization of the gas molecules (Table 3.3) then it is capable of ionizing, that is, ejecting an electron from the neutral molecule, it thus leaves behind a positive ion. The new electron thus ejected along with the primary electron continues the process of ionization. Since a molecule is much heavier compared to an electron, it can be considered relatively stationary, making no contribution to the ionization process. On the contrary, the electrons move very fast under the influence of the applied electric field and continue to release further electrons from the gas molecules in a systematic way developing an "electron avalanche". The "avalanche" of electrons thus formed finally reach the anode as shown in Figure 3.5.

The field intensity at which impact ionization occurs, the value of drift velocity of electron in air is usually ~10^7 cm/s, while the value of positive ions is about 150 times lower ~10^5 cm/s. Accordingly, the transit time required by electrons and ions to cross a gap differs by about 150 times.

As early as 1901, the great physicist Townsend realized that the electrical breakdown of a gas depends upon the behavior of electron and ion swarms. The

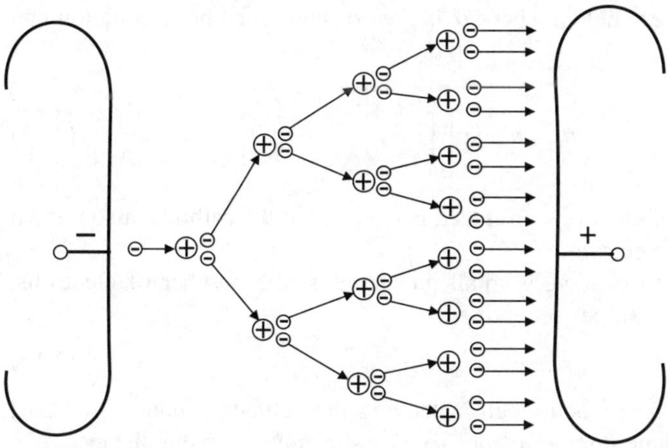

Figure 3.5 Development of an electron avalanche in uniform field.

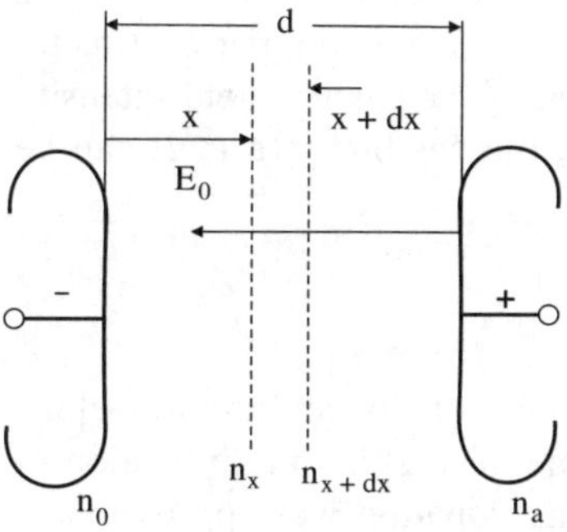

Figure 3.6 Electrons in a uniform field.

significant parameters in such swarms are the electric field intensity E, and the number of gas molecules per unit volume [2.3, 2.4]. Townsend first described the process of charge carrier multiplication in the form of an avalanche. Later he also gave the mathematical derivation of its development.

If only the process of electron multiplication by electron collision is considered in uniform field between two plates, Figure 3.6, then neglecting other processes (recombination and diffusion), the number of electrons produced by collision t an element dx, at distance x from the cathode is given by,

$$dn_x = n_x \alpha dx \tag{3.10}$$

where x is the distance form the cathode, α the Townsend's ionization coefficient, and nx the number of electrons at distance x from the cathode.

In a uniform field where the field intensity E is constant, the ionization coefficient α can be considered constant. By integrating equation (3.9) and applying the initial condition, $n_x = n_0$ at $x = 0$, the following equation is derived for a uniform field,

$$n_x = n_0 e^{\alpha x} \tag{3.11}$$

and for weakly nonuniform fields, where α is not constant, the above equation can be rewritten as,

$$n_x = n_0 \exp\left[\int_0^x \alpha dx\right] \tag{3.12}$$

where n_0 is the number of electrons produced per second at the cathode, also known as the initial number of electrons.

Therefore, in the case of very small gap lengths, the number of electrons, striking the anode (at $x = d$), per second is,

$$n_d = n_0 e^{\alpha d} \tag{3.13}$$

This means that on average, each electron leaving the cathode produces $(e^{\alpha d} - 1)$ new electrons and the same number of positive ions in traversing the distance d.

The expressions (3.11–3.13) show distinctly the exponential or an avalanche form of growth of the number of charge carriers by primary or α process.

Townsend's first ionization coefficient α (for electropositive gases) is defined as the mean number of ionizing collisions made by a single electron per centimeter drift across the gap in a uniform field. It is a function of the electric field intensity E, and at constant temperature it is dependent upon the gas pressure P. It can be proved that,

$$\frac{\alpha}{p} = f\left(\frac{E}{P}\right)$$

The coefficient α can be calculated with the help of molecular parameters. However, α is usually obtained experimentally by measuring the multiplication of electrons in high electric fields. For air, the following equation was approximated by Engelmann [3.5].

$$\frac{\alpha}{p} \approx 1.11 \times 10^{-4}\left(\frac{E}{p} - 25.1\right)^2 \tag{3.14}$$

where E is in *V/cm* in Torr and α in cm^{-1}. Equation 3.14 is plotted for α/p and E/p in Figure 3.7 at a constant temperature.

With the development of electronic circuitry and detectors capable of resolving short duration pulses, time resolved studies of transient phenomena occurring in gases between a pair of electrodes due to the development of charge carriers only became possible around 1940. Raether was able to take the first photographs of the trace of an electron avalanche in 1939 [3.6]. He used the so called "Wilson Cloud Chamber", which caused condensation of water vapour droplets on the charge carriers of an avalanche at appropriate gas pressure.

In his experiment, a short duration voltage pulse (a few tens of *ns* duration) was applied on the electrode system shown in Figure 3.8 (a,c). When the applied voltage magnitude reached its desired peak value, just sufficient to develop an avalanche, it was maintained at this level. However, a breakdown was prevented from taking place. Just at this stage, primary electrons were produced in between the

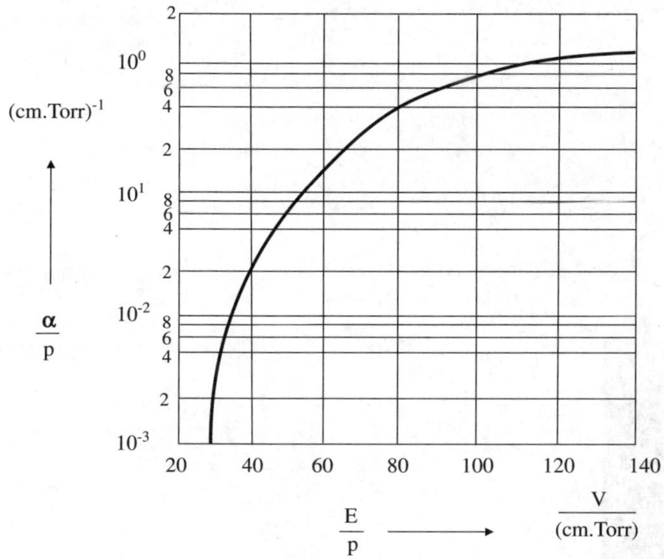

Figure 3.7 Apparent ionization coefficient α/p as function of E/p for air.

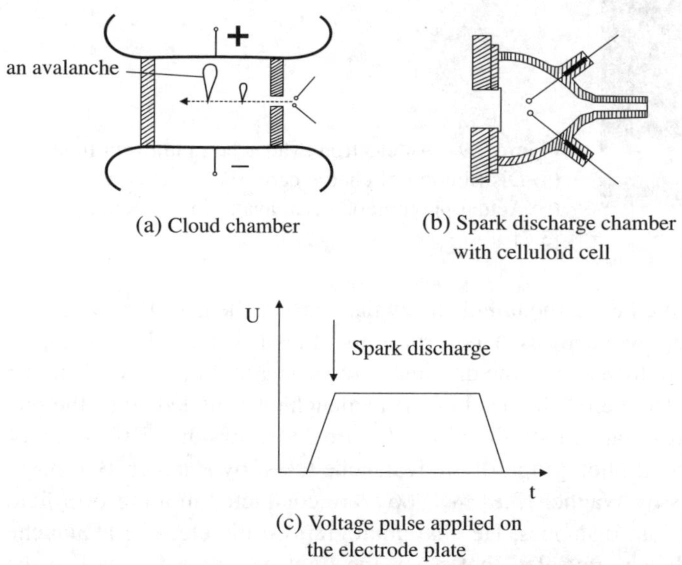

Figure 3.8 Experimental arrangement to produce an avalanche.

electrode system with the help of an external spark discharge source, as shown in Figure 3.8 (b). This gave rise to the development of an electron avalanche in the gap. In order to avoid a breakdown, the process was controlled by a steep reduction in the applied voltage to zero after the duration of a few tens of nanoseconds.

The drift velocity of electrons is about 150 times greater that of the ions. Hence as soon as the avalanche is formed, the positive ions remain practically stationary

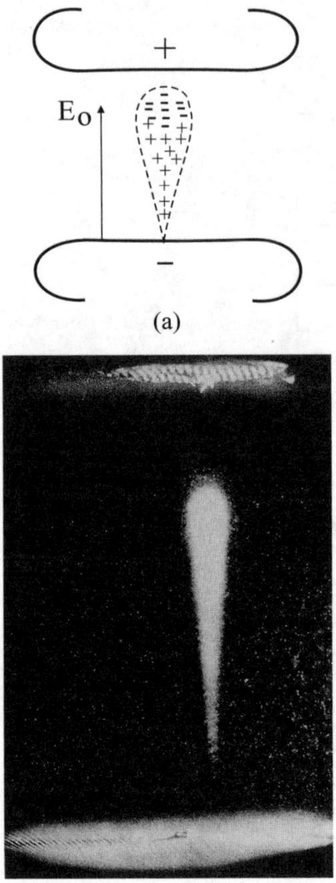

(a)

(b)

Figure 3.9 An electron avalanche in uniform field. (a) Distribution of charge carrier and the shape, (b) Actual photograph of an avalanche by Raether [3.7].

where they are produced, i.e., at the tail of the avalanche. The head of the avalanche is consequently built up by electrons. The form of the track is wedge shaped, apparently due to the thermal diffusion of the drifting electron swarm having acceleration in the direction of electric field. The head of the avalanche is rounded since the diffusion of electrons takes place in all directions. Figure 3.9 shows the distribution of charge carriers and actual photograph of an avalanche taken by Raether [3.7, 3.8].

The experiments by Wagner [3.8] in 1966 were conducted in a uniform field having slightly longer gap distances. He also photographed the electron avalanche in a cloud chamber. It was revealed that when the light was first detected in the chamber, the number of charge carriers in the avalanche, that is the space charge, was insufficient to cause a distortion in the applied electric field E_0. The center of the electron cloud moved at the electron drift velocity corresponding to the field E_0. The first light thus detected was described as "primary avalanche", shown in Figure 3.9 (b). The distribution of electrons in the gap is caused by diffusion. Through these experiments, it could be concluded that only beyond a particular development of the avalanche, when the total number of electrons reach ~10^7 to 10^8, the space charges of ions and electrons in the avalanche become strong enough to distort the applied electric field E_0.

The form of an electron avalanche reveals that positive and negative space charges are built within the avalanche itself. It is also evident from Figure 3.9 and equation (3.11) that the space charge concentration increases exponentially with the avalanche length.

The electric field E_a, developed due to space charges in an avalanche in free space, is shown in Figure 3.10 (a). According to the concept explained in Section 3.2.1 for uniform field with space charge, a field intensity enhancement is expected to develop at the border of the opposite polarity space charges. When such an avalanche is produced between two electrode plates having a uniform field E_0 applied externally, the field at the avalanche head is diverging, directed radially towards the electrode as shown in Figure 3.10 (b). This figure also shows the resultant field distribution including the field distortions caused to the existing uniform field E_0 due to an electron avalanche.

Similar to the case of a uniform field with space charge explained earlier, the following conclusions can be drawn here;

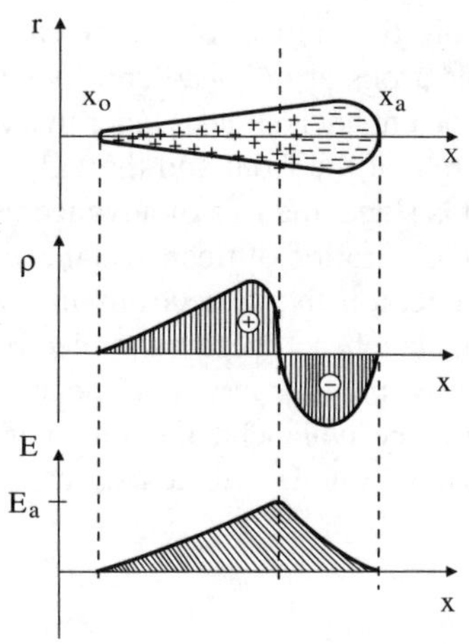

(a) Space charges and field due to electron avalanche in free space.

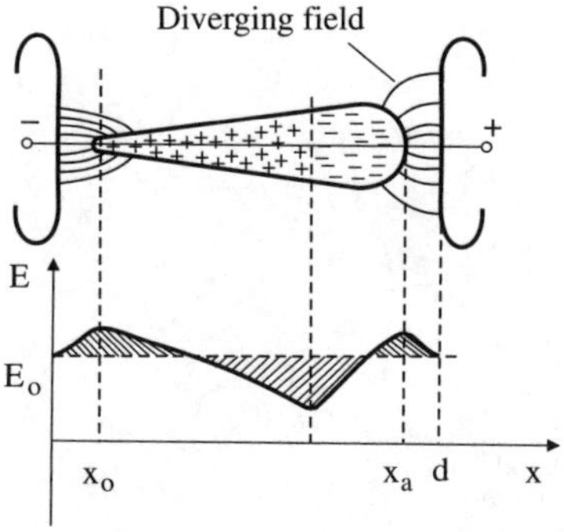

(b) Distortion of field by avalanche.

Figure 3.10 Effect of electron avalanche in uniform field.

a. Compared with the applied field E_0, the field intensity at the head and the tail of the avalanche is increased due to the presence of space charges.

b. Within the avalanche, the field intensity is reduced between the opposite polarity electron and ion clouds.

The enhancement of field intensity at the head and the tail of an avalanche can initiate a new type of discharge process, known as streamer or Kanal mechanism, which will be dealt with in the next section.

3.2.3 Development of Streamer or "Kanal Discharge"

The growth of charge carriers in an avalanche in uniform field E_0 is exponential, as shown by equation (3.11) and (3.12). However, this hypothesis is valid only as long as the electrical fields due to the space charges of electrons and ions are negligible compared to the applied external field E_0. This is true as long as the gap distance "d" between the electrodes is very small, in mm range. If the gap distance is increased, the avalanches are able to acquire bigger sizes, developing stronger space charges. When the space charge concentration (the number of charge carriers in an avalanche n_0e^{ax}) exceeds ~10^7 to 10^8, the electrons at the head and ions at the tail of an avalanche find themselves in a state of an enhanced field. Under this condition, the ionization process proceeds at a higher rate at the front and the tail of the avalanche, but at a lower rate at the center. At this stage, the rate of advancement of the avalanche towards the anode increases by the generation of fresh avalanches because of the enhanced field intensity in the upper region. New electrons in the vicinity of the primary avalanche produce fresh avalanches. These new electrons could be produced by photo ionization, caused by the primary avalanche. The new avalanches thus formed at the head of the primary avalanche are known as "secondary avalanches". This development is known as "anode-directed streamer", shown in Figure 3.11.

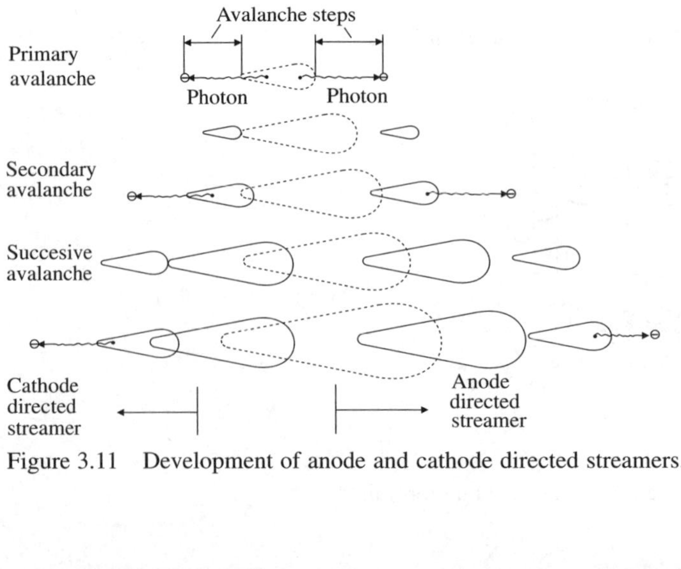

Figure 3.11 Development of anode and cathode directed streamers.

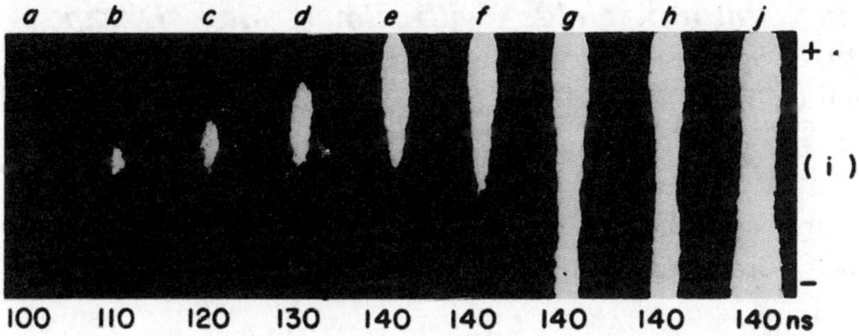

Figure 3.12 Development of streamer in nitrogen at 200 Torr, Chalmer [3.12].

Simultaneously, at the tail of the primary avalanche, a favourable condition arising due to positive ion space charge extends the ionization process backwards, giving rise to the "cathode-directed streamer". Obviously, this requires a sufficient number of electrons also at the cathode side of the primary avalanche. The earlier theories by Raether and Loeb [3.6, 3.9 and 3.10] described that the rapid development of primary avalanches required production of electrons by a process such as photo ionization. However, detailed quantitative investigations by Davies et al. [3.11] in 1971 revealed that the secondary electrons produced from the cathode by "photo-electric effect" can account adequately for the development of cathode directed streamer velocities that have been measured experimentally.

Details about the development of streamer discharge were obtained by the measurement of the light output from a developing discharge, supplemented by observations on avalanches in cloud chambers by Raether [3.7] shown in Figure 3.9(b). Similar photographs of space charge accelerated by the development of streamer avalanche process in nitrogen at 200 Torr were taken by Chalmers et al. [3.12], shown in Figure 3.12. These photographs were taken at short intervals of time between 100 and 140 ns after the application of the voltage.

Experiments performed by Chalmers showed that the light fronts, which could be photographed, move towards both the anode and the cathode with a very high velocity (>>10^7 cm/sec, the drift velocity of electrons at which the impact ionization occurs) [3.13]. The region from which the illumination originates has a form of a narrow filamentary channel. This rapidly developing filamentary luminous channel was given the name "streamer".

3.2.4 Breakdown Mechanisms

The electrical breakdown of a gas is possible only when a highly conductive channel is formed between the two electrodes. In order to achieve this, vigorous ionization processes take place to produce a large number of charge carriers in the gas. The breakdown in uniform and weakly nonuniform fields can take place involving avalanche process of below or above critical amplification, depending upon the given conditions of electrode shapes and gap distances. The breakdown mechanisms are accordingly distinguished as the Townsend and the Streamer mechanisms. The avalanche process described above remains, however, the base for both of the mechanisms.

3.2.4.1 Breakdown in Uniform Fields with Small Gap Distances (Townsend Mechanism) When the gap distance d between two electrodes in a uniform field is very small (in mm range), α the Townsend's first ionization coefficient, which is a function of field intensity E, may still have quite a low value even at the breakdown field intensity. Under these conditions, the avalanche space charge concentration may not be able to acquire its critical amplification (the total number of electrons $\approx 10^8$). In other words, only a sub-critical amplification of avalanche is able to take place.

During an electrical breakdown, a conducting discharge channel through the insulating gas bridges the electrodes. The charge carriers required in order to build this discharge channel are not only produced within the gaseous dielectric across the gap (primary or α process), but are also released from the electrode surfaces (secondary or γ-process).

As described by Meek, Craggs and Alston [3.13, 3.14], the secondary ionization processes are a consequence of the transfer of energy gained by the electrons in an avalanche. These secondary processes are ionization of the gas caused by positive ions, photons and the excited molecules. Ejection of electrons from the cathode is explained by the following effects:

a. **Positive ion effect "γ_{ion}":** While the positive ions produced in the primary avalanche cannot gain enough kinetic energy in the electric field to ionize molecules, they may have sufficient potential energy to cause ejection of electrons upon striking the cathode.
b. **Photon effect "γ_p":** Excited molecules in the avalanche may emit photons on returning back to their ground state of the energy level. This radiation falling on cathode may produce photo-emission of electrons.
c. **Metastable effect "γ_m":** Molecules in metastable condition may diffuse back to the cathode and cause electron emission on striking.

At low pressures of gas, the contributions made by positive ions and photons to the process of ionization are believed to be insignificant. The three processes of cathode effect are described quantitatively by a coefficient "γ" as follows,

$$\gamma = \gamma_{ion} + \gamma_p + \gamma_m$$

"γ" is known as Townsend's secondary ionization coefficient. It is defined as the number of secondary electrons on an average produced at the cathode per electron generated by the primary process, that is, per ionizing collision in the gap. "γ" strongly depends upon the cathode material and it is a function of field intensity and pressure of the gas,

$$\gamma = f\left(\frac{E}{p}\right)$$

Like α, γ also represents a probability process. If the mean number of secondary electrons, per avalanche produced, are μ, then considering the equation (3.13),

$$\mu = \gamma\left(e^{\alpha d} - 1\right) \tag{3.15}$$

If the primary electron generation process begins with n_0 number of electrons, the second generation begins with μn_0 number of electrons.

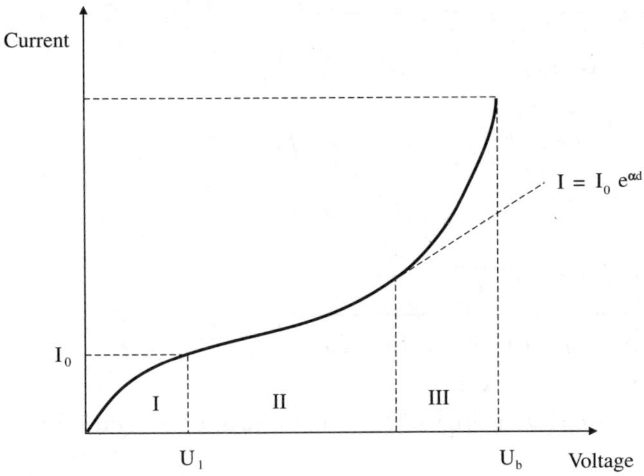

Figure 3.13 A general schema of the development of conduction current up to breakdown.

At this stage, it would be interesting to learn the growth of pre-breakdown average conduction currents in small gaps within uniform fields. Townsend first performed such experiments as early as 1914 [3.15] and later similar experiments were conducted by many physicists such as Rees, Kluckow, Bandel and Hoger etc. [3.16, 3.17, 3.18, 3.19]. These experiments have been performed in different gases like air, oxygen, hydrogen, nitrogen and its mixture with CH_4 under varying conditions, Raether [3.7]. From the equations developed with theoretical considerations and confirmed by experimental measurements, average current growth through the gap with respect to the applied voltage and time are shown in Figures 3.13–3.15.

In order to measure the U-I characteristic, Townsend's original experimental arrangement had uniform field electrodes enclosed in a glass vessel. This vessel was provided with a quartz window for irradiating the cathode with ultraviolet light to emit photo electrons [3.14]. As the voltage applied is raised, the initial current through the gap increases slowly to a value I_0. The magnitude of this current depends upon the ultraviolet illumination level of the cathode, region I, shown in Figure 3.13. The electrons emitted from the cathode move through the gas with an average velocity determined by their mobility at the existing field intensity in the gap. The initial increase in current is followed by an approach to saturation because some of the electrons emitted from the cathode return to it by diffusion. The proportion of electrons that diffuse back decreases as the voltage is increased, but not all the electrons emitted reach the anode, even at the voltages at which impact ionization in the gas begins to occur. Thus, in general, there is no well-defined plateau in the U-I characteristic, and the current eventually increases rapidly through the regions II and III with increasing voltage until a breakdown occurs at some well defined voltage $U = U_b$ [3.13]. Whatever may be the level of initial illumination of cathode, the voltage U_b at which the breakdown occurs remains unaltered.

The increase in current in region II is derived from the process of field intensified impact ionization by primary or α-process. The secondary or γ-process accounts for the sharper increase in current in the region III and also for the eventual spark breakdown of the gas. Figure 3.14 shows the U-I characteristic in helium at a

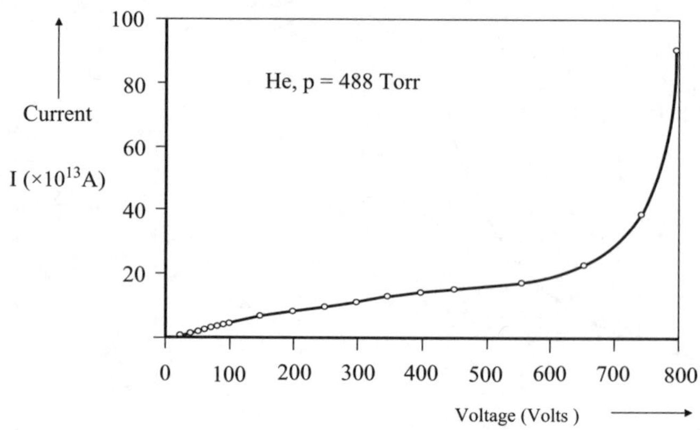

Figure 3.14 Voltage-current (U-I) characteristic in helium measured by Rees [3.16] in 1963.

pressure of 488 Torr, measured by Rees [3.16]. It is evident from Figure 3.13 and 3.14 that the characteristics measured by Townsend and Rees are very similar.

It is a distinguishing feature of breakdown that the voltage across the gap drops in the process that produces a high conductivity channel between the electrodes. This takes place in a very short time (in μ secs). Investigation of the mechanisms through which this high conductivity is built has been a challenging problem for physicists working on gas discharge. The mechanism of breakdown described by Townsend is based upon the space charge of the slow positive ions, accumulated in the gap by photo-effect at the cathode. This space charge produced by the secondary electrons and the rapidly succeeding avalanches favors in general the ionization conditions for the electrons as well as the rapid growth of current. Experimental investigations using an oscilloscope of the growth of current to breakdown with time have given a good insight into this development and also clarified the mechanism.

If α is the primary ionization coefficient for the applied uniform field E_0, the amplification of ionizing collisions of the electrons is given by $e^{\alpha x}$. In the event of positive ion space charge distorting the field, the amplification of α which increases with distance and time is given as,

$$\exp\left[\int_0^d \alpha(x,t)\,dx\right]$$

For this case, $\mu(t)$ can be written as,

$$\mu(t) = \gamma\left\{\exp\left[\int_0^d \alpha(x,t)\,dx\right] - 1\right\}$$

which grows continuously above 1, until breakdown occurs.

Experiments have been performed to study the current growth in uniform fields in air and other gases initiated by a single electron ($n_0 = 1$) with the help of a light flash. The process with higher number of electrons ($n_0 >> 1$) is achieved by illuminating the cathode with constant intense light [3.18–3.20].

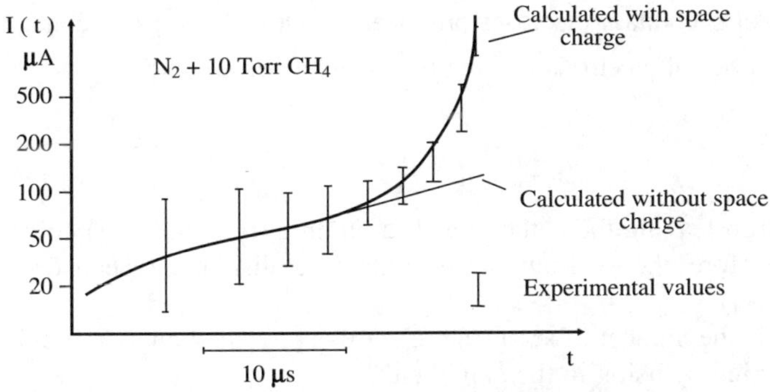

Figure 3.15 Current growth initiated by a single electron, (E/p_{20} = 44.8 V/cm.Torr, d = 2 cm, αd = 12.3, μ = 1) on semilogarithmic plot by Hoger [3.19] 1963.

Hoger [3.19] developed one such method of measurement in nitrogen. With the help of theoretical considerations confirmed with experimental results, he computed the current growth started by a single and more number of electrons as shown in Figure 3.15. The vertical lines indicate the measured values. At lower values of current, the statistical distribution scatters more compared to the distribution at higher current values. Mielke [3.20] and Schröder [3.21] also measured similar results in air.

In the same manner as shown in Figure 3.13, the growth of current I_0 corresponding to a constant illumination can be evaluated by assuming reasonable values of μ in various regions of the current plot. It can be interpreted that at low values of current, the growth is almost free from the influence of space charge (only α-process), whereas at higher values, the space charge effect of positive ions (γ-process) is responsible for the development of current with time, Raether [3.7].

The movement of charge carriers (electrons and positive ions) in the gap is responsible for the growth of circuit current. The saturation level of the curve in Figure 3.13 (region II) is also called the steady state region. If it is assumed that the number of positive ions diffusing per second at the cathode are just equal to the number of newly-formed electrons arriving at the anode, the steady state growth of electronic current in this region can be given by,

$$I = I_0 e^{\alpha d} \tag{3.16}$$

where I_0 is the initial photo-electric current at the cathode.

Although this expression represents the combined effect of a number of avalanches, it fails to signify the breakdown.

The current in region III has a much steeper rise until breakdown occurs, which is rendered to the γ or secondary process. The Townsend's current growth equation in this region is derived in [3.15] as follows.

Let

n_0 = the number of electrons emitted by primary process from the cathode (at $x = 0$) per second. In other words, n_0 avalanche develop in uniform field at the cathode with initial electrons.

n_0' = the number of secondary electrons produced at the cathode per second.

n_0'' = the total number of electrons leaving the cathode per second.

Thus

$$n_0'' = n_0 + n_0' \tag{3.17}$$

Since each electron leaving the cathode makes on an average $(e^{\alpha d} - 1)$ collisions in the gap d, therefore, the total number of ionizing collisions per second in the gap will be $n_0''(e^{\alpha d} - 1)$.

By definition, γ is the number of secondary electrons produced on an average at the cathode per ionizing collision in the gap, then,

$$n_0' = \gamma n_0''(e^{\alpha d} - 1) \tag{3.18}$$

Submitting equation (3.18) in (3.17), we have

$$n_0'' = n_0 + \gamma n_0''(e^{\alpha d} - 1)$$

or

$$n_0'' = \frac{n_0}{1 - \gamma(e^{\alpha d} - 1)} \tag{3.19}$$

From equation (3.13), the number of electrons arriving at the anode is given by,

$$n_d = n_0'' e^{\alpha d}$$

By putting the value of n_0'' in the above equation, we have

$$n_d = \frac{n_0 e^{\alpha d}}{1 - \gamma(e^{\alpha d} - 1)} \tag{3.20}$$

Under the steady state conditions, the total current in the gap can therefore be given by,

$$I = \frac{I_0 e^{\alpha d}}{1 - \gamma(e^{\alpha d} - 1)} \tag{3.21}$$

This equation describes the growth of average current in the gap before the breakdown. According to equation (3.15), $\gamma(e^{\alpha d} - 1)$ in the denominator of equation (3.21) represents μ, the mean number of secondary electrons produced per avalanche. For $\mu << 1$, the secondary ionization, or γ-process is insignificant. Then equation (3.21) reduces to $I \approx I_0 e^{\alpha d}$, which represents the region II in Figure 3.14. The applied voltage and hence the field intensity is low in this region. This condition is also described as "non-self-sustaining conduction", under which a breakdown would not be able to develop by itself.

As the applied voltage, and thus the field intensity E is increased, the value of μ approaches 1. Then the denominator of this equation approaches zero and, therefore, the current I tends to rise unlimitedly. At this stage, the current is however limited by the impedance offered by the power supply and by the gas itself. Under these conditions, the conduction process becomes self-sustained to maintain the level of required charge carriers, described as a "self-sustaining process". The quantitative condition for breakdown can be expressed as,

$$\mu = \gamma(e^{\alpha d} - 1) = 1 \tag{3.22}$$

This equation is known as the "Townsend Criterion" for spark breakdown of electropositive gases in uniform field. At the final stage of breakdown, the electron amplification is normally much greater than one ($e^{\alpha d} >> 1$), hence the criterion reduces to,

$$\gamma e^{\alpha d} = 1 \tag{3.23}$$

Under the condition when $\mu > 1$, a strong concentration of charge carriers grow in the subsequent "generations" of electron production. Under the effect of space charge, a highly conductive discharge channel is able to form which ultimately leads to breakdown.

These equations and conclusions are also valid for weakly nonuniform fields where the μ is defined in slightly different way as follows,

$$\mu = \gamma\left[\exp\left(\int_0^d \alpha dx\right) - 1\right] \tag{3.24}$$

Following the examples of the basic gas-discharge concepts by Townsend, Pedersen [3.22] derived breakdown criterion for weakly nonuniform fields.

In order to nullify the effect of space charge, that is, to achieve Townsend criterion ($\mu = 1$), experiments were performed by many physicists on electrodes with very small gap separations of *mm* range, [3.7]. In their experiments, avalanches with high αd were produced while maintaining a very low value of γ (of the order of $\sim 10^{-7}$).

In this way, series of avalanches with $e^{\alpha d}$ approaching a value 10^7 could be produced by photosuccessors in the gas by an alpha-particle at $\mu \approx 1$. At this stage, no effect of space charge could be observed for many (about four) generations of electron production. The space charge was found to be effective only after around ten generations of electron production.

The Townsend's mechanism of spark breakdown can, therefore, be explained on the basis of the observation that the ionization process begins with the production of a number of "series of avalanches". It extends all over the area of the gap and ultimately constricts into a spark, flashover or breakdown by giving rise to a highly conductive channel in which both electrons and ions flow to achieve the required magnitude of current for breakdown. This leads to the complete loss of insulating properties of the gas. A conceptual schematic of the breakdown mechanism is shown in Figure 3.16.

3.2.4.2 Breakdown with Streamer (Streamer or Kanal Mechanism)

Whereas the Townsend mechanism of spark breakdown depends upon many generations of avalanches, a single electron avalanche mechanism is the fundamental characteristic of the streamer concept of breakdown, Raether, Loeb [3.7, 3.23].

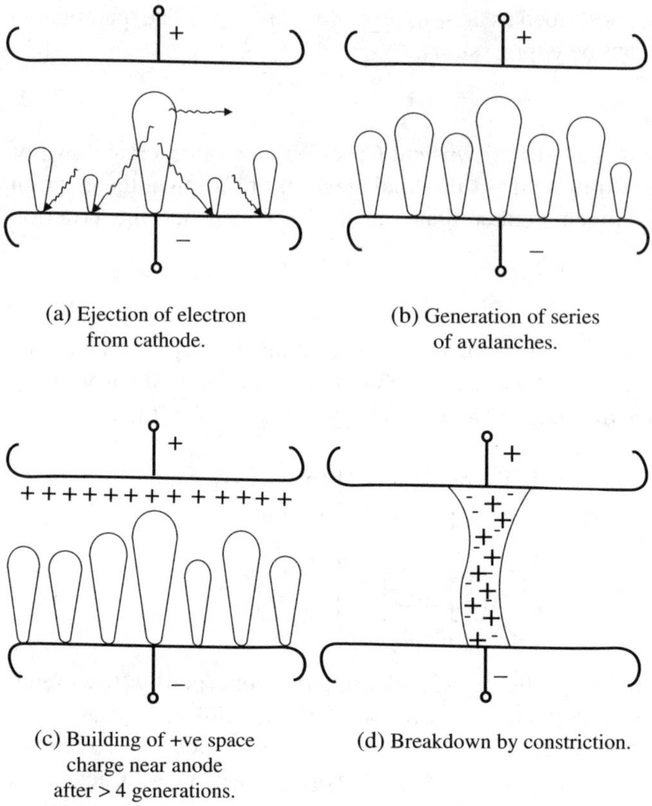

Figure 3.16 Townsend breakdown mechanism schematic.

According to Townsend's generation mechanism, the drift velocity of electrons determines the time required for breakdown by normal avalanche propagation in a few generations. The estimated time required for breakdown in comparatively longer gap distances by this theory was too long, contrary to the actual time measured experimentally. With the development in electrical and optical measurement techniques, Raether confirmed this as early as 1939 [3.6]. This necessitated a novel approach for breakdown mechanism suitable for longer gap distances in uniform fields that could not be explained by the Townsend mechanism.

As described by Raether, it is the "eigen space charge" of the avalanche itself that produces the instability. The streamer breakdown mechanism describes the development of spark breakdown directly from a single avalanche. The stark space charge development within the avalanche due to rapid growth of charge carriers transforms it into a conducting channel.

The expression "streamer discharge" derives its name from the literal meaning of streamer. The word "streamer" means a ribbon attached at one end and floating or waving at the other in wind. When a number of such long narrow strips wave together, they appear to be a set of waves or ripples moving forward. Streamer also means a column of light shooting up in an aurora. This kind of a visual display can be seen due to movement of ionized particles in pockets when partial breakdown takes place in gaseous dielectrics. The term "Kanal" has been taken from German language which means a canal or a channel.

Experimental observations were made for the development of space charge in cloud chambers by Loeb and Meek [3.10] and Raether [3.6, 3.9]. Thereupon, the analytical estimation of the conditions for the space charge field of avalanche E_a to be able to acquire a magnitude of the order of the externally applied field E_0, confirmed that the transformation from avalanche to streamer processes began to develop from the head of an electron avalanche when the number of charge carriers increased near a critical value,

$$n_0 e^{\alpha x_c} \approx 10^8$$

For an avalanche initiated by a single electron ($n_0 = 1$) in a uniform field, this corresponds to a value,

$$\begin{aligned} \alpha x_c = \alpha d_c &= \ln 10^8 \\ &= 18.4 \end{aligned}$$

where x_c is the length of an avalanche in the field direction when it amplifies to its critical size.

In other words, x_c is the critical length of the electrode gap d_c. It implies that the streamer mechanism is possible only when $d \geq x_c$. If x_c is longer than the gap length d, ($x_c > d$), then the initiation of streamer is unlikely as shown in Figure 3.17.

On the basis of experimental results and some simple assumptions, Raether developed the following empirical formula for the "streamer breakdown criterion" [3.7],

$$\alpha x_c = 17.7 + \ln x_c + \ln \frac{E_a}{E_0} \tag{3.25}$$

The interaction between the space charges and the polarities of the electrodes result in distortion of the uniform field. Field intensities toward the head and the tail of avalanche acquire a magnitude ($E_0 + E_a$), while above the positive ion region, just behind the head, the field is reduced to a value ($E_0 - E_a$), Figure 3.17. It is

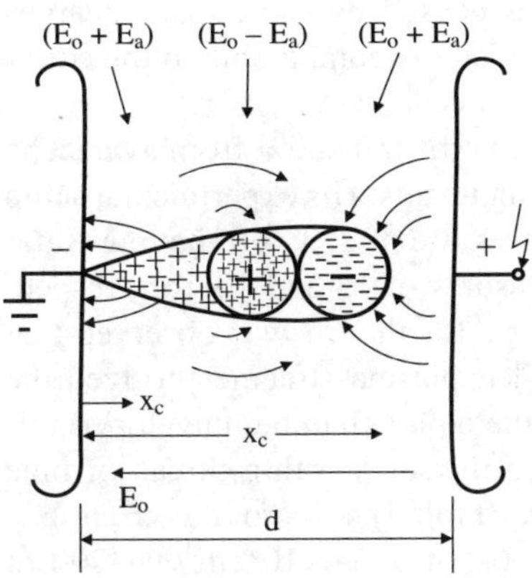

Figure 3.17 Effect of space charge field E_a of an avalanche of critical amplification on the applied uniform field E_0.

evident that the space charges within an avalanche strengthen as the length of the avalanche increases.

Raether [3.7] has also given a relation to estimate the magnitude of eigen charge field. If the cluster of any like polarity charge particles acquires a shape close to a sphere of radius r_a, then the field at the surface of this space charge in the form of a sphere can be estimated as,

$$E_a = \frac{e\exp(\alpha x)}{4\pi\varepsilon_0 r_a^2}$$

where e is the elementary charge of an electron and ε_0 the absolute permittivity constant.

The condition for transition from series of avalanche to streamer breakdown assumes that this eigen space charge field approaches nearly equal to the externally applied field $(E_a \approx E_0)$. Hence the criterion for breakdown becomes,

$$\alpha x_c = 17.7 + \ln x_c$$

The minimum value of αx_c required for breakdown in a uniform field gap by streamer mechanism is obtained on the assumption that the transition from avalanche to streamer occurs when an avalanche of critical size just extends across the gap d_c. By incorporating this condition, the criterion for streamer breakdown takes the form,

$$\begin{aligned} \alpha d_c &= 17.7 + \ln x_c \\ &= 10^8 \\ &= 18.4 \end{aligned} \tag{3.26}$$

Thus the condition $x_c = d_c$ gives the smallest value of α to produce streamer breakdown, where d_c is given in cm. For $\alpha x_c = \ln 10^8$, x_c works out to be equal to 2.01 cm. Hence, a gap distance of 2 cm between the uniform field electrodes can be considered to be minimum or critical gap distance d_c for streamer phenomenon to take place in atmospheric air. However, to make this analytical criterion figure round, $\alpha d_c = \alpha x_c = 18$ or even 20 is described in literature, Pedersen [3.22], Mosch [2.1].

The "streamer breakdown criterion" can therefore be interpreted as a condition for the development of significant field distortions caused by stark space charges within a single avalanche so that its eigen field intensity is comparable to the externally applied field.

Cloud chamber experiments performed to study the transition from avalanche to streamer by Raether [3.7] revealed very interesting results. His experimental setup consisted of a 3.6 cm uniform field gap in air at a pressure of 273 Torr, having water and alcohol vapors for condensation at partial pressures of 13 and 10 Torr, respectively. On applying a voltage pulse, as described by Raether, it was observed that when αx nearly approached a value 20, a so-called plasma streamer bridged the electrodes after about 10 to 15 ns. A dense fog in the cloud chamber enveloped this plasma streamer. Observations in the dark showed that only a thin thread of blue light was visible. Hence the streamer needed only 10 to 15 ns to cross a 3.6 cm gap distance. This amounts to a velocity of propagation of a few 10^8 cm/sec (200 to 400 cm/μsec). In this time, a normal avalanche breakdown as described by Townsend would have propagated only a few millimeters. A streamer was thus measured to grow much faster (of the order of 300 cm/μsec, that is, about 1/100 of the velocity of light), compared to the drift velocity of electrons, which is only 10 to 20 cm/μsec. Later, the initial velocity of streamer propagation was also measured between 200 to 500 cm/μsec by Dawson and Winn in 1965 [3.24]. However, it must be pointed out that such experiments require precision measurement techniques. A slight variation in the conditions may lead to a large difference in measured values.

The description of the cathode-directed streamer breakdown mechanism given by Meek [3.25] and Loeb [3.26] for uniform field is described below.

When the avalanche extends across the gap, the electrons are swept into the anode, and the positive ions in the tail of the avalanche stretch out across the gap as shown in Figure 3.18. A highly localized space charge field due to positive ions

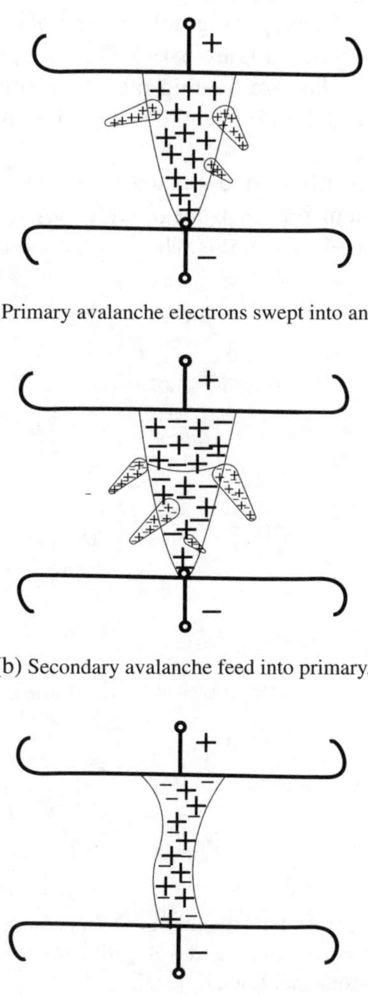

(a) Primary avalanche electrons swept into anode

(b) Secondary avalanche feed into primary.

(c) Self propagating streamer breakdown.

Figure 3.18 Schematic of a cathode directed streamer breakdown showing development stages.

is produced near the anode but since the ion density elsewhere is low, it does not constitute a breakdown in the gap. In the gas surrounding the avalanche, secondary electrons are produced by photons and photoelectric effect at the cathode. The secondary electrons initiate the secondary avalanches, which are directed toward the stem of the main avalanche. If the space charge field developed by the main avalanche is of the order of the applied field intensity, the secondary electrons are produced in the gas due to field enhancement. Thus the secondary avalanches feed into the primary avalanche as shown in Figure 3.18 (a), (b). The positive ions left behind by the secondary avalanches effectively lengthen and intensify the space charge of the main avalanche in the direction of the cathode and the process develops in the form of a self-propagating streamer breakdown shown in Figure 3.18 (c).

Significant development in high-speed photographic techniques, strengthened by the incorporation of image converters and image intensifiers have made it possible to record the progress of the discharge light ouput at earlier stages than before.

Typical results of streamer discharge progression leading to breakdown are shown in Figure 3.12 taken by Chalmer [3.12]. These photographs were obtained from discharge development in a 3 cm gap in nitrogen. Figure 3.19 shows the photograph of an avalanche where secondary avalanches are feeding into the primary avalanche. It was taken in a gap of 3.6 cm in air at 270 Torr and at a field intensity of about 12.2 kV/cm by Raether [3.7].

Pederson [3.22] has attempted to theoretically derive streamer criterion for breakdown, considering the attachment coefficient for electronegative gases (SF_6) and the nonuniformity of the weakly nonuniform electric fields. However, the quan-

Figure 3.19 Photograph of a cathode directed streamer developed from an avalanche, Raether [3.7].

titative figure ($\alpha x_c = \alpha d_c = 18$) accepted for the criterion for electropositive gases and uniform field remains unchanged.

The fundamental criteria for the electrical breakdown of gaseous dielectrics, expressed in terms of growth of ionization process, are of limited use for high voltage apparatus design/practice engineers. This, however, gives an indepth knowledge of the behavior of gaseous dielectrics.

3.2.5 Breakdown Voltage Characteristics in Uniform Fields (Paschen's Law)

In uniform fields, Townsend's criterion for breakdown in electropositive gases in given by equation (3.22), as

$$\gamma(e^{\alpha d} - 1) = 1$$

or

$$\alpha d = \ln\left(\frac{1}{\gamma} + 1\right)$$

where the coefficients α and λ are functions of E/p and are given as follows,

$$\alpha = pf_1\left(\frac{E_0}{p}\right)$$

and

$$\gamma = f_2\left(\frac{E_0}{p}\right)$$

where E_0 is the applied electric field and p the gas pressure. In a uniform field electrode system of gap distance d,

$$E_b = \frac{U_b}{d}$$

where U_b is the breakdown voltage and E_b the corresponding field intensity. E_b is equal to the electric strength of the dielectric under given conditions. When the applied field intensity $E_0 = E_b$, Townsend's criterion for breakdown in electropositive gases in uniform field can be represented in terms of the product of the gas pressure and the electrode gap separation "pd" as,

$$f_2\left(\frac{U_b}{pd}\right)\left\{\exp\left[pdf_1\left(\frac{U_b}{pd}\right)\right] - 1\right\} = 1 \qquad (3.27)$$

or

$$U_b = f(pd) \qquad (3.28)$$

This is known as Paschen's law, one of the important laws in high voltage engineering. It states that the breakdown voltage, U_b of gaseous dielectrics in uniform field electrode gap is a unique function of the product of pressure and the gap distance, pd, between the electrodes. It is derived from Townsend's theory of breakdown in gaseous dielectrics applicable to uniform electric fields. The scientist, Paschen, established it experimentally in 1889 from the measurement of breakdown voltage in air, carbondioxide, and hydrogen—all basically electropositive gases. Perhaps the property of electronegativity of gases was not known at that time. It could be shown later that this law is also applicable to electronegative gases [3.14, 3.22].

Most of the experimental work performed to verify Paschen's Law has been carried out under controlled conditions and for small gap distances between the electrodes where the degree of field uniformity, the Schwaiger factor "η" could easily be maintained at unity, a condition for uniform field. The transition from uniform field to extremely nonuniform field has a range known as "weakly nonuniform field", for which the Schwaiger factor η is ≥ 0.25 and <1 for air at atmospheric pressure. The behavior of dielectrics has been found to be very similar in uniform as well as in weakly nonuniform field configurations.

The value of α depends upon the properties of particular gas and that of γ upon the electrode material. At a low value of E/p, that is at high pressures, where a steady state can be achieved, experiments have been performed on the spatial growth of ionization in a large number of both electropositive and electronegative gases. In these experiments the conduction currents in the gap were maintained below 0.1 μA, so that the distortion of uniform fields due to space charge remained at a minimum. The results have shown that this generalized Townsend's theory of breakdown is applicable over a vide range of physical conditions. The theoretically estimated values of the breakdown voltages U_b have been found in agreement with those observed experimentally.

A schematic of the variation of U_b with respect to "pd" is shown in Figure 3.20. Equation (3.27) does not imply that the breakdown voltage varies linearly with "pd", although in practice it is found to be nearly linear over certain regions. As evident from Figure 3.20, the breakdown voltage attains a minimum value $U_{b\min}$ around a particular value of the product $(pd)_{\min}$.

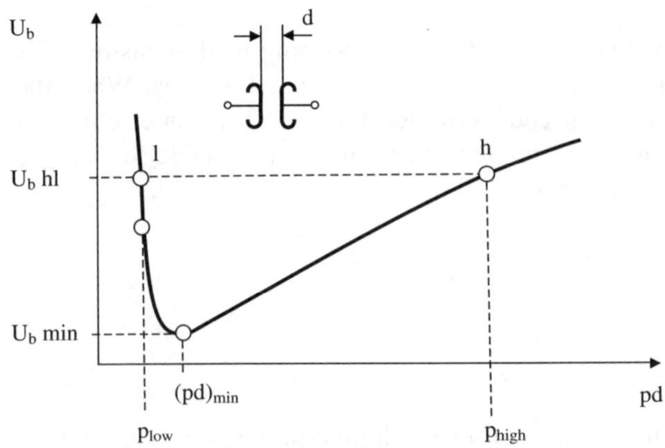

Figure 3.20 Breakdown voltage vs *pd* characteristic in uniform field (Paschen's curve).

To explain the shape of this curve, it is convenient to consider a gap with fixed spacing (d = constant), and let the pressure be decreased from a point P_{high} on the curve at the right of the minimum. As the pressure is decreased, the density of the gas decreases; consequently the probability decreases for an electron colliding with the molecules as it travels towards the anode. Since each collision results in loss of energy, a lower electric field intensity or a lower voltage suffices to provide electrons with the kinetic energy required for ionization by collision to achieve breakdown.

When the minimum of the breakdown voltage is reached and the pressure still continues to decrease, the density of the gas becomes so low that relatively fewer collisions take place. Under such conditions, an electron may not necessarily ionize a molecule on colliding, even if the kinetic energy of the electron is more than the energy required for ionization. The probability considerably reduces for making successful impact ionization by an electron. In other words, an electron has a finite chance of ionizing that depends upon its energy. The breakdown can occur only if the probability of ionization is made greater by increasing the field intensity. This explains the increase in breakdown voltage to the left of the minimum. At low pressures, P_{low}, high vacuum conditions exist; thus this phenomenon is applicable in high voltage vacuum tubes and switchgears. Under these conditions, the effect of electrode materials on the breakdown voltage also plays an important role. Paschen's law is not valid to the left of the minimum.

Many scientists have measured the characteristic values of minimum breakdown voltage $U_{b\min}$ and $(pd)_{\min}$ of Paschen's curve experimentally in different gases. These measured values for some common gases used in practice are given in Table 3.4. As explained in the literature, the measured values conform well to the theoretically calculated values so long as the field emission caused by field distortions is prevented, since Paschen's law is valid only for the uniform fields. For example, a minimum breakdown voltage as low as 64 V was measured by Cueilleron [3.27] in neon between caesium coated electrodes at a gas pressure of 26 Torr. However, on leaving the range of validity of Paschen's law, that is, at extreme

TABLE 3.4 Characteristic values of minimum breakdown voltages and the product pd

Gas	$(pd)_{\min}$ Torr.cm	$U_{b\min}$ Volts
Air	0.55	352
Nitrogen (N_2)	0.65	240
Sulphurhexafluoride (SF_6)	0.26	507
Hydrogen (H_2)	1.05	230
Oxygen (O_2)	0.7	450
Carbondioxide (CO_2)	0.57	420
Neon (Ne)	4.0	245
Helium (He)	4.0	155
Sodium Vapour (Na)	0.05	320

pressures or values of pd, the breakdown voltages strongly depend upon the electrode materials and their conditions. In other words, the breakdown voltages under such conditions mainly depend upon the work function of the cathode materials.

The validity of Paschen's Law for air in weakly nonuniform fields was experimentally investigated by Banwari [3.28]. In his experiments, the degree of uniformity of electric field, "η", between two identical spheres and calottes was kept at about 0.9, a magnitude well above its lower limit of 0.25 for weakly nonuniform fields. The experiments were carried out at atmospheric and sub-atmospheric pressures in order to also investigate the minimum point of validity of Paschen's Law in the curve, the Pachen's minimum.

The measured ac breakdown voltage U_b vs pd characteristic is shown in Figure 3.21. Beginning with atmospheric pressure at the extreme right of the curve it can be observed that the spark breakdown voltages reduce linearly up to a minimum value of "pd". This verifies the applicability of Pashen's Law.

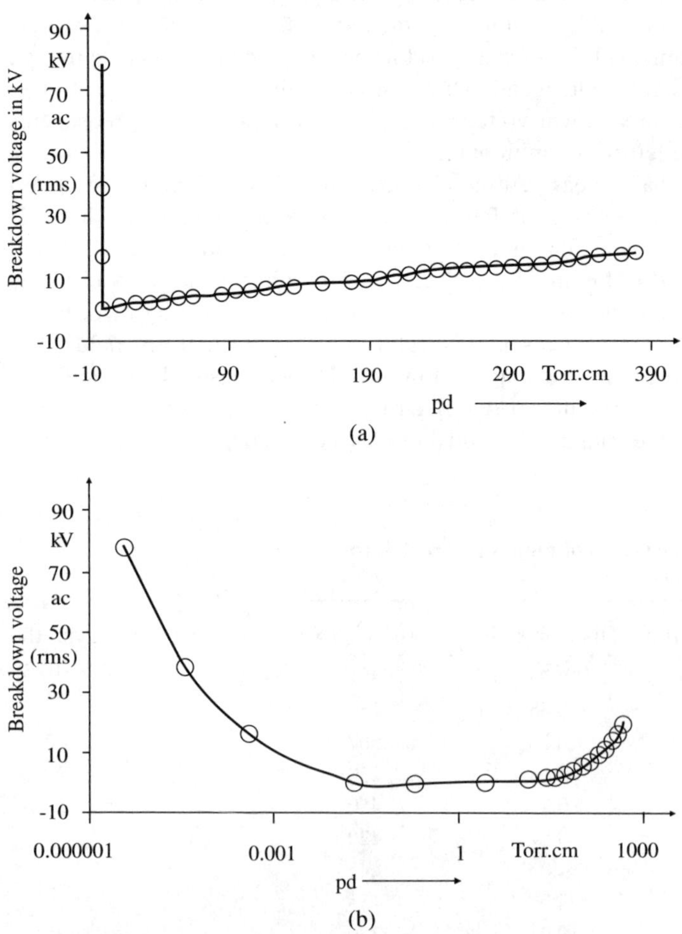

Figure 3.21 Paschen's curve measured in weakly nonuniform field (a) on linear scale for "pd" (b) on logarithmic scale for "pd".

Figure 3.22 Faraday Glow Discharge with ac power frequency voltage.

In the experiments conducted by Banwari [3.28], the normal spark breakdown ceased to occur abruptly around a pressure of 25 Torr. At pressures of 22 Torr and lower (up to 1×10^{-4} Torr), Faraday glow discharge incepted in the tube on applying the ac power frequency voltage shown in Figure 3.22. The occurrence of glow discharge is not dependent on the product "*pd*"; rather it was a function of pressure alone. Hence Paschen's Law is not valid as soon as glow discharge begins to occur while lowering the pressure.

The Faraday Glow Discharge is known to occur when cathode rays, the stream of high-energy electrons, travel through the gas gaining energy from applied electric field and losing it in collisions with the molecules. Apart from ionization these collisions also result in excitation of the gas molecules that give away light radiation (photons) when they return to their ground energy level. This results in ambient-plasma state of the gas having very high conductivity. Auroras, the spectacular displays of luminous radiation in the sky near Polar Regions, are due to ambient plasma state of atmospheric air at high altitudes. Aurora lights are emitted when atoms in the ionosphere are struck by high energy electrons coming from the sun get directed towards the Polar Regions due to the earth's magnetic field. This phenomenon can be described as Faraday glow discharge in nature.

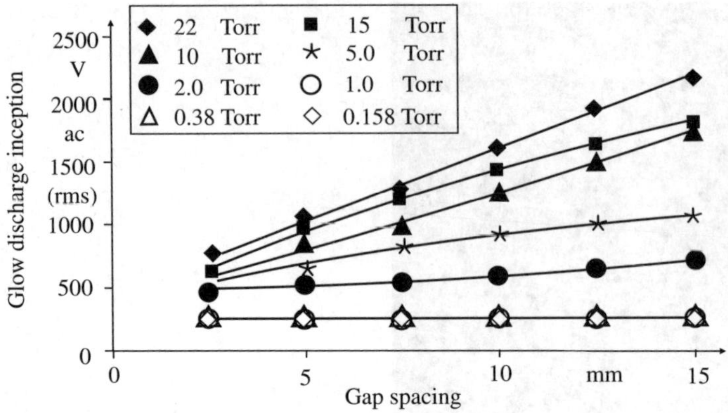

Figure 3.23 Glow inception voltage measured for decreasing pressure and increasing gap distance.

Depending upon the physical conditions, the glow discharge inception voltages could be very low, in the range of a few tens of volts to a few hundred volts. Glow discharge inception voltages measured by Banwari [3.28] for atmospheric air at different pressures are shown in Figure 3.23.

Since the transition from spark breakdown to glow discharge in a gas is a function of pressure alone, the failing point of Paschen's Law, the Paschen's Minimum, should be related to pressure alone, not to the product "*pd*", as given in Table 3.4 by Cueilleron [3.27]. The measured values of $U_{b\min}$ given in this table for various gases are not very clearly mentioned if these are for spark breakdown or for inception of glow discharge.

For a wide range of pressure and gap length, the breakdown voltage in uniform fields in air can be calculated by combining the relation given by Schumann [3.29] and the Townsends's criterion for breakdown (equation 3.22). According to Schumann, over a wide but restricted range of E/p, the ratio α/p may be expressed as,

$$\frac{\alpha}{p} = C\left[\frac{E}{p} - \frac{E_c}{p}\right]^2 \tag{3.29}$$

where E and E_c are field intensities, E_c being the limiting value of E at which an effective impact ionization begins; p is the pressure and C a constant.

The breakdown criterion $\gamma(e^{\alpha d} - 1) = 1$ can also be written as,

$$\alpha d = \ln\left(\frac{1}{\gamma} + 1\right) = K$$

or

$$\frac{\alpha}{p} = \frac{1}{pd}\ln\left(\frac{1}{\gamma} + 1\right) = \frac{K}{pd} \tag{3.30}$$

where K can be treated as being constant because γ, which is affected by cathode surface and gas pressure, has a very small value ($<10^{-2} - 10^{-3}$), and thus $1/\gamma$ is quite a large number. Therefore, K does not change much even if there is considerable change in γ. For breakdown with Townsend mechanism, K is of the order of 8 to 10. However, the exact value of K is of minor importance and may be considered to be constant for a large range of values of p and E. By combining equations (3.29) and (3.30), the following expression is obtained,

$$\frac{K}{pd} = C\left[\frac{E}{p} - \frac{E_c}{p}\right]^2$$

or

$$\frac{E}{p} = \frac{E_c}{p} + \sqrt{\frac{K/C}{pd}}$$

Thus the expression for breakdown voltage in uniform fields can be given as,

$$U_b = \left(\frac{E_c}{p}\right)pd + \sqrt{\frac{K}{C}}\sqrt{pd} \qquad (3.31)$$

The values of the constants E_c and K/C at $p = 1$ bar and temperature 20°C (normal temperature and pressure, n.t.p.) were determined by Sohst [3.30] and Schröder [3.21] as following,

$$\widehat{E_c} = 24.36 \text{ kV/cm}$$
$$K/C = 45.16 \text{ (kV)}^2/\text{cm} \cdot \text{bar}$$

Inserting these values, equation (3.31) becomes

$$\widehat{U_b} = 6.72\sqrt{pd} + 24.36(pd) \quad \text{kV} \qquad (3.32)$$

where p is given in bar and d in cm, therefore pd in bar.cm.

Calculated values of breakdown voltages using equation (3.32), for uniform field in air, were compared with the available experimental results from various authors, Dakin et al. [3.31], as shown in Figure 3.24. The peak values of the

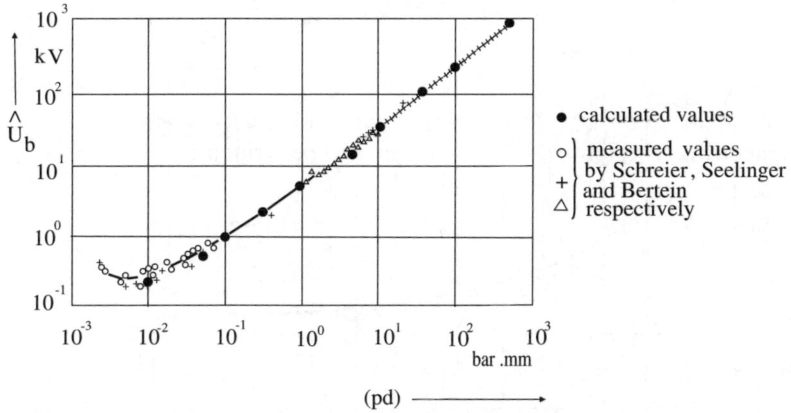

Figure 3.24 Paschen's curve for air in uniform field at temperature 20°C Dakin et al. [3.31].

breakdown voltages in kV and pd in bar.mm are plotted on a double logarithmic scale paper. As seen in this figure, the calculated and the measured results agree with each other well, except for the very low values of pd. In this region, where the E/p values are quite high due to low pressures, Schumann's quadratic equation (3.29) no longer holds good. This is the region where the air acquires ambient plasma state and the Paschen's Law does not hold good as explained above.

It is more convenient to introduce the "relative gas density", δ, a dimensionless quantity, in place of gas pressure p. The relative gas density takes care of the effect of temperature on the mean free path of electrons in the gas at constant pressure. The number of collisions made by an electron in crossing a gap is, therefore, proportional to the product δd and γ. Correction for the variation in ambient conditions of air can be made by introducing the factor of "relative air density", δ, defined as,

$$\delta = \frac{p}{760} \cdot \frac{293}{273+t} = 0.386 \frac{p}{273+t} \tag{3.33}$$

where p is in Torr and t in °C.

At normal temperature ($t = 20^o C$) and pressure ($p = 760$ Torr), δ is equal to one. By introducing δ in place of p in equation (3.32), this equation can be rewritten for normal temperature and pressure as,

$$\hat{U}_b = 6.72\sqrt{d} + 24.36\,d$$

or

$$\hat{E}_b = 24.36 + \frac{6.72}{\sqrt{d}} \quad \text{kV/cm} \tag{3.34}$$

It is interesting to note that even in uniform field at constant pressure and temperature, the electric strength of air is not constant. It tends to ≈ 24 kV/cm for long gap distances of the order of about 10 cm and greater. The value ≈ 31 kV/cm is applicable only for $\delta d = 1$, that is for one cm gap at 760 Torr and 20°C. For gap distances of a few mm, the electric strength is much higher than 31 kV/cm. It has been measured to be about 92 kV/cm for a gap distance of 0.1 mm. A plot of equation

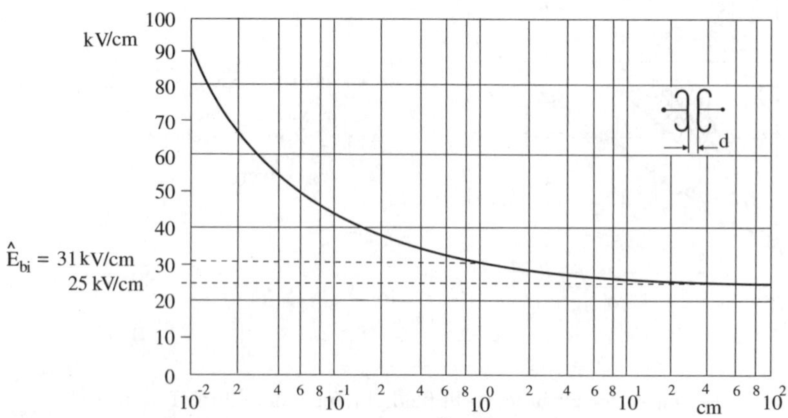

Figure 3.25 Breakdown voltage characteristic of atmospheric air in uniform fields.

(3.34) is shown in Figure 3.25. This represents the breakdown characteristic of atmospheric air at normal temperature and pressure in uniform fields. The electric strength value 31 kV/cm, measured across one cm gap in uniform field at normal temperature and pressure, is known as the "inherent" or "intrinsic strength" of air E_{bi}. For longer gap lengths in uniform fields, the electric strength of air reduces to about 25 kV/cm as shown in the figure. A breakdown with streamer mechanism is more likely to develop for such gap distances.

Paschen's law is found to be applicable over a wide range of *pd* values (of the order of 2000 Torr.cm). At extremely high and also at very low values of *pd*, that is, to the left of the minimum where a high vacuum exists, deviations from Paschen's law are observed.

At or near atmospheric pressures, the electrode material does not affect the breakdown. However, at extremely high and also at very low pressures, considerable influence of electrode materials, their state and even their surface area is reported, Trump, Cookson and Dokopoulusm [3.32–3.34]. A typical example of the influence of electrode material on high field intensity breakdown in air, measured by Trump et al. [3.32] in 1950, is shown in Figure 3.26. Significantly higher breakdown voltages are obtained between stainless steel and aluminium electrodes. These measurements were performed in uniform fields at constant gap distance, varying the air pressure. The difference in measured characteristic could be attributed to the roughness of the two materials. This needs further investigation.

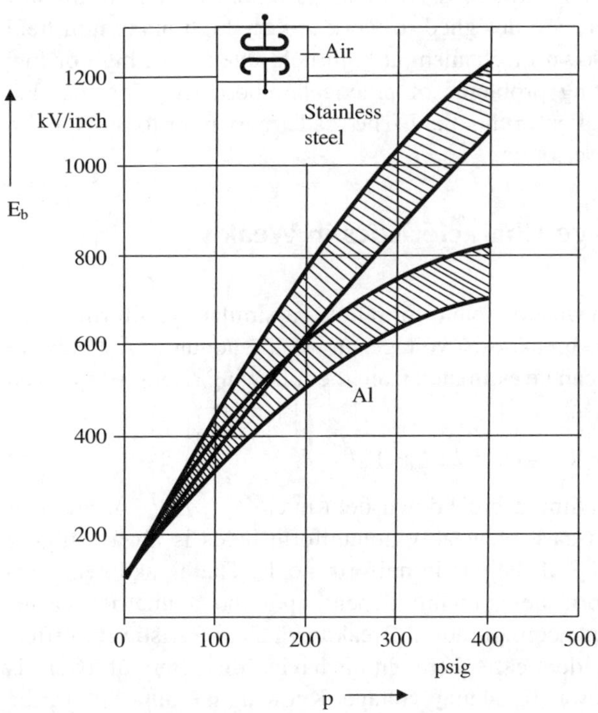

Figure 3.26 Effect of electrode material on breakdown voltage at high pressure and high field intensity, Trump [3.32].

It has been widely accepted that in the extreme pressure conditions, some cathode dependent processes dominate the breakdown mechanism. These additional processes are introduced mainly due to high field intensity, at which considerable field emission of electrons from the cathode takes place. That is why at extremely high pressures the departure of breakdown voltage from Paschen's curve is associated with the transition from Townsend to Streamer breakdown, as Paschen's law is applicable only with the Townsend's mechanism. At very low pressures also, deviation from Paschen's curve is observed where the breakdown mechanism ceases to be influenced by the gas particles and once again becomes electrode material/surface condition dominated (as in vacuum).

With regard to the effect of temperature, the validity of Paschen's law has been experimentally confirmed up to 1100°C, Alston (1958) [3.35]. Further increase in temperature ultimately results in failure of Paschen's law. It could be because of thermal ionization (above 2000 K), thermionic emission, or distortion of the electrodes by heat.

Large number of investigations have been performed to study the effect of various parameters on the breakdown voltage in compressed gases such as; the nature of gas, presence of dust or foreign particles, electrode material, state of electrode surface, oxide layers, area of electrode surface and also on the phenomenon of conditioning of electrodes. A review of the work on this subject is summarized by Cookson [3.33].

The use of SF_6 gas at high pressures and of vacuum as insulants in high voltage equipment having weakly nonuniform fields has increased considerably in the last three decades. Such equipment are designed to work at very high maximum field intensities. Hence, the breakdown mechanism at high field intensities, high or low pressures are posing interesting problems of practical importance. Considerable research work has been carried on in this field. The reader may refer to Section 3.5 and Chapter 5 for detailed description.

3.2.6 Breakdown Voltage Characteristics in Weakly Nonuniform Fields

The breakdown mechanism in weakly nonuniform fields is similar to uniform fields. Since the PB inception and the breakdown voltages in weakly nonuniform fields are equal, the breakdown voltage can be estimated from the following relationship given by Schwaiger, equation (2.4),

$$U_i = U_b \approx E_{b\max} \cdot \eta \cdot d \tag{3.35}$$

At breakdown, the "maximum breakdown field intensity", $E_{b\max}$ of an electrode system or equipment in case of weakly nonuniform fields is always higher than the "electric strength" E_{bi}, of the gas in uniform fields. The breakdown characteristics in weakly nonuniform fields mainly depend upon the geometrical factor of uniformity η of the electrode configuration. Breakdown characteristic for different gap lengths between two identical spheres in air having diameters of 10 cm is given in Figure 3.27 for ac power frequency voltage. Knowing the dimensions, the factor η can be found out from the Schwaiger curves given in Figure 2.6. If the

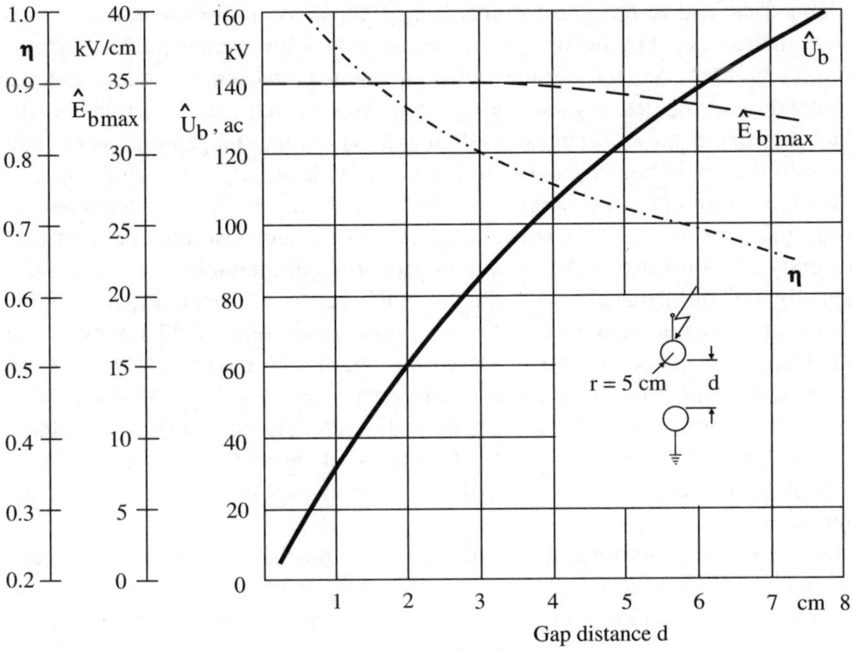

Figure 3.27 Variation in ac breakdown voltage in air, maximum breakdown field intensity at the electrode and the Schwaiger factor with increasing gap distance in weakly nonuniform fields.

measured values of $\hat{U}_b - d$ characteristic are known $\hat{E}_{b\max} - d$ characteristic can be calculated. From these curves it can be observed that $\hat{E}_{b\max}$ does not change much within a certain range of gap distance d, as also shown by the linear part of the $\hat{U}_b - d$ characteristic. For this particular electrode configuration, the linearity in characteristics can be considered to be limited up to a gap distance equal to the radius of the spheres. For a rough estimation of breakdown voltage of air by equation (3.35), one may take, therefore, a mean value of $\hat{E}_{b\max}$ from the curve in Figure 3.27. It works out to be equal to 34.5 kV/cm. As the gap distance d is increased, the field remains to be weakly nonuniform but the value of Schwaiger factor η decreases.

3.3 BREAKDOWN IN EXTREMELY NONUNIFORM FIELDS AND CORONA

Electrode configurations in atmospheric air having extremely nonuniform fields are more common than those having weakly nonuniform fields. This is because considerable efforts are needed to achieve weakly nonuniform fields and may greatly increase the dimensions of the electrodes. Secondly, for the majority of the high voltage equipment installed in open air, the space for providing clearance is not a problem. The presence of undesirable foreign particles and electrode protrusions in solid and liquid dielectrics lead to field distortions. Hence, these too introduce extreme nonuniformity in the fields. For the high voltage apparatus design, certain field distortions are taken for granted so that their unavoidable presence may not lead to unexpected lower breakdown voltages. As a result of intensive research and development, the modern compressed gas switchgear systems (GIS) using SF_6 gas and very high voltage solid dielectric power cables are produced with weakly nonuniform fields.

As the difference between maximum and mean field intensities in equipment increases (smaller η), the field characteristic becomes more nonuniform. Consequently, a poorer use of the insulating properties of the dielectric takes place. In extremely nonuniform fields at voltages much below the breakdown, a stable breakdown process confined locally to the region of extreme field intensity can be maintained. This phenomenon is known as "Partial Breakdown" (PB). When it occurs at free electrodes in gaseous dielectrics, it is called "Corona". For rated voltages above 100 kV, it is often very difficult and economically unviable to produce open air apparatus free of corona at normal working voltages. Understanding physical implications of the damage caused by corona, is therefore, very important.

As in the case of uniform fields, the discharge mechanism in extremely nonuniform fields can also be described with the development of basic avalanche and streamer processes. The most suitable electrode combinations to simulate extremely nonuniform fields are needle-plane or rod-plane electrode configurations. These asymmetrical electrode configurations have a highly localized region of extreme field intensity at the tip of the needle/rod electrode. Since the fundamental process of corona depends upon the polarity of the applied voltage, these electrode systems are found suitable for the proposed distinction between the positive and negative corona.

3.3.1 Development of Avalanche Discharge

Consider a needle-plane electrode configuration in air, Figure 3.28, with a +ve dc voltage applied to the needle. Above a certain magnitude of the applied voltage, U_l it gives rise to the impact of ionization process that develops in the form of avalanches.

On increasing the voltage, the field intensity at the tip of the electrode increases. Above a certain magnitude E_i of field intensity, the air loses its insulating properties locally in this region. PB or corona begins to take place in that region. The corona in turn produces space charge. Under the influence of space charge, the field intensity situation may get completely changed. Due to considerable differences in the movement of electrons and positive ions, the interaction between the fields developed due to space charge and the applied voltage is dependent upon the polarity of the electrode. Both the positive and the negative polarities are, therefore, considered separately. PB produced on applying ac has a periodic variation of the polarity, which makes the process rather complicated.

3.3.1.1 Positive Needle-Plane Electrode Configuration (Positive or Anode Star Corona) The process through which an avalanche is formed at a

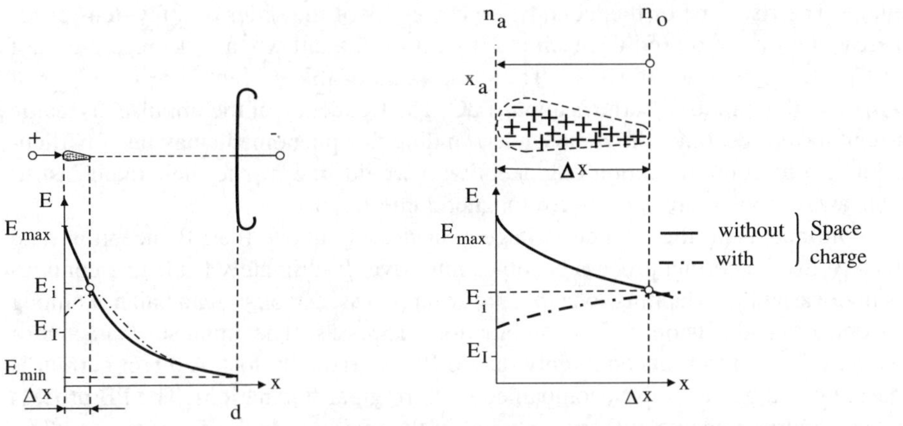

(a) Field in the gap (b) Field in the region of ionization.

Figure 3.28 An electron avalanche in front of positive point electrode (a) Field in the gap, (b) field in the region of ionization.

positive needle electrode is analogous to the one in uniform field. However, the applied field intensity E in this case falls sharply close to the needle electrode tip. Beyond a short distance, Δx, depending upon the applied voltage, the field intensity may fall below the minimum field intensity required for impact ionization E_I, Figure 3.28 (a). Thus, the avalanches are not able to extend themselves beyond a maximum length of Δx in the ionization region. It is generally much shorter than the critical amplification length x_c of the avalanche in this case. This is shown again in Figure 3.28 (b) on a lower scale ratio diagram for x.

In this case, the magnitude of field intensity E strongly depends upon the location. Since α is a function of E, the number of electrons arriving at the anode n_a can be given according to equation (3.11) as,

$$n_a = n_o \exp\left[\int_0^{x_a} \alpha(E_x)dx\right] \tag{3.36}$$

When a positive *dc* voltage, just sufficient to incept an avalanche, is applied to the needle (the anode), the electrons at the head of the avalanche are quickly absorbed by the positive electrode. A positive space charge due to the heavy and slow ions remains at the back, having a very slow movement, especially because of rapidly decreasing applied field at the anode tip. On increasing the applied voltage the ionization process is enhanced, hence stronger space charges build up. The two like polarity charges coming close to each other result in weakening the electric field in the region in front of the tip, as shown by dotted lines in Figure 3.28. Inception of further avalanche discharge is possible only when there is a drift of space charge away from the anode with time accompanied by radial diffusion, possibly toward the cathode. This type of discontinuous process gives rise to an impulse form of discharge current at voltages above the PB inception, U_i, in spite of applying a *dc* voltage. The rise time of this conduction current is of the order of a few tens of *ns*, whereas the time required to reach the instant on the tail when it decreases to half of its peak value is about 100 *ns*. The peak value of this current is of the order of 50 μA and the impulse charge is about 1 pC. The frequency of the impulse discharge current may go beyond 400 MHz. Understanding this phenomena may be of particular interest to communication engineers who would like to site their radio instruments away from sources of electromagnetic interference.

On increasing the applied voltage considerably above the PB inception field intensity E_i, the corona process becomes intensive. It ultimately leads to a continuous time variant overlapping local breakdown process, establishing a balance among the ionization, diffusion and recombination processes. The impulse character of conduction current produced mainly due to PB is gradually lost. A direct current is measured that, however, is accompanied with irregular fluctuations. The PB process becomes more intensive and concentrated in the region near the tip of the needle.

The optical impression of this quasi-stationary local breakdown process is a weak, bluish light like a "star in the sky", adjacent to the needle tip. This light phenomenon is explained by the excited state of gas molecules due to electron impact. The atoms at higher energy levels emit a quantum of light and fall back to the original state of lower energy level. This process is known as positive or anode "star corona" (in German Glimmentladung) or popularly known as simply "corona". The word "corona" is taken from Greek where it means the halo of light seen around a heavenly body. In Latin it means a crown. The "star corona" usually takes place at extremely sharp and pointed electrodes, for example, at needles, thin wires and sharp edges etc., on applying either *dc* or the relatively slower changing power frequency *ac* voltages. The situation demands, on the one hand, a steep fall in potential gradient so that the avalanches do not acquire their critical amplification length, and on the other hand, that the charge carriers get enough time to build the space charge. The audible noise produced by star corona is a continuous "hissing" sound. The magnitude of mean potential gradient required for breakdown in air with stable positive corona preceding the breakdown lies around 15 kV/cm and below. It reduces as the gap distance is increased.

3.3.1.2 Negative Needle-Plane Electrode Configuration (Negative or Cathode Star Corona) When negative polarity *dc* voltage, just sufficient for the inception of impact ionization is applied to a needle (the cathode), the condition is similar to the one discussed in the previous section only to some extent. In this case as well, the possibility of impact ionization is limited to a short distance Δx because of a steep fall in potential gradient at the needle towards the plane. The ionization process started with a single electron adjacent to the needle tip is able to extend the avalanche to a maximum length of Δx only, analogous to the conditions shown in Figure 3.28. Like in case of positive polarity voltage, here too the avalanches do not acquire their critical length of amplification. Hence the avalanche process limits itself to a short region adjacent to the needle tip and it is not able to expand farther.

However, in this case the avalanche develops in the opposite direction, that is, the avalanche head is towards the plane (anode), as shown in Figure 3.29(b). Due to the high field, the electrons nearer to the needle electrode drift away quickly;

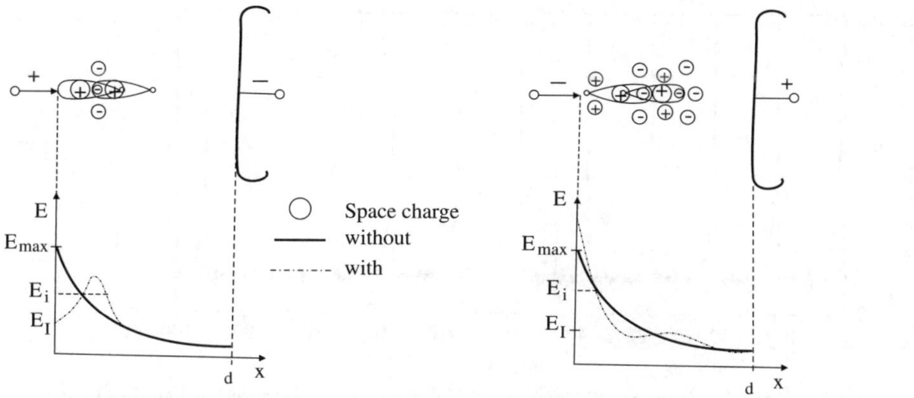

Figure 3.29 Field intensity distribution between needle plane electrodes with stable star corona.

however, because of the steep fall in field intensity they slow down beyond this region in the dielectric. Since oxygen in the air is an electronegative gas, oxygen molecules absorb the relatively slower electrons in front of the avalanche, forming negative ions. Thus negative space charge is developed by these heavy and slow ions leading to weakening of the field at a short distance away from the tip of the negative electrode. It prevents the avalanche process to develop farther. In the mean time, the positive space charge left behind by the avalanche process shifts towards the negative electrode tip, increasing the field intensity considerably as shown in Figure 3.29 (b). After a certain time, even the negative space charge shifts away from the vicinity of the needle tip, diminishing the field weakening effect. New avalanche formation is then possible. Under the influence of these space charges, lower field intensity results away from the tip of the electrode, leading to a much higher voltage required for breakdown in this case.

In short, it can be concluded that as in the case of positive polarity, in the case of negative polarity, a discontinuous process of charge carrier production and their migration in the dielectrics also takes place. The magnitudes of conduction currents in gaseous dielectrics without PB are fairly low. Since PB in dielectrics involve high discharge currents, the conduction current increases considerably with PB. Under static conditions above the PB inception level, an impulse form of PB current is conducted through the dielectric in a very regular and repetitive pulse form, as shown in Figure 3.30. The frequency of these current pulses may vary from a few kHz to MHz. This pulse current was first measured and studied in detail by Trichel [3.36] in 1939, and hence are named after him as "TRICHEL Pulses". Many researchers have measured these later on, Woboditsch [3.37]. An oscillogram of measured Trichel pulses is shown in Figure 3.30.

As in the case of PB inception voltage, the inception voltage of Trichel pulses is also practically independent of the gap length. The frequency of pulse increases with increasing voltage. It depends upon the geometrical radius of the cathode, the gap length and the gas pressure. On raising the voltage, the mode of these pulses

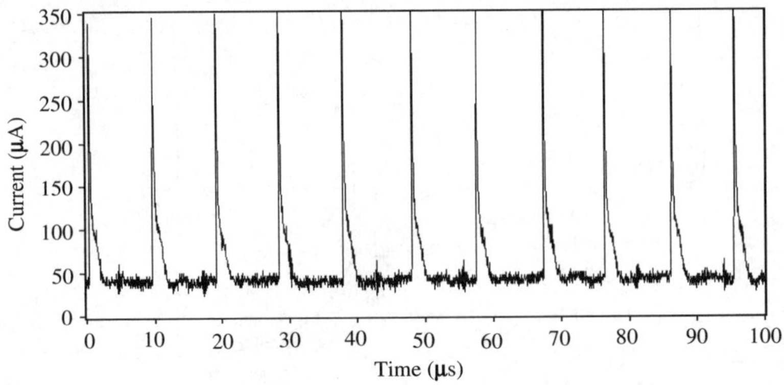

Figure 3.30 Impulse form of PB current , "Trichel Pulses", measured at needle-plane electrode system on applying negative polarity dc voltage.

does not change over a wide range. Eventually, at much higher voltages, the impulse character of the PB current is lost due to overlapping of individual pulses. A direct current is then measured, which is accompanied with certain fluctuations. A steady PB or star like "cathode corona", similar to the anode corona is also observed in this case but it is located slightly away from the needle tip. The cathode corona appears reddish, as compared to the bluish hue of an anode corona.

The mean potential gradient in the gap required for breakdown of air with stable negative corona preceding the breakdown is ~20 kV/cm for gap lengths up to a few *cm*. It reduces for longer gap distances but the reduction is less compared to positive corona preceding the breakdown.

3.3.2 Development of Streamer or Kanal Discharge

Let the field distribution between the needle-plane electrode system be modified in such a way that the fall in field intensity (the potential gradient) at the tip of the electrode is no longer as steep. It can be achieved, for example, by increasing the radius of curvature of the needle, taking most suitably a hemispherical rod in place of needle electrode. This increases the depth of the region Δx in the dielectric, where the impact ionization is possible. With this result, the avalanches developed in this region can acquire critical length of amplification, x_c, leading to a transformation from avalanche to streamer discharge (section 3.2.4.2). As a consequence of the interaction between field produced by the stronger space charge and the applied field, an extended effect of polarity is observed in this case of the development of avalanche of above critical amplification. Both the polarities are therefore discussed separately. Practical experience has shown that PB with streamer discharge may be produced on applying *dc*, *ac*, as well as switching impulse (*si*) type of voltages to a hemispherical rod-plane electrode system for gap lengths beginning with a few *cm* and above. The magnitude of voltage required for PB inception U_i in this case is comparatively higher than needed to produce avalanche discharge of below critical amplification (star corona).

3.3.2.1 Positive Rod-Plane Electrode (Positive Streamer Corona)

Consider a situation after the inception of avalanche discharge in the region next to the positive rod. When an avalanche grows to its critical amplification, the following equation holds true;

$$\int_0^{x_c} \alpha dx = \ln \frac{n_{xc}}{n_0} \approx 18 \tag{3.37}$$

where x_c is the length of the avalanche when it acquires its critical amplification as explained in Section 3.2.4.2. The avalanche process is able to develop only under the condition that the field intensity in the region is above E_I, the minimum field intensity required for impact ionization, as shown in Figure 3.28.

The electrons at the head of an avalanche are immediately absorbed by the positive rod electrode (anode). Due to two like polarity charges coming close to each other, the positive space charge left behind can lead to weakening the field next to the anode to such an extent that further ionization may not be possible in this region. Since the avalanches acquire their critical amplification, there is a strong concentration of positive space charge towards the tail of the avalanche, shown in a circular form for the purpose of illustration in Figure 3.31. An increase in field intensity towards the tail of these avalanches results due to the presence of negative space charge, developed at the front of the next generation of avalanches. It is formed by photon emission from the avalanche heads and also by the formation of negative ions of oxygen molecules, an electro-negative gas. The resultant enhanced field intensity magnitude may, therefore, be much higher than the minimum intensity required for PB inception E_i (curve 1 in Figure 3.31). In this process, the avalanches in the second generation grow very close to the first. Because of similar reasons explained above, an increase in field intensity is caused again due to the positive space charge of the second and the negative space charge at the heads of the third stage of avalanche process. The development of one stage of avalanche process grows into the other shown by the curves 1, 2, 3, 4, in Figure 3.31. The space charge effect in this case is very strong. The ionization as well as the PB process continues in the gap so long as a subsequent avalanche started by photo-ionization falls in the field region above E_I and the enhanced resultant field intensity is above PB inception, E_i. In spite of the actual potential gradient in the dielectric due to applied voltage decreasing below these required levels, the discharge process may continue to extend in the gap.

The positive and the negative space charges involved in the growth of the streamer string compensate for each other. The conductivity of this nearly neutral charge carrier channel is, therefore, quite low. However, the basic condition for the development of streamer discharge $\left(\int_0^{x_c} \alpha x_a \approx 18\right)$ must be fulfilled at each point. Since the positive charge carriers, being heavier particles, do not move themselves much, the process is comparable to the movement of charges like a wave. In fact, numerous avalanches begin together and the whole process is a continuous development of a large number of streamer PB trajectories. The cathode directed Streamer Corona grows in the space from the rod electrode (anode) in the main field direction like a "shower of discharge", as shown in the photograph in Figure 3.32.

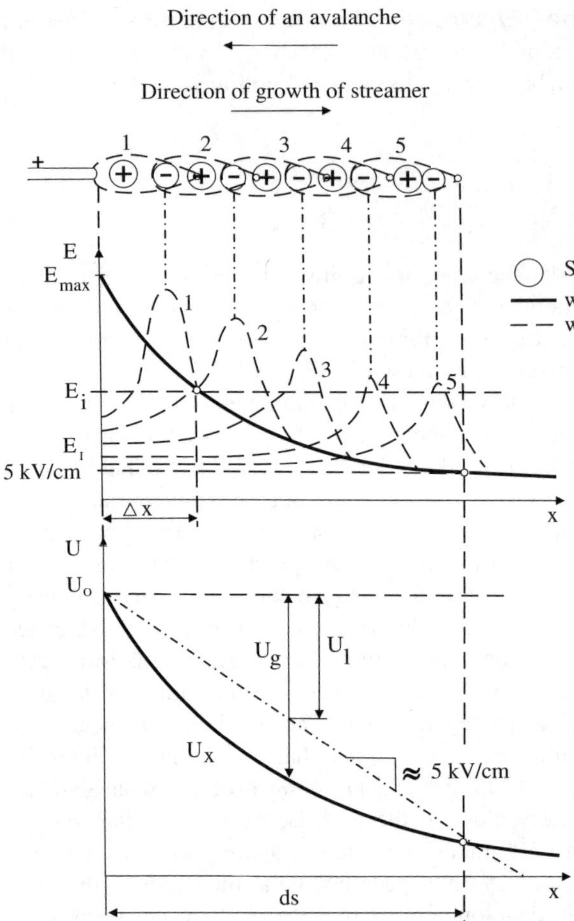

Figure 3.31 Schematic of the development of streamer corona in front of a positive rod electrode with variation in field and potential as a consequence of space charge.

This photograph was taken by Lemke [3.38] on applying a long duration pulse of 1/5000 μs and 100 kV positive peak-voltage on a needle-plane electrode system having a gap distance of 20 cm. Similar photographs were also taken by Nasser [3.40]. The experiments conducted by Nasser revealed that, on applying a single impulse voltage, even 200 partial breakdown branches may develop in a gap length of just 2.5 cm. The number of these branches grows exponentially in the dielectric with increasing gap length. It is an interesting characteristic to note that these large number of branches never cross each other.

The streamer corona is also accompanied with an impulse form of current. Depending upon the extent of streamer, this current may acquire its maximum magnitude of a few milliampere to an ampere within nano-seconds. The streamer current decays to its 50% value within about 100 ns. The impulse discharge is generally in the range of 100 pC to 100 nC but may be even higher in some cases. The next

Figure 3.32 Photograph of a positive streamer corona, Lemke [3.38, 3.39].

process of displacement and formation of space charge is introduced with a time lag. The current discharge process accompanied with streamer corona, is therefore, repetitive having irregular frequency. This is the reason that the audible noise produced by streamer corona is a flutter sound, unlike hissing in case of star corona.

The integrated optical light appearance of "Streamer Corona" is like a weakly illuminated "bunch" or more appropriately a "shower" of discharge. Hence, it is also known as "bunch discharge", also mentioned sometimes as "illuminated string discharge", (in German "Büschel" and "Leuchtfaden" discharge respectively). However, it is popularly known as "streamer corona".

Referring to Figure 3.31, it can be seen that, for a known potential distribution U_x in the gap, the range of growth of streamer corona can be determined, if the potential gradient requirement for the streamer process to grow is known. Let the process be analysed through the involvement of energy in the space between the electrodes. The increase in potential due to space charge U_g is represented by its corresponding energy gain (where $U_g = U_0 - U_x$). Because of continuous production of new charge carriers, some energy is consumed simultaneously. Hence a loss of energy in the gap is caused. Let this loss be represented by a potential loss U_l. The PB process ceases to continue when this loss of energy is no more compensated by the gain, that is, when $U_l \approx U_g$. This occurs at $x = d_s$, the extent of streamer as shown in Figure 3.31. Measurements by Lemke [3.38] in 1967 revealed that the mean potential drop per centimeter length in a positive streamer corona trajectory is 4.4 kV. In other words, it can be said that, for a positive streamer corona to grow, a mean potential gradient above 4.5 kV/cm is required in the gap.

Since the whole process is dynamic, it is very difficult to determine such figures for a nonlinear process. A decade later in 1977, Les Renardieres research results [3.50] confirmed Lemke's findings. According to them, the average potential gradient along the streamer PB channels is 4–5 kV/cm over a wide range of gap lengths when breakdown with dc positive polarity is accomplished.

The propagation of streamer corona through a gap depends upon the electrode geometry and the magnitude and type of applied voltage. It may also depend upon the external circuit parameters. Hence the velocities of propagation of streamer trajectories measured by different researchers vary to some extent. Both Nasser [3.40] and Lemke [3.38] have explained the development of positive streamer corona.

Figure 3.33 shows a schematic development of positive streamer given by Nasser. His experiments were performed on a 2.5 cm gap between needle-plane electrode configurations. The needle had a radius of 0.5 mm. A 1/50 μs impulse voltage of magnitude 25 kV was applied on the needle electrode. In the beginning, the streamer velocity was measured to be about 300 cm/μs. The streamer propagation velocity reduced continuously as the discharge grew, both in axial and radial directions. The main streamer trajectories arrived at the cathode after 0.02 μs of voltage application with a velocity of about 100 cm/μs. The streamer channels on the sides grew at much slower speed, depending upon their location from the main field axis. After about 1 μs, they all acquired their final magnitudes.

The photograph in Figure 3.34 shows the propagation of streamer corona between positive needle-plane electrodes having a gap distance of 20 cm on applying a near standard switching surge type of impulse voltage (Lemke [3.38]). The oscillogram below shows the variation of voltage and current with time. An interesting observation was made during these measurements that the first streamer discharges were always able to extend themselves up to opposite electrode, when started with a peak voltage of 90 kV. This confirmed that an average potential gradient of

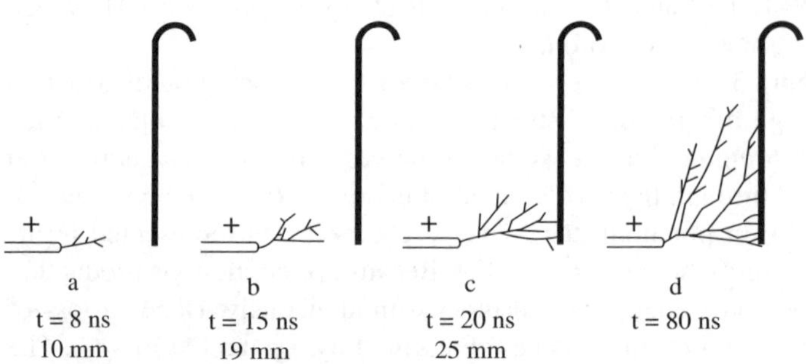

Figure 3.33 Schematic of the development of positive streamer, Nasser [3.40].

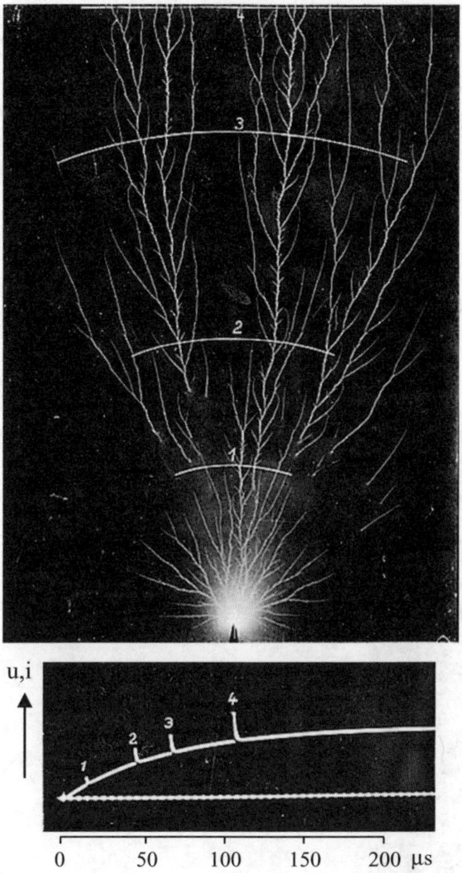

Figure 3.34 Development of positive streamer corona with time, Lemke [3.38].

4.5 kV/cm is required for the propagation of streamer corona. Higher magnitudes of PB pulse currents are measured with increasing voltage as the streamer trajectories grow towards the cathode.

3.3.2.2 *Negative Rod-Plane Electrode (Negative Streamer Corona)*

As with positive rod-plane electrodes, the development of a streamer discharge in negative rods also begin with an avalanche of above critical amplification. The negative streamer mechanism is in principle comparable to the positive streamer except for the location of the formation of space charge. The direction of the avalanches in this case is, however, opposite; that is, their heads are away from the rod. A strong positive space charge is therefore built in front of the rod in the dielectric, increasing the field intensity right at the tip of the electrode. The electrons form a negative space charge of their own and that of negative oxygen ions at the head of the avalanche slightly away from the tip of the electrode. The two like polarity charges weaken the field in the vicinity. Higher basic field is required to be applied in this situation to enable the impact and the photo-ionizations as well as PB to continue farther in the gap.

When the PB process develops farther, a sort of "scattering" of negative space charge takes place by radial diffusion because of the high mobility of electrons. Consequently, weakening of the negative space charge takes place. The field intensity, which is affected by the concentration of the space charge, again increases to some extent. Because of this space charge effect on the field, the anode directed streamer corona is not able to grow in the gap to the extent compared to cathode directed corona at the same potential. The radial diffusion of electrons is also responsible for a comparatively lesser number of distinct trajectories of streamer corona able to develop at the rod. This phenomenon is very clearly seen in Figure 3.35. Toepler in Technische Hochschule, Dresden, originally took these photographs,

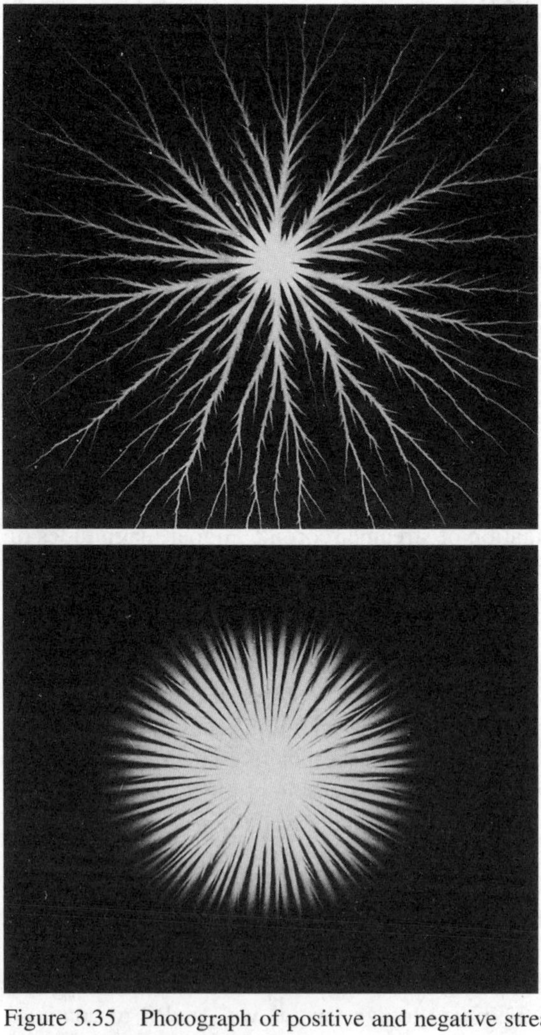

Figure 3.35 Photograph of positive and negative streamer corona, "Lichtenberg Figures" taken by Toepler, TU Dresden.

commonly known as "Lichtenberg figures". Lichtenberg, a German scientist, studied this phenomenon in 1777.

While the average potential gradient requirement for the propagation of positive streamer corona is estimated to be about 5 kV/cm, it is of the order of 10 to 15 kV/cm for negative streamer corona. This has also been confirmed by Les Renardieres [3.51].

3.3.2.3 Symmetrical Positive and Negative Electrode Configurations in Extremely Nonuniform Fields In case of unsymmetrical electrode configurations, the cathode or the anode directed streamer corona grows only from the tip of the rod or blunt needle electrode towards the plane. This happens because the field intensity at the sharp electrode is much higher compared to the plane. If a symmetrical electrode configuration of blunt needle-needle or rod-rod is applied to a bipolar *dc* voltage, the situation is very interesting to analyze.

Under this condition, both the cathode as well as anode directed streamer corona initiate simultaneously. They grow in the air from the two electrodes towards the opposite electrode as shown in the Figure 3.36. As explained in the previous sections, the extents of the growth of the cathode and anode directed streamers in the gas are quite different, evident from the Lichtenberg figures shown in Figure 3.35. The positive streamer corona grows deeper in the gap as compared to negative streamer corona under the influence of the development of "specific space charges". Hence, in this case the zero potential line may not fall exactly in the middle of the two electrodes in spite of the geometrical symmetry of the electrodes and a symmetrical bipolar voltage being applied. It may shift towards the negative electrode since the anode corona has a tendency to extend itself more in the gap, Figure 3.36.

It would be interesting to measure the bipolar *dc* breakdown voltage for the symmetrical electrode system. The positive and the negative streamer corona

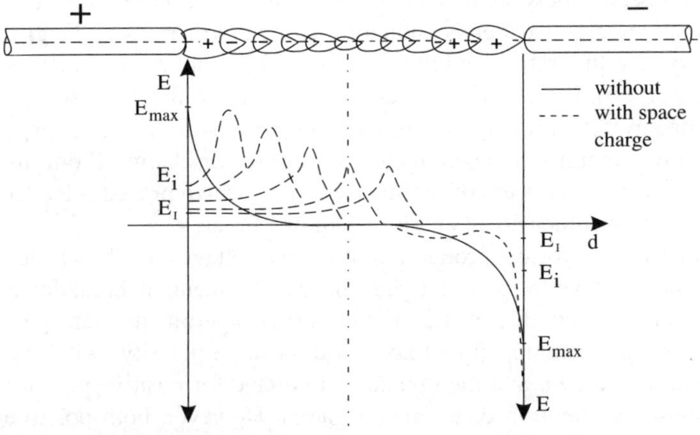

Figure 3.36 Growth of streamer and field development between symmetrical rod-rod electrodes.

trajectories have entirely different (even up to 2.5 times) potential gradient across them. Hence the voltage magnitude required for breakdown would strongly depend upon the individual extent of growth of the two coronas in the gap, which in turn depends upon the geometrical shape of the electrodes.

Instead of applying a bipolar symmetrical *dc* voltage, if mono-polar voltage is applied on one electrode and the other is grounded, the situation would be again entirely different. As explained in Section 2.3.1.1, grounding of one of the electrode introduces more nonuniformity in the field. On applying a positive polarity voltage on one electrode, it induces opposite, that is, the negative polarity voltage on the other and vice versa. If the other electrode is not grounded, a symmetrical field pattern across the center line between the two electrodes would appear, Figure 2.5. The grounding of one of the two electrodes is another cause of introduction of asymmetry in the field between the symmetrical electrodes. The breakdown voltages for symmetrical electrode system measured with positive and negative polarity *dc* voltages separately while grounding one of the electrode may not have much difference as compared to the breakdown voltages measured on an asymmetrical electrode system, such as rod-plane, and so forth.

3.3.3 Development of Stem and Leader Corona

Even when the streamer corona, described above, extends itself all through the gap between the electrodes, spark breakdown of the medium may still not take place. This is because the conductivity of streamer trajectories is low for accomplishing a breakdown. In order to achieve a sufficiently conductive channel required for breakdown, an extended phenomenon of PB activity is necessary for strengthening the discharge process. Study of the development of breakdown mechanisms with extended streamer process requires investigations in longer gap distances, of the order of a meter and above.

The phenomenon of the development of spark breakdown in both positive as well as negative rod plane electrode configurations for relatively long gap lengths (a meter and above) has been widely studied and reported in technical literature. The development of breakdown largely depends upon the electrical properties of the gas, the gap length, electrode configuration and the type of voltage applied. As early as in 1906 Toepler (1870–1960), Professor in TH Dresden suggested for the first time that the streamer corona trajectories may undergo an abrupt transformation (Gänger [3.41]). Toepler was investigating the phenomenon of surface discharge. From his findings, Toepler introduced the concept of "stem discharge", which precedes leader corona, to describe the three-dimensional spark breakdown in air.

Development of facilities for the production of higher voltages in laboratories in years to come made it possible to investigate the development of breakdown mechanism in longer air gap lengths. At Les Renardieres, experiments were performed for breakdown in gap lengths up to 13.5 m with positive polarity switching impulse voltage. Since it is known that the breakdown voltage for positive polarity is lower, it has drawn more attention of the investigators. However, both positive and negative stable leader PB development have been investigated by many individuals as well as teams of researchers [3.38–3.51].

As seen in the photograph of a positive streamer, Figure 3.32, a concentration of streamer trajectories develops at the immediate front of the electrode. A large number of streamer channels begin in a quick succession, causing high density of current in the region in front of the tip of the electrode. In the next stage of discharge development, this may lead to excessive heating, and thus thermal ionization in the immediate vicinity of the electrode. A sudden "constriction" phenomenon of the streamer trajectories takes place when the voltage is increased. With constriction, substantially brighter than the streamer corona, a single "stem"-like discharge appears at the electrode. This may be a single or multiple branch trajectory discharge, depending upon the electrode configuration. This type of a highly conductive channel is known as "stem" or "stembunch", and "Stielbüschel" discharge in German, a terminology also introduced by Toepler.

Both positive as well as negative stembunch discharges in the gap have a stepped structure. At the end of each step, streamer corona precedes the formation of the following step. Figure 3.37 shows a photograph of the warming up of a stem

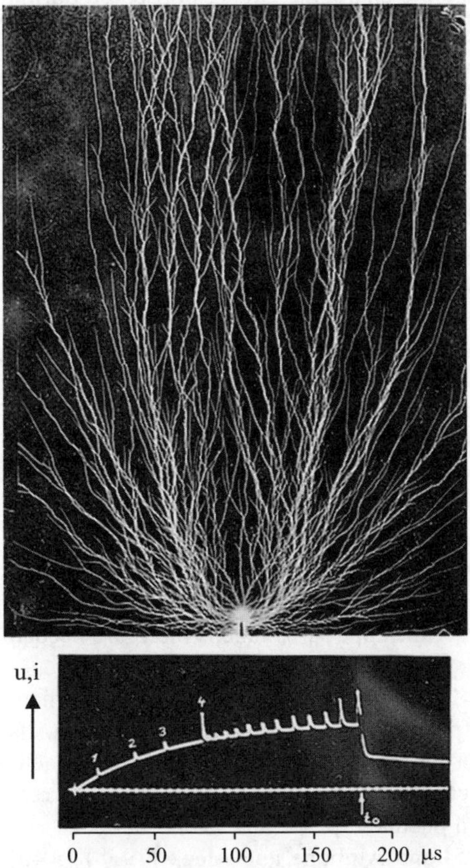

Figure 3.37 Formation of "stem" discharge on a positive needle-plane electrode gap of 20 cm, Lemke [3.38].

discharge on a positive needle-plane electrode system having a rather short gap length of 20 cm on applying a near switching surge type of a voltage, Lemke [3.38]. The oscillogram shows the variation in applied voltage and PB current pulses with time.

The development of stem discharge has also been studied by Petropoulus and Menemenlis [3.42–3.44]. A 1/50 μs positive polarity, 100% breakdown voltage magnitude was applied to the sphere forming an extremely nonuniform field in air between sphere and plane with a gap distance of 18 cm. Spheres of 5 to 15 mm diameter were chosen as anode and a disc of 30 cm in diameter as cathode. The lengths of stem steps measured in this study generally varied from 5 to 15 mm. Longer steps were rarely observed, but they could acquire lengths up to 40 mm. For gap lengths smaller than 18 cm, the stepped structure of the stem discharge has been reported to be obscure. The average velocity of propagation of this discharge was found to be between 20 and 150 mm/μs under the experimental conditions described.

When the stembunch discharge becomes stronger, it develops into a narrow and intensively brighter channel known as "leader". This terminology was originally proposed by Schonland in 1934 to describe the analogous process observed during lightning in the atmosphere, Meek and Craggs [3.25]. The initial stage of the formation of a positive leader corona as an extension of stembunch, is shown in Figure 3.38. Here too, the oscillogram below shows the voltage profile with respect to time, superimposed with the PB current pulses.

The leader channel is accompanied with a less bright, diffused, voluminous conical shower like streamer corona at its tip. As described by Meek and Craggs, this has a filamentary structure. After a passage traced by leader the luminosity of the filaments of streamer corona falls. The main leader channel follows the tortuous path traced out by the streamer corona, ahead of the leader channel. As these fine streamers of leader corona approach the cathode, they converge into a less number of thicker channels. On bridging the gap between the two electrodes, the initial narrow channel towards the anode is re-illuminated by the main conduction current. This flow of current through the leader channel neutralizes the space charges that had developed in the gap during the growth of the corona.

On applying a sufficiently high voltage to the electrode, the PB Process initiates after a time-delay. During this delay electron space charges are generated at suitable locations. This is known as "statistical time-lag". This is the reason that, if the time to reach the crest of the applied voltage is very short (of the order of 1 μs, as in the case of a lightning impulse) the PB with stable leader is not able to develop in the gap. A longer time to reach the crest of the applied voltage is therefore needed to produce a stable leader corona. The general aspect of the development of breakdown mechanisms is quasi continuous and the different stages are linked together.

A leader formation alone is not sufficient criterion for spark breakdown. Its growth over a wide range of voltage may terminate spontaneously. Only on approaching sufficiently close to the opposite electrode, its velocity and discharge current increase appreciably and spark breakdown occurs.

Allibone and Schonland 1934 [3.45] and Allibone and Meek (1938) [3.46] have reported the earliest investigations on the development of leader corona. Breakdown mechanisms in long air gaps were examined by them for both positive

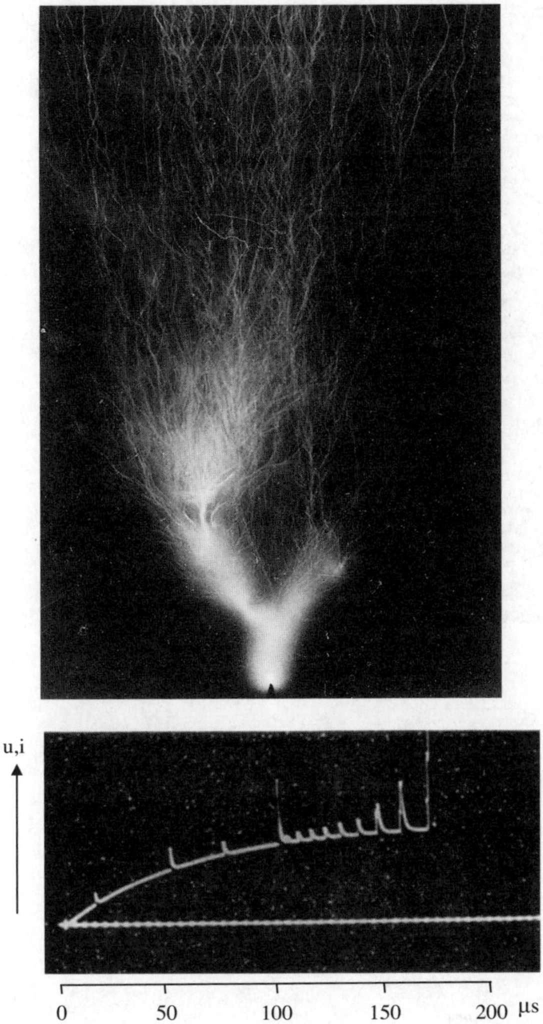

Figure 3.38 Beginning of a positive leader corona at a blunt needle electrode in 20 cm gap length just before its final jump for breakdown, Lemke [3.38].

and negative polarity voltages. These experiments were conducted by means of an especially developed rotating mirror streak camera (high speed photographic techniques) and published by Waters, Jones, Dyson and Hemmings in 1961 [3.47, 3.48]. Lemke made commendable contributions in 1967 [3.38, 3.39]. Les Renardieres group worked as a team on very long air gaps and published reports in 1977 and 1981 [3.49–3.51].

3.3.3.1 Development and Propagation of Positive Leader Corona On applying a positive impulse voltage on the rod-plane air gap, first a dense, filamentary streamer corona discharge occurs. Figure 3.39 (a) shows an actual time resolved photograph and Figure 3.39 (b) a diagrammatic representation of the same event

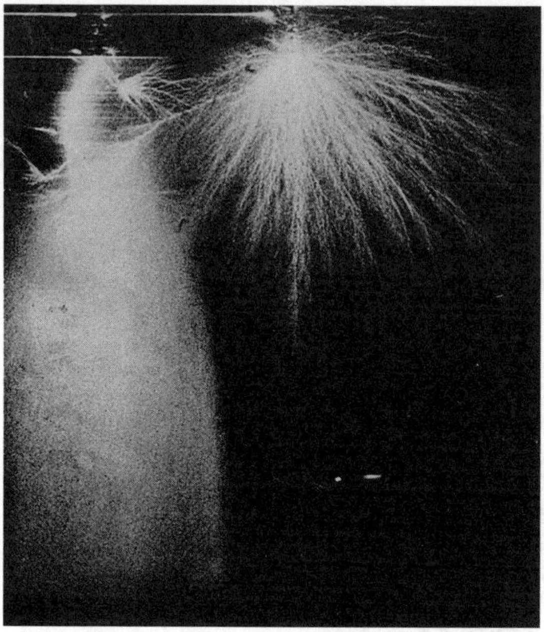

(a)

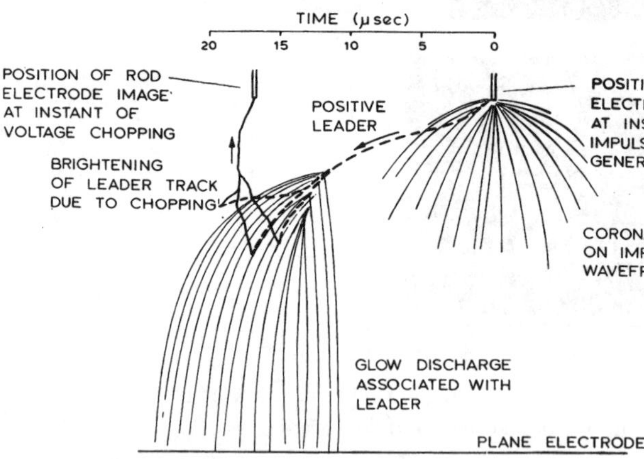

(b)

Figure 3.39 (a) Time resolved photograph of arrested breakdown in 200 cm positive rod-earthed plane gap, on applying 1 MV 0.5/2000 μs impulse. The horizontal light streaks are time markers [3.47, 3.48]. (b) A diagrammatic representation of Figure 3.39 (a), also given in [3.47, 3.48].

taken by Watters et al. [3.47, 3.48]. This phenomenon is described as the "first corona" or the "corona burst". It is in the form of a large conical or roughly hemispherical volume, centered at the electrode. In the beginning, several embryo-discharge channels, formed at the anode surface, may propagate only a few *cm* and not grow further.

The electrode separation in these experiments was chosen so that approximately 90% times breakdown (U_{b-90}) occurred on applying 1 MV, 0.5/2000 μs impulse voltage of both the polarities. This gave a gap distance of 200 cm for positive and 92 cm for negative polarity impulse voltages. However, the breakdown could be arrested at any desired stage before completion by means of a parallel diverter gap system in order to prevent fogging of the photographic records. The photographic techniques, especially developed for this purpose by Dyson et al. [3.48] comprised a camera possessing an aperture of f/1, a focal length 61 mm and resolving power 50 lines/mm. At a rotating mirror speed of 500 rps, the camera produced a tracing speed of 0.38 mm/μs, and a maximum time resolution of 20 ns was obtained. The velocity of propagation of discharge channels was measured both by slit technique and by measurement of the length of the filaments under rapid chopping of voltage.

In their experiments, the positive leader commenced to propagate after an average delay of 0.4 μs, that is, when the "first corona" had already extended to about 40% of the gap distance. A uniform velocity of 200 cm/μs for the propagation of first corona was measured. The initial velocity of propagation of the stable leader channel was recorded as approximately 20 cm/μs. If the flashover failed to occur, these leader channels terminated nearly at this distance. In case the leader succeeded in progressing further, a gradual acceleration in its propagation velocity developed and breakdown was assured. When a leader length developed to about 50% of the gap distance, it had a discontinuous stepped structure. Beyond this length it underwent a sudden increase in velocity, losing the stepped structure. No leader channel development originating at the cathode was recorded in this case. Figure 3.40 shows the mean velocity of propagation of positive leader, measured by Waters et al. [3.47]. However, as confirmed later [3.50], the leader propagation velocity depends upon

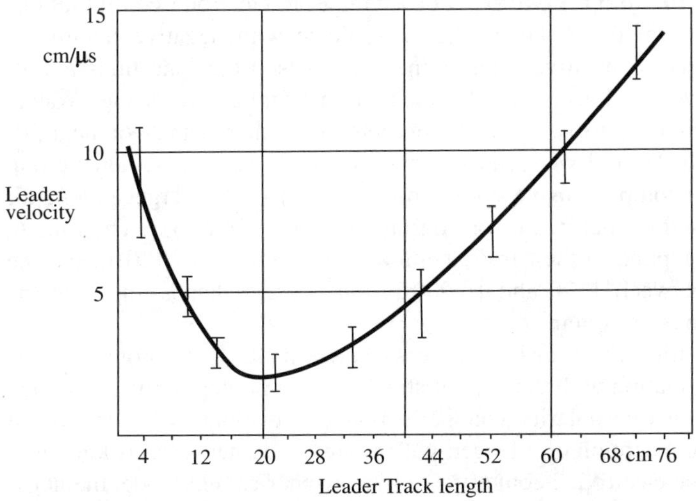

Figure 3.40 Mean stable leader channel propagation velocity for 1 MV, 0.5/2000 μs positive voltage application on a 200 cm rod-plane gap [3.47–3.48].

the electrode configurations and characteristics of the applied voltage, that is, its crest magnitude and time taken to reach the crest.

A very clear conclusion could be drawn from these experiments that a leader channel is always formed by high current density in one of the first (streamer) corona filaments. It later undergoes a critical stage in its development when the leader tip progresses to about 20 cm from the rod. The capability of the leader to cross the gap is determined at this critical stage of its development.

The spatial propagation characteristics of the leader channels are those of a narrow irregular path, often in a conducive direction at an angle to the main field. The luminosity of these channels is generally high. The voluminous streamer corona at the leader tip, also having a weak, lesser bright filamentary structure, ionizes the air ahead of the leader channel. It is described as "leader corona" by some authors but it is in fact a streamer corona. It is responsible for maintaining the higher level of current flow in the leader propagating channels. It is also responsible for modifying the field distribution pattern in the gap by developing space charges.

In case the discharge is chopped before completion, the track established by the leader undergoes a rapid increase in brightness, much stronger than that produced during the leader growth itself. This is known as a "return stroke", which has a very high velocity as well, beyond the limits of measurements as reported by Waters and Jones [3.47]. This phenomenon is also described as "re-illumination" or "re-strike" [3.50]. It could happen when the propagating downward and the upward channels meet to establish least resistance path through the dielectric enabling very high current to flow.

The experimental investigations by Waters and Jones also reveal that the velocity of propagation of unstable leader at its final jump is much higher as compared to the stable leader propagation velocity shown in Figure 3.40.

3.3.3.2 Development and Propagation of Negative Leader Corona and the Phenomenon of Space Leader For asymmetric electrode configurations in extremely nonuniform fields the breakdown voltage with negative polarity is much higher compared to positive. This is the main reason that one finds a very limited number of investigations reported in the literature for negative leader. Waters et al. [3.47, 3.48] published excellent photographs and other details of negative leader development in 1961. Two decades later in 1981, the Les Renardieres group in France published a comprehensive study report on negative discharges in long air gaps up to 7 m, a result of their teamwork during 1971 to 1978 [3.51]. The time to crest of the voltage applied in their investigations varied from 6 to 320 μs and the time to half value between 1600 and 10,000 μs. Such wave shapes represent the switching surges in power systems.

While investigating the transition process between the first-corona and the subsequent leader formation with the help of streak camera photography, it revealed that in the case of negative polarity voltage, it is not so distinct as in the case of positive polarity. The inception of a leader, both positive and negative, is known to be dependent upon the electrode geometry. For a hemispherical electrode, the negative leader inception time may coincide with the first corona. As in the case of positive polarity, in this case the first corona is also a filamentary streamer.

One may distinguish three types of leader channels formed with negative polarity voltage before the final breakdown:

- the "negative leader", which develops from the cathode toward anode
- the "positive leader", from a weak point on the anode toward the cathode just before the final stage of breakdown
- the "space leader" (in gap lengths above 1 m), which incepts somewhere in the gap and propagates toward both the electrodes, irrespective of any affinity. The space leader causes an extension of the negative leader by re-illumination.

Waters et al. studied the negative leader propagation on a 92 cm gap; hence they did not observe occurrence of the space leader. Figure 3.41 shows the development of a negative leader.

The typical phenomenon of the development of "space leader" is reported by Les Renardieres for all investigations in long gaps of more than two meters on applying a negative polarity impulse voltage [3.51]. After the first corona occurrence, an "ionized zone", having a bright luminous nucleus at the tip originates at any suitable location in the gap. This is described as "space stem" discharge. The whole process of positive and negative streamers with their common space stem repeats itself at a more advanced stage in the gap. It appears to be like a "bi-directional" corona on a fictitious electrode. Like the first streamer corona, the negative space stem also has slightly branched individual filaments. On the contrary, the

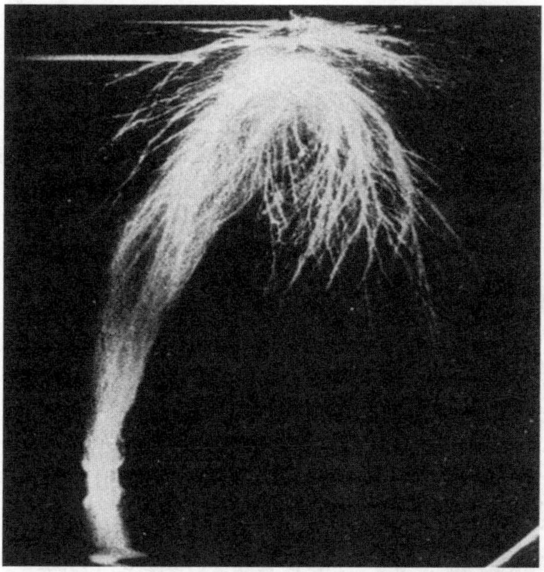

Figure 3.41 Time resolved photograph of an arrested breakdown in 92 cm gap between high voltage rod and earthed plane on applying a negative polarity impulse peak of 1 MV [3.47, 3.48].

positive space stem forms dense strings, connecting the space stem with the negative leader. The velocity of propagation of these streamers with their common space stems is of the same order as typical streamer velocities (a few m/μs).

The "space leader" generally originates and elongates from a space stem. They initiate at a distance from the main leader, which increases as the leader progresses into the gap. During the experiments at les Renardieres [3.51] space leaders have been observed in all the gap lengths between 2 and 7 m, on applying impulse voltages with time to crest varying from 6 to 320 μs. However, the characteristics of the space leaders, that is, number per impulse, their length, position, instance of appearance etc. depend upon the gap distance and the shape of the impulse voltage. A still photograph of a space leader is shown in Figure 3.42. This space leader was formed off the main leader channel course in a gap length of 7 m between a 300 mm radius hemispherical rod and plane electrode system. The space leader profile is marked on the photograph with an arrow.

Some important conclusions regarding the propagation of both positive and negative leader coronas can be drawn from the experimental results given in [3.51] as follows:

- the leader velocity during its continuous propagation depends to a great extent upon the gap length
- it is very difficult to estimate the average leader velocity because of re-illumination process accompanied with instantaneous elongation of leader. The direction of propagation is highly random and formed typically in steps. However, values of about 10 cm/μs may be accepted with reasonable approxi-

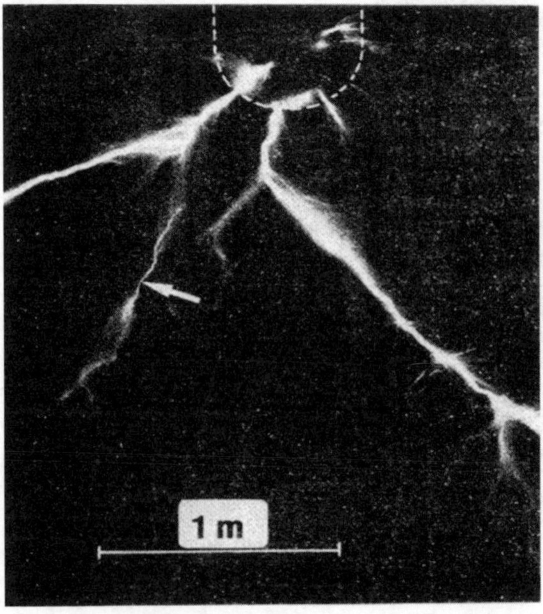

Figure 3.42 Profile of a "space leader" marked on a still photograph, Les Renardieres [3.51].

mations as average propagation velocity of leader for 50% breakdown voltage level ($U_{b\text{-}50}$), independent of the gap distance.

- there is a clear evidence of the dependence of the velocity of negative leader propagation upon the voltage characteristics.

On increasing the mean rate of rise of the applied voltage, the average leader velocity increases. The average conduction current also increases with the voltage slope, having the same trend as the velocity.

Despite greater complexity of the negative discharge compared to the positive, following similarities do exist as described in [3.51]

- the first streamer corona inception field has the same minimum value in both, but it has a smaller statistical spread in case of negative polarity. The magnitudes of streamer extension and light output in a particular phase of discharge are also smaller for negative polarity.
- positive streamers seem to have the same nature of discharge for both polarities. The comparison of a positive leader in positive discharge with a negative leader in negative discharge indicates that the nature of both the channels is similar.
- with both positive as well as negative polarities, the inception of leader channel depends upon the geometry of the electrodes and characteristics of the applied voltage.

The whole process of breakdown with leader discharge is dynamic. To be able to estimate the potential gradients inside the leader discharge is very difficult, especially when the mean potential gradient along the leader channel decreases during its propagation through different stages. The potential gradient requirement is estimated either by the breakdown characteristics of a gap or by computation with different methods of charge simulations for estimating the field on leader models. Conflicting numerical values are thus available from the literature.

The mean axial potential gradient in a stable positive leader is estimated by Lemke [3.39] to be higher than 1 kV/cm. As the leader channel extends, the injection of current in the leader by the corona at its tip increases the discharge current, bringing down the potential gradient to slightly less than 1 kV/cm. For negative leader, Les Renardieres [3.51] report estimates the mean potential gradient of 2 to 3 kV/cm, irrespective of the gap lengths. Unlike the hissing and flutter sounds produced by star and streamer corona respectively, the leader corona produces loud cracking audible noise (AN). It also quite intensively generates ozone.

3.3.3.3 Electromagnetic Interference (EMI) Produced by Corona In real life situations, the extremely nonuniform fields prevail most in the power system in atmospheric air. Corona activity is therefore a common feature. Corona or PB produces impulse current discharge giving rise to electromagnetic waves at a very wide frequency range. The star and the streamer coronas are normally permitted to a certain extent in the power system at free electrodes in atmospheric air. Efforts are made to avoid the leader corona due to its severity. However, the occurrence of leader corona in the form of "surface discharge" or "tracking" at the interface of some solid and liquid dielectrics with gaseous mediums is very common, especially under polluted conditions. All these coronas appear different, and produce different kind of audible noise and Electromagnetic Interference (EMI). They also generate ozone and cause electric power loss.

The individual characteristics of the EMI, caused by the three types of coronas, were investigated by Arora et al. [3.52, 3.53]. In their experimental investigations, three different electrode systems were selected to produce three different type of coronas; star, streamer and leader. A Spectrum Analyzer was used to measure the EMI level associated with the coronas with the help of two broadband antennae namely; "Stoddart Biconical" and "Stoddart Conical Logarithmic Spiral". Measurements were made at one-meter distance away from the source in dB above one μV per MHz.

For star corona, an EMI magnitude at its inception was measured to be −35 dBm in a frequency range of 30 to 50 MHz on applying only 10 kV ac, rms voltage to a needle electrode. On increasing the voltage to 45 kV, the highest level of EMI of −18 dBm was measured between 30 to 40 MHz. The EMI at higher frequencies was present but at lower intensity. An EMI level of −38 dBm was measured around 180 MHz, i.e. the video transmission frequency. Measurements at higher frequencies revealed that low intensity EMI due to star corona was present even up to 500 MHz.

In case of streamer corona at a 2.5 cm diameter hemispherical rod, the EMI inception could be detected on applying *ac rms* voltage of 30 kV at around 40 MHz. The maximum interference of −24 dBm could be measured on applying 70 kV but at around 40 MHz. These experiments revealed a very interesting fact that EMI produced at higher frequencies by streamer corona were of considerably low intensity, only 5 to 10 dBm above the system noise. No interference could be recorded beyond 160 MHz.

Unlike star and streamer coronas, the EMI by leader corona, produced as surface discharge in these experiments, was of much higher intensity. A maximum intensity of −13 dBm at 29 MHz could be detected. Higher intensity EMI was found to be present throughout the frequency range up to 400 MHz and intermittently in the range from 400 to 1000 MHz. The EMI was in fact found to be present even beyond 1 GHz but could not be measured due to constraints of the suitable antenna.

It can be concluded that the leader corona has the highest nuisance value causing maximum EMI. The star corona causes comparatively much lower interference, while the streamer corona causes minimal interference. The level of EMI increases, with increase in applied voltage/field conditions for all types of coronas, irrespective of their inception levels.

3.3.4 Summary of the Development of Breakdown in Extremely Nonuniform Fields

When a free electron in gaseous dielectric acquires enough kinetic energy from the applied electric field, it makes a series of successful impact ionizations colliding with neutral molecules. This process develops in a very systematic manner and leads to the formation of innumerable number of avalanches. In the case of extremely nonuniform field electrode configurations in gaseous dielectrics, PB, a local breakdown phenomenon known as "corona" takes place. It incepts at the location of highest field intensity on increasing the applied voltage. The complete breakdown in extremely nonuniform fields is always preceded with stable PB or corona. The type of corona determines the magnitude of breakdown voltage or the electric strength of the gaseous dielectrics.

Depending upon the electrode configurations, three types of coronas are distinguished. These are:

- **Star Corona:** It takes place in the gas at very sharp electrodes, for example, needle, protrusions, edges, brim etc. The avalanche formation at such electrodes is not able to acquire its critical amplification due to steep fall in field intensity. Since it is not able to grow in the gap, in the dark it appears like a "star" in the sky. It produces a hissing sound and the EMI due to this type of corona could extend up to 500 MHz [3.52 and 3.53].
- **Streamer Corona:** At larger and smoother electrodes, e.g., rod, sphere etc., where the gradient of the potential gradient is not very steep, the avalanche formation is able to acquire its critical amplification. Hence a streamer like PB, named as "streamer corona," takes place. It grows much deeper in the gap between the two electrodes. A clear "flutter" sound is produced by streamer corona. The EMI produced by this type of corona is found to limit itself up to only 40 MHz. The potential gradient across these channels is estimated to be ~4–5 kV/cm for positive and 10–15 kV/cm for negative polarity voltages.
- **Leader Corona:** On increasing the gap distance between the rod or sphere and plane electrodes to 1 *m* and above and increasing the applied voltage, streamer corona trajectories constrict. A "stem" like discharge is bunched together that grows into a bright stepped structure with branches known as "leader corona". At the tip of each branch, the dense streamer corona produces the charge particles for its high conductivity. The streamer corona at the tip also determines the tortuous path of the leader corona. A distinct "cracking" audible noise (AN) can be heard. The EMI produced by leader corona is measured up to 1 GHz and above [3.52 and 3.53]. The production of ozone by leader corona is so high that it can be smelled from a distance. The potential gradient across stable leader corona channels is estimated to be 1 kV/cm for positive and 2–3 kV/cm for negative polarity voltages.

Figure 3.43 shows the streamer, stembunch, and leader coronas all in one photograph, produced by a rotating Tesla coil set above a car by Terren [3.54]. Peter Terren, a physician and free lance researcher is attracted to anything having to do with high voltage. He has a very interesting site; http://tesladownunder.com of such pictures of stable PB in atmospheric air. The least bright, dense filamentary discharge in this picture is the streamer corona. At some locations constriction of these filaments and formation of stembunch can be seen. The very bright channels, less in number, are the leader corona channels.

In order to facilitate a better understanding of the subject, Table 3.5 complies the essential features of the development of PB in extremely nonuniform fields.

Figure 3.43 Streamer, stembunch and leader coronas produced by rotating Tesla coil set above a car, Terren [3.54].

3.3.5 Breakdown Voltage Characteristics of Air in Extremely Nonuniform Fields

For a stable partial breakdown process to be able to establish a fairly good conductive path between the electrodes, it requires a vigorous production of charge carrier particles. When this is accomplished, the PB process is rendered "unstable", leading to a "spark breakdown" of the dielectric between the two electrodes. Before the breakdown, depending upon electrode configuration and gap distance, three main types of stable PB may develop. These could grow to provide sufficiently conductive unstable breakdown channel. The three possibilities are:

1. A stable star corona leading to spark breakdown by forming an unstable leader, and finally an arc.
2. A stable streamer corona extended nearly up to opposite electrode, followed by a spark breakdown producing an unstable leader, and ultimately an arc.
3. A dense filamentary streamer corona is followed by stembunch discharge. Stable leader corona channels are then formed, accomplishing the spark breakdown at their advance stage with a final jump of unstable leader followed by an arc.

The breakdown voltage characteristics are distinguished accordingly. Figure 3.44 shows a schematic of the possibilities of accomplishing spark breakdown through these mechanisms for different types of voltages. It must be made clear that in all the cases, an "unstable leader" gives rise to a highly conductive breakdown channel, finally forming an arc. In other words, for achieving the minimum required conductivity of the breakdown channel, it is the leader discharge that acquires the plasma state of the gas transforming into an arc towards its final jump. Besides many other parameters, the magnitude of the breakdown voltage strongly depends upon

TABLE 3.5 Stable Partial Breakdown in air (corona) in extremely nonuniform fields

Type of PB Characteristics	Star Corona (below critical amplification of avalanche)	Streamer or Kanal Corona (above critical amplification of avalanche)	Stem and Leader Corona
Present with (Type of Voltages)	dc and ac	dc, ac Switching surge Lightning impulse	dc, ac Switching surge
Accompanied physical process	Impact ionozation	Impact and Photo ionizations	Impact, photo and Thermal ionizations
Audible Noise	Hissing (normal)	Flutter (low intensity)	Cracking (loud)
EMI	up to 500 MHz High intensity: ≈30–40 MHz	up to 160 MHz High intensity: ≈40 MHz	up to >1 GHz High intensity: ≈29 MHz
Average potential gradient (peak) required in the gap for growth/propagation	≤15 . . . kV/cm (+) 15 . . . 20 kV/cm (–)	4 to 5 kV/cm (+) 10 to 15 kV/cm (–) ≈7 kV/cm (~) peak	0.75–1.25 kV/cm (+) 2.....3 kV/cm (–) ≈ 2.....2.5 kV/cm (~) peak
List of associated nomenclature in the literature	—Star corona (+, –) —Glow corona (+, –) (Glimmentladung (+, –) —Corona (+, –) —Trichel (–) discharge	—Streamer Corona (+, –) —Bunch discharge (+, –) (Büschel Entladung) —Illuminated string corona (+, –) (Leuchtfaden Entladung) —Filamentary (+, –) glow discharge (dense streamer)	—Stembunch (+, –) discharge —Stepped leader (+, –) —Leader corona (+, –) —Space stem (–) —Space leader (–)

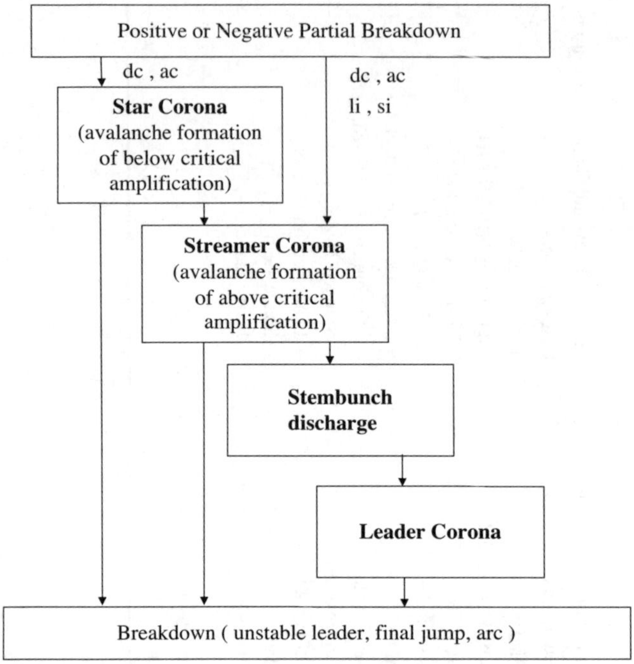

Figure 3.44 Schematic of breakdown initiated with stable PB by different types of voltages.

the type of stable corona that precedes the spark breakdown that in turn depends upon the electrode shapes and the gap distance.

In extremely nonuniform field configurations, as discussed, there is a significant effect of polarity of the applied voltage on breakdown. However, in case of *ac* power frequency voltage, where the polarity reverses every 10 ms, the breakdown always occurs at the positive polarity as the electric strength of air is lower for this polarity. The average breakdown voltage/potential gradients in this case are, therefore, measured between the *dc* positive and negative polarities.

In the case of impulse voltages (lightning "li" and switching "si" impulses), not only the polarity but also the time required to reach the crest of the voltage "T_{cr}" affects the breakdown. Beginning with the lightning impulse, lower breakdown voltages are measured as the time to crest of the applied voltage is increased. Depending upon the electrode configuration and gap length, a minimum of breakdown voltage is measured for a particular time to crest of the applied voltage. On increasing the time to crest further, higher breakdown voltages are measured again. A sort of a U curve characteristic is obtained as explained later.

3.3.5.1 *Breakdown Preceded with Stable Star Corona* Unlike uniform and weakly nonuniform fields, in the case of extremely nonuniform fields, the avalanches on a sharp electrode are not able to grow deeper in the gap towards the opposite electrode because of steep fall in potential gradient at the tip of the electrode. The breakdown process in this case begins at the sharp electrode. Breakdown with stable star corona is, therefore, only possible in very small gap lengths of a few *mm* to a few *cm* range, having a steep potential gradient at one or both the electrodes. Stable star corona is possible to be produced only with static *dc* or slow changing *ac* power frequency voltages. These provide sufficient time to build a steady "space charge" field. Under these conditions alone, the avalanche formation even at advance stages of PB is not able to achieve its critical amplification before the complete breakdown. With this result, the mean breakdown voltage in the electrode gap acquires comparatively higher values of the order of 15 to 20 kV/cm (peak) for the two polarity *dc* voltages.

Uhlmann in his dissertation in 1929 [3.55] measured breakdown voltage characteristics between a 30° needle-plane electrode configuration for different gap distances separately with positive and negative polarity *dc* voltages, shown in Figure 3.45. A strong effect of polarity on the breakdown voltage magnitudes is observed in these cases of extremely nonuniform fields. The *ac* (peak) breakdown voltage characteristic would fall slightly above the positive polarity *dc*.

As seen in Figure 3.45, the average potential gradient required for breakdown decreases as the electrode gap distance is increased. This is an indication of the development of transition process. One may conclude that for this electrode configuration, breakdown with stable pure star corona is possible only up to a gap length of about 1.5 *cm*. Under these conditions, it can be seen that the average potential gradient measured initially for breakdown in the gap are about 15 kV/cm for positive polarity and 20 kV/cm for negative polarity voltages, as described earlier. At longer

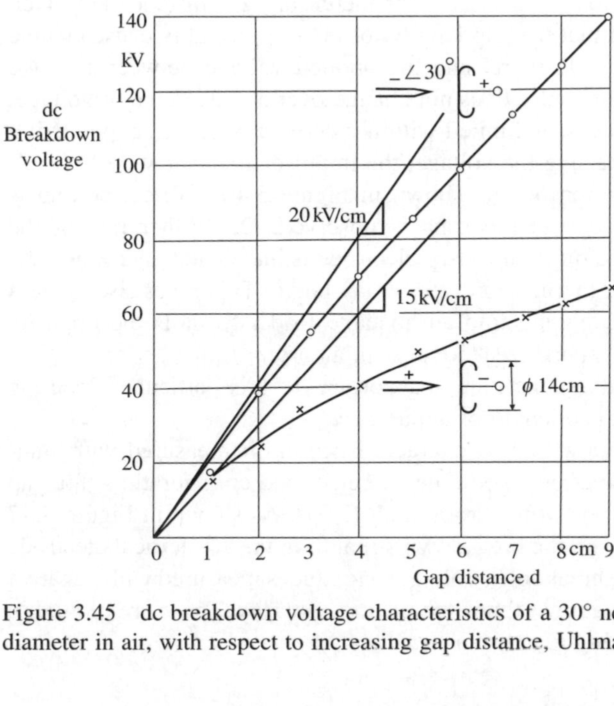

Figure 3.45 dc breakdown voltage characteristics of a 30° needle and a plane of 14 cm diameter in air, with respect to increasing gap distance, Uhlmann [3.55].

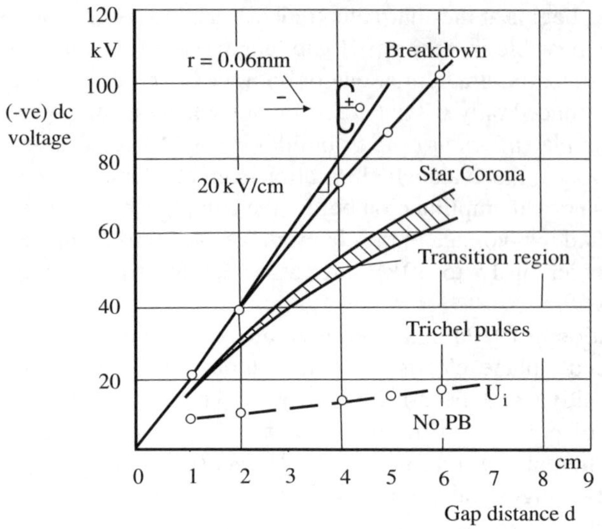

Figure 3.46 dc negative needle-plane breakdown and corona discharge characteristics in atmospheric air, Kuffel [3.56].

gap lengths, where a lower average breakdown potential gradient is measured, streamer corona may have commenced towards the advance stage of breakdown.

Similar characteristics were measured with *dc* voltage by Kuffel and Zaengl [3.56] for negative polarity needle-plane electrode configuration at different gap lengths. The very sharp needle electrode (cathode) is reported to have a radius of 0.06 mm, Figure 3.46. Below the lowest curve, no PB current is measured. The PB and the Trichel pulse inception voltages, U_i, for increasing gap distance are given by this curve. It can be seen that the gap length does not appreciably cause change in the PB inception voltage. On increasing the applied voltage between the two electrodes, the mode to Trichel pulse does not change over a wide range of voltage. The avalanche formation process is limited within its critical stage, as explained in Section 3.3.1.2. On further raising the voltage, the impulse characteristic of the PB current is lost. Over a transition region, shown in Figure 3.46, a direct current is measured and a steady (stable) star corona is observed. On further raising the voltage, the PB process at the tip of the sharp electrode is intensified, increasing the current density so much that thermal ionization may begin. This gives rise to stem discharge instantaneously extending into an unstable leader towards the opposite electrode and accomplishing spark breakdown with an arc.

A breakdown only with stable pure star corona for this particular electrode configuration is limited to a gap length of about 2 *cm*.

Impulse voltage breakdown, U_{b-50} characteristics of air measured with standard *li* and *si* for needle-plane and needle-needle electrode configurations for gap distances up to 15 cm are reported by Arora et al. [3.57]. As shown in Figure 3.47 (a), strong effects of polarity on the breakdown strength in unsymmetrical electrode system can be observed. The breakdown voltage magnitudes measured with negative polarity *li* and *si* are more than two times the positive polarity. These are plotted in Figure 3.47 (b) with increasing gap distances.

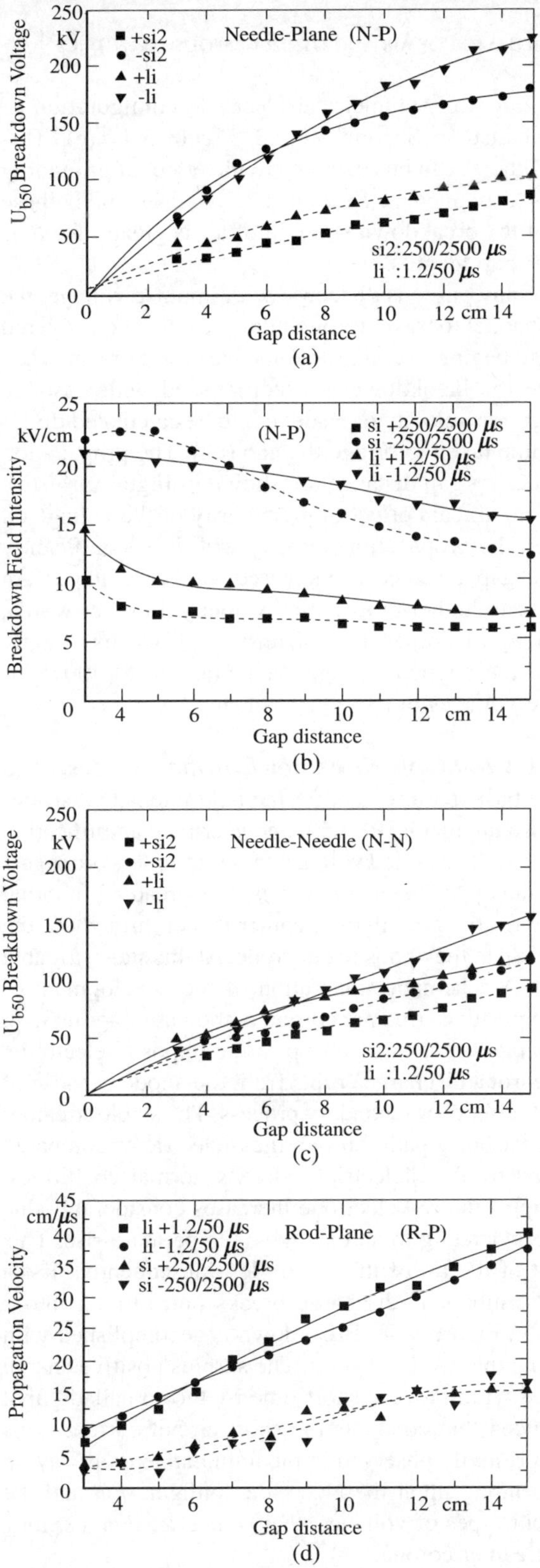

Figure 3.47 Breakdown of air in short gap lengths with li and si impulse voltages, Arora et al. [3.57]. (a) Breakdown characteristics for needle-plane. (b) Average breakdown strength with increasing gap distance. (c) Breakdown characteristics for needle-needle. (d) Measured propagation velocity of thc breakdown channel with increasing gap distance.

Effect of polarity of the voltage on symmetrical electrode configuration of needle-needle is marginal as explained in Section 3.3.2.3, Figure 3.47 (c). This marginal difference in measured voltages can be attributed to the effect of grounding of the lower needle electrode, as explained in Section 2.3.1.1. From both these figures it can also be observed that the breakdown strength of air is greater for *li* as compared to *si* even for such short gap lengths.

In the event of breakdown of any gaseous dielectric with impulse voltage, the time required for the breakdown channel to travel through the gap can be considered from the instant the applied voltage begins to collapse to the earliest moment when it acquires zero magnitude. Since the breakdown is accomplished with unstable leader bridging the gap, the propagation velocity of leader could be calculated dividing the gap distance (cm) by the time taken to bridge the gap (μs). The propagation velocities thus calculated for increasing gap distance are shown in Figure 3.47 (d). It can be observed that there is no significant effect of the polarity of the voltage on the propagation velocity. However, the propagation velocity with *li* is significantly higher as compared to *si*. As the gap distance is increased, higher voltages are required for breakdown, giving rise to higher propagation velocity. In other words, it can be concluded that the velocity of propagation of unstable leader is directly proportional to the rate of rise of the applied voltage. Velocities up to 400 m/ms could be measured with *li* impulse voltages in these experiments.

3.3.5.2 Breakdown Preceded with Stable Streamer Corona As described in detail in Section 3.3.2, when the basic requirement for the field intensity distribution in the gap is met, streamer or Kanal, that is, PB with above critical amplification of avalanche is able to grow. On raising the applied voltage, these streamers propagate in the main field direction toward the opposite electrode, besides spreading in radial direction. If the condition required for the growth of streamer is met throughout the gap length, the PB is able to extend up to the opposite electrode. At this stage, a stable streamer PB is rendered unstable. A schematic illustration of the development of breakdown mechanism with stable positive streamer corona is shown in Figure 3.48.

As soon as the streamer is able to extend itself up to the opposite electrode, say the cathode, denser streamer corona discharge erupts from the anode because of "cathode effect", also known as "γ-effect" or secondary process. The stable streamer corona establishes a much lower resistance path through the dielectric as compared to the insulation resistance offered by the dielectric under its normal conditions. Hence the current density at the tip of the rod electrode increases considerably due to the conduction of charge carriers, leading to an excessive temperature rise. This causes thermal ionization in front of the tip of the electrode. Subsequently, first a short bright "stem" and then a "stembunch" discharge breaks out, turning into a thriving unstable leader as shown in Figure 3.48. Breakdown is accomplished with a "final jump" of the leader bridging the two electrodes. The surplus positive charge in the leader channel is neutralized within a very short time by the so-called "final jump". Ultimately an arc is produced that conducts the short circuit current. This type of breakdown mechanism is primarily observed in medium gap lengths, say up to about 1 or maximum 2 *m*, depending upon the electrode configuration and the type of applied voltage. All the four types of voltages, that is, *ac*, *dc*, *li* and *si* may produce breakdown with stable streamer corona.

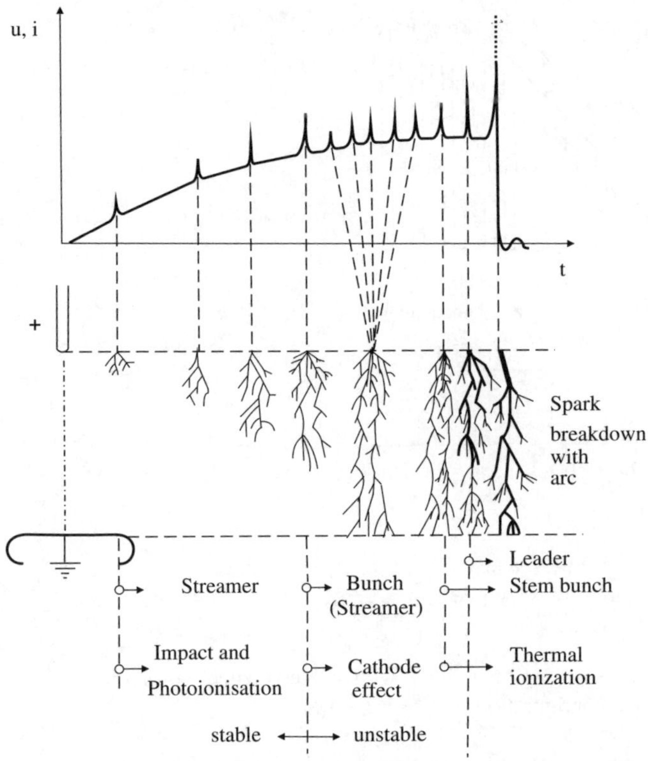

Figure 3.48 Breakdown mechanism development with stable streamer corona.

An analytical explanation of the mechanism described above is difficult. However, a distinction between breakdown with stable and unstable streamers can be made in terms of the degree of uniformity of the field. Consider a positive sphere-plane electrode configuration. For small gap distances, a weakly nonuniform field exists in the gap. On applying sufficient magnitude of voltage to the sphere to achieve the required maximum field intensity for breakdown $E_{b\max}$, the inception of streamer immediately leads to an abrupt breakdown, region A in Figure 3.49. No stable PB occurs in this region. On increasing the gap distance, the field becomes more nonuniform. Beyond the region A of the gap, a condition is reached when the streamer corona inception at the sphere is not able to extend immediately up to the opposite electrode. Hence, a stable streamer partial breakdown activity takes place in the gap as shown in the region B.

The borderline between these two phenomena is described in terms of a limited value of η, the "η_{lim}". In the region A, η is larger than η_{lim} whereas it is smaller in the region B. The threshold degree of uniformity "η_{lim}" is defined at the point where the weakly nonuniform field turns over to extremely nonuniform field. The field intensity at the sphere under these conditions must acquire the value $E_{b\max}$, required for breakdown. As illustrated in this figure, stable streamer corona may occur only in the region B. The PB inception voltage U_i in terms of the electrode geometry

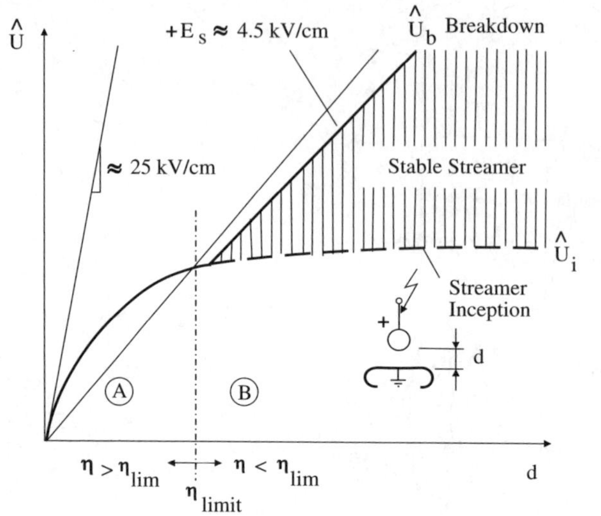

Region Ⓐ : $U_i = U_b \approx E_{b\,max} \bullet d \bullet \eta$

Region Ⓑ : $U_i \approx E_{b\,max} \bullet d \bullet \eta$, $U_b \neq f(\eta)$,

$U_b \approx E_s \bullet d$

Figure 3.49 Threshold curves showing breakdown with stable streamer corona for positive sphere-plane electrode configuration.

can also be estimated for the region B, but not the breakdown voltage. The magnitude of the breakdown voltage in the region B primarily depends upon the average potential gradient required for the growth of streamer corona "E_s" and the actual extent of its growth in the gap. The borderline between region A and B represented by η_{lim} also represents the distinction between the weakly and extremely nonuniform fields. The position of the line representing η_{lim}, the transition from weakly to extremely nonuniform field, depends upon the electrode configuration.

Consider a breakdown when η is equal to η_{lim}. The breakdown voltage for the region A is given by equation (3.35),

$$U_b = E_{b\max} \cdot d \cdot \eta_{\text{lim}} \tag{3.38}$$

The breakdown voltage for the region B is given by the equation,

$$U_b \approx E_s \cdot d \tag{3.39}$$

where E_s is the average potential gradient or the potential drop per unit length across the stable streamer corona trajectory, which is about 4.5 kV/cm for positive polarity voltage.

Equating the two equations given above and considering a value of 25 kV/cm for $E_{b\max}$ (the maximum breakdown field intensity for air in weakly nonuniform fields), the value of η_{lim} for air can be calculated as follows:

$$E_{b\max} \cdot d \cdot \eta_{\text{lim}} \approx E_s \cdot d$$

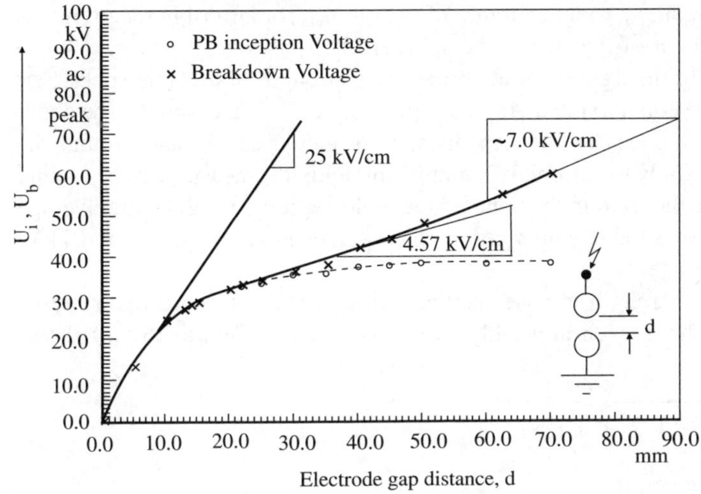

Figure 3.50 PB inception U_i and breakdown voltage U_b characteristics for 9.92 mm diameter sphere-sphere electrodes, Prem [3.59].

or

$$\eta_{\lim} \approx \frac{E_s}{E_{b\max}}$$
$$\approx \frac{4.5}{25} \approx 0.2$$

For estimating the value of $\eta_{\lim}$ in atmospheric air, experimental investigations were made by Arora et al. [3.58, 3.59] with *ac* power frequency voltage. Measurements were carried out on three sets of identical sphere-sphere electrode system of diameters 9.92, 15.0 and 20.0 mm. The upper electrode was applied *ac* voltage and the lower was grounded. The PB inception voltage, U_i, and the breakdown voltage, U_b, were simultaneously measured with increasing gap distance. Figure 3.50 shows the *ac* peak PB inception and breakdown voltages with increasing gap distance for a 9.92 mm diameter sphere-sphere electrode system. Identical curves were also measured for the other two pairs of spheres.

The measured results revealed that for the gap distance up to about twice the diameter of the spheres, no PB could be detected. The field can therefore be considered to be weakly nonuniform up to the gap distance of twice the diameter of the identical spheres when the lower sphere is grounded. Above this gap distance the PB inception U_i and the breakdown U_b voltages could be measured separately; hence, the field is rendered extremely nonuniform. Stable streamer corona occurred in the range between the two characteristics shown in Figure 3.50.

The value of Schwaiger Factor limit, $\eta_{\lim}$, at the gap distance where the transition occurred was estimated taking in to consideration the effect of grounding of the lower electrode as well as the effect of the connecting shanks of the spheres. The carefully estimated value of $\eta_{\lim}$ was found to vary between 0.24 and 0.25. It would therefore be safe to assign a rounded figure of 0.25 to η_{lim} for atmospheric air. It was found to be independent of the size of the spheres [3.59].

It can be seen in the figure that the breakdown strength of the atmospheric air up to the gap distance equal to the diameter of the sphere, D, is measured to be about 25 kV/cm (peak). It reduces as the gap distance is increased. At and around the transition from the weakly to extremely nonuniform field, it is measured to be around 15 kV/cm (peak). In the extremely nonuniform field region (d > 2D), the average potential gradient across the streamer corona has been estimated to be around 7 kV/cm for *ac* (peak) voltage.

The breakdown strength of gaseous dielectrics strongly depends upon degree of uniformity of the field, even in weakly nonuniform fields. Figure 3.51 (a) shows

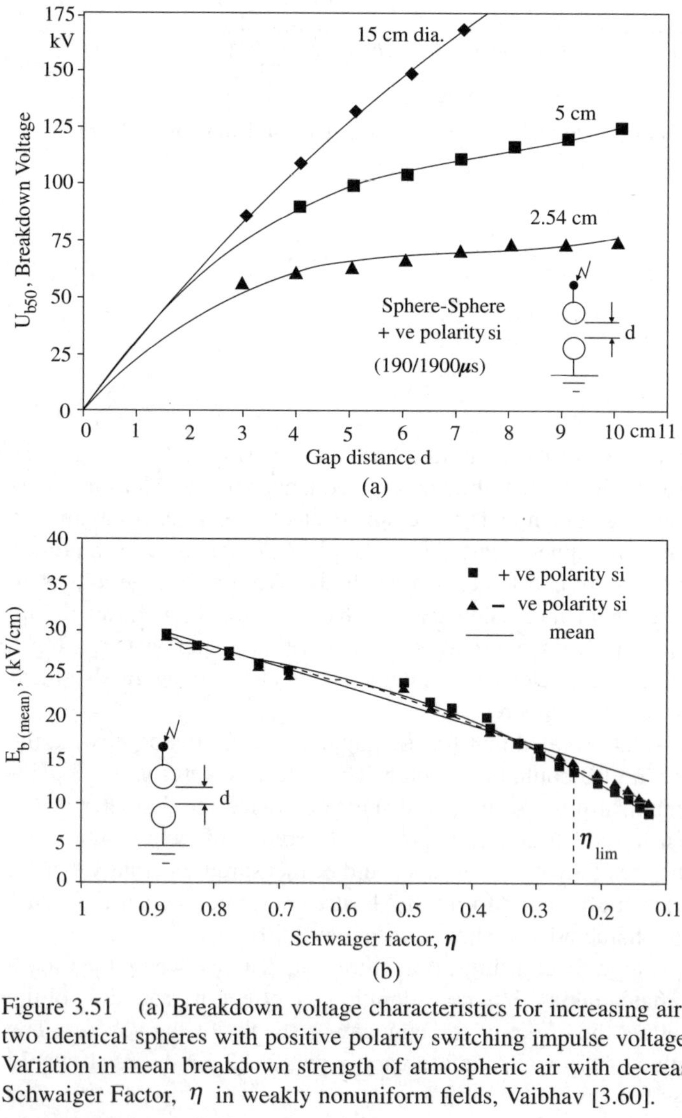

Figure 3.51 (a) Breakdown voltage characteristics for increasing air gap distance between two identical spheres with positive polarity switching impulse voltage, Vaibhav [3.60]. (b) Variation in mean breakdown strength of atmospheric air with decreasing value of Schwaiger Factor, η in weakly nonuniform fields, Vaibhav [3.60].

the variation in positive polarity switching breakdown voltage, $U_{b\text{-}50}$ for air with increasing gap distance between two identical spheres of diameters of 2.54, 5 and 15 cm measured by Vaibhav [3.60]. Since the field is more uniform between 15 cm diameter spheres, much higher strength of air is measured. Figure 3.51 (b) shows the variation in mean breakdown strength of atmospheric air with decreasing value of Schwaiger Factor, η measured between sphere-sphere electrodes with both polarities of *si* voltages of 190/1900 μs shape also measured by Vaibhav [3.60]. Similar but slightly higher breakdown voltage characteristics were measured with *li* impulse voltages of shape 1/50 μs. It was revealed by these measurements that the effect of polarity on the breakdown strength of air is negligible for the values of η between 1.00 to 0.65. In this range of η in which weakly nonuniform field exists, a linear characteristic of breakdown strength was measured. A transition in breakdown phenomenon is evident from the reducing breakdown strength below the η value of 0.65. Below the value of η_{lim} (0.24) the mean breakdown strength reduces considerably.

Characteristics of breakdown with *dc* voltage for increasing air gap up to 2.5 m are shown in Figure 3.52 with both positive and negative polarities for sphere-sphere and rod-plane electrodes. The curve number 1 for a large size sphere-sphere configuration represents weakly nonuniform field even beyond 50 cm of gap distance. Since no partial breakdown activity takes place before the breakdown, space charges of significant magnitude are unable to develop in this field configuration. No effect of polarity is therefore measured on the breakdown voltage characteristic. The curve number 2, measured with negative polarity voltage on a rod-plane gap, a case of extremely nonuniform field, is accompanied with stable negative streamer corona before the breakdown. The mean potential gradient requirement of 15 kV/cm for short gap distances and 10 kV/cm for longer gap distances for negative streamer can

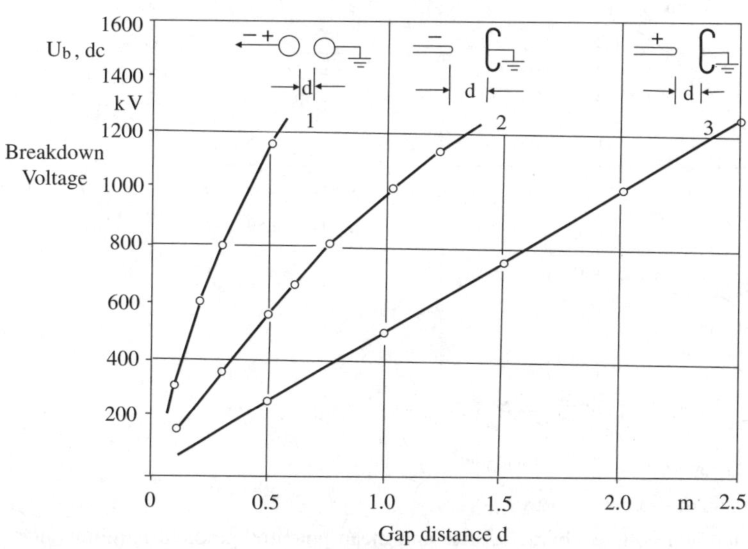

Figure 3.52 dc breakdown voltage characteristics with both polarities for sphere-sphere and rod-plane long air gaps.

be observed to have met in this curve. The last curve, number 3, accompanied with stable positive streamer, represents an average potential gradient requirement of about 5 kV/cm in the gap. A significant effect of polarity on the breakdown voltage on asymmetric electrodes can be observed. The breakdown characteristics with *ac* (peak) power frequency voltage for rod-plane electrode configuration would almost fall in between the positive and negative polarity *dc*, curves number 2 and 3.

The average potential gradients measured in these cases are slightly lower because of fairly long air gaps. The respective streamer coronas begin to transform into their next stage of development before the breakdown in gap distances above 1 m.

3.3.5.3 Breakdown Preceded with Stable Streamer and Leader Coronas (Long Air Gaps) The breakdown characteristics described so far suggest that the magnitude of breakdown voltages strongly depend upon the types of stable partial breakdown, which are able to develop in the gap before the final jump. The mean potential gradient requirement for the pre-breakdown stages in gap distances up to 6 m were estimated by Lemke [3.38]. The breakdown voltage characteristic for a positive rod-plane gap, Figure 3.53, on applying 60/2500 μs shape of positive polarity *si* impulse voltages, represent breakdown with stable streamer corona up to a

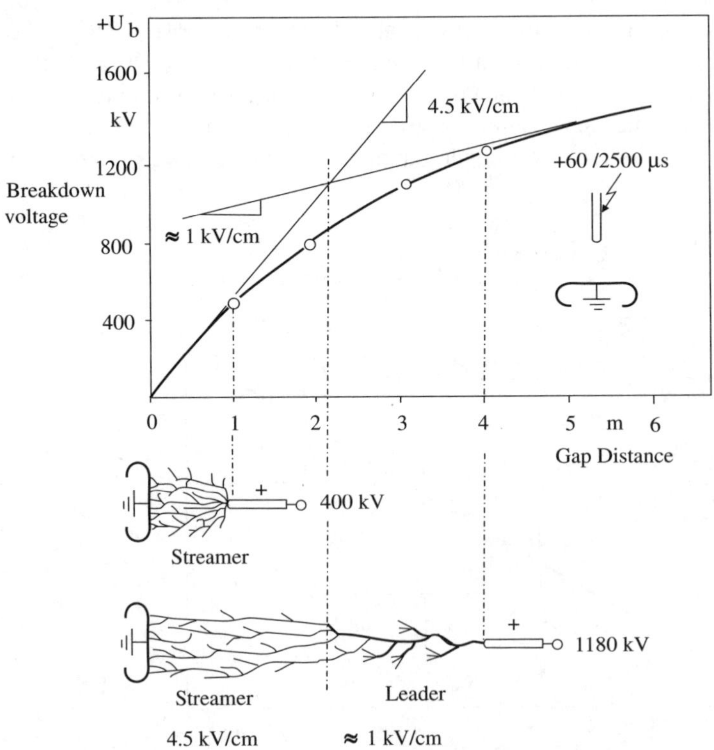

Figure 3.53 Breakdown voltage characteristics and mean potential gradient requirement for the propagation of streamer and leader coronas, Lemke [3.38].

relatively small gap distance of 1 m. The average potential gradient for breakdown in this region was measured to be 4.5 kV/cm. On increasing the gap distance, a lower potential gradient requirement for breakdown was measured continuously in the transition range of gap distance. Finally, an average potential gradient of 1 kV/cm was required for breakdown for the gap distances above 4 m in this case. Stable leader corona could be observed for gap lengths above 2 m for the given electrode system. As illustrated in Figure 3.53, the breakdown voltage characteristic falls between the two tangents drawn on the curve, representing the potential gradients of circa 4.5 and 1 kV/cm. The region with stable streamer extends to a maximum of 2 m. The breakdown characteristic in the region beyond this length is determined by stable leader corona before the complete breakdown.

Breakdown mechanism in very long gap distances in air preceded with stable leader corona is described in Figure 3.54. Consider a long gap distance, above 2 m, between a positive rod or sphere and plane electrode configuration. On applying a preferably slow changing voltage, that is, *ac* or *si* of sufficiently large magnitude, at first a very strong and dense filamentary streamer discharge appears at the high voltage electrode. This is known as "first corona" shown in the schematic of

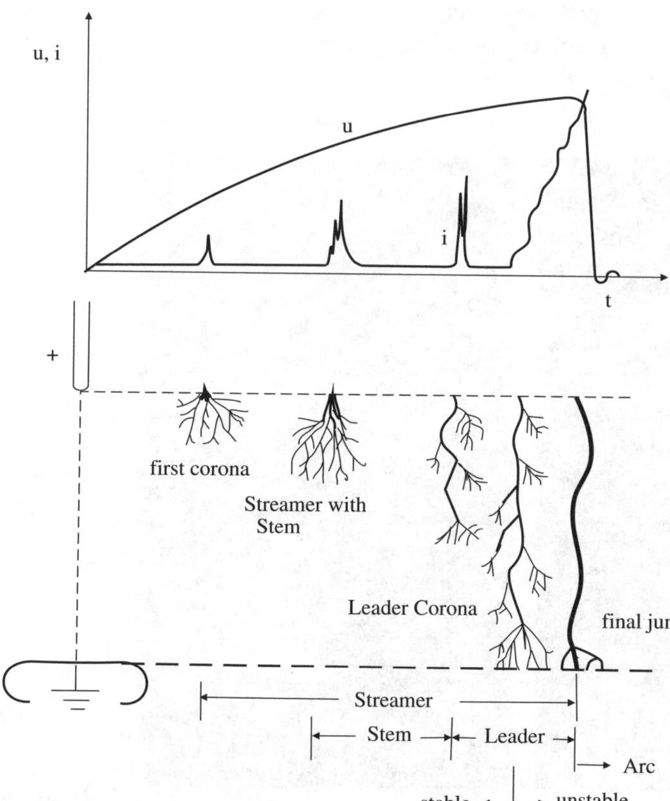

Figure 3.54 Breakdown mechanism schematic with stable leader discharge.

breakdown, Figure 3.54. Unlike cases of short gap lengths, in this case the minimum collective current density in discharge channel required to produce a "leader" (about 200 mA) may be achieved even without the cathode effect. The first corona is then followed by stable stem at the rod electrode and then the leader extending in the gap. The stepped form of leader branches is accompanied with streamer corona at their tips. The leader corona propagates, traversing the path laid down by streamer corona as it provides the leader the required current density.

Thus a stable leader corona may prevail in the gap extending close to the opposite electrode, as illustrated in Figure 3.54. On increasing the applied voltage, breakdown of air between the electrodes is accomplished with the "final jump" by an unstable leader bridging the gap, followed by an arc. Photographs of a stable leader and a breakdown with its final jump on a 7 m gap length between rod-plane, applying positive polarity *si* were taken by Lemke [3.38], shown in Figure 3.55.

The breakdown mechanism with negative polarity voltages in very long gap lengths is slightly different. It is described in detail in Section 3.3.3.2. compared to the positive polarity. The difference occurs in two ways. Firstly, there is an eruption of a positive leader from any weak point on the anode. Secondly, a typical phenomenon of the occurrence of "space leader" somewhere in the gap takes place, extending in either direction. The latter phenomenon is observed only in gap lengths above 2 m.

Breakdown voltage characteristics for very long gap distances in air with different types of voltages have received considerable attention all over the world since

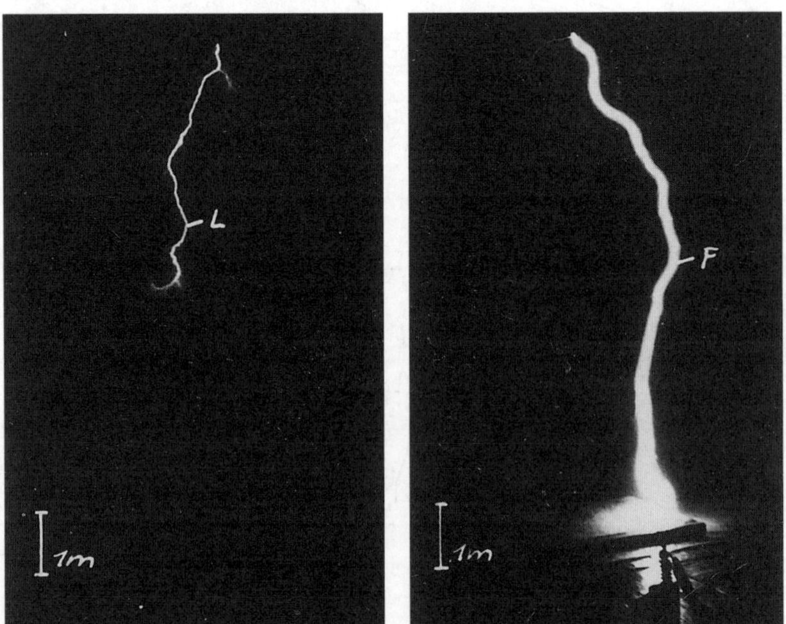

Figure 3.55 Stable leader and breakdown with final jump for a 7 m gap with positive polarity si voltage, Lemke [3.38].

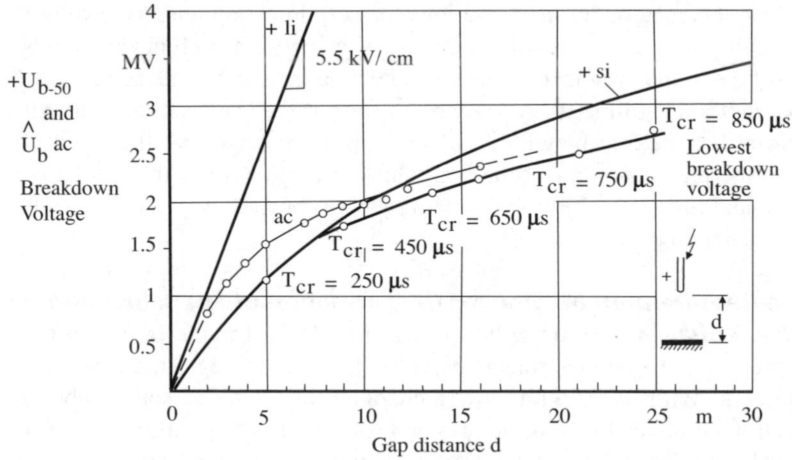

Figure 3.56 Breakdown voltage characteristics for positive polarity rod-plane very long gap lengths in air (1013 mm bar, 20°C) for standard li, si and si of critical time to crest, Kind [3.61], and for ac power frequency voltage, Alexandrov et al. [3.62].

the 1960s. The necessity to achieve higher transmission voltages inspired the engineers to investigate the breakdown mechanisms of very long air gaps. Gap distances up to 27 m have been investigated with various types of voltages.

Breakdown characteristics with increasing gap distance d, between rod-plane electrodes in air are brought together in Figure 3.56 for standard positive polarity lightning (1.2/50 µs) and switching (250/2500 µs) impulse 50% breakdown voltages as well as for *ac* power frequency voltage. These characteristics were published by Kind et al. [3.61] and Alexandrov et al. [3.62]. The curve with *ac* power frequency voltage in this figure was calculated and confirmed by measurements. The instant of breakdown in this case is obviously expected to occur at the positive peak value of the alternating cycle. A sort of saturation can be observed in this curve above 5 m gap distance. As seen in Figure 3.56, considerably higher magnitudes of positive polarity *li* voltages are required for the breakdown as compared to *ac* and *si*. The negative polarity *li* breakdown voltages in this case of asymmetric electrodes will be much higher than even the positive polarity. High breakdown voltages with *li* impulse are measured because stable PB activity is not able to develop in the gap due to its short duration. The breakdown voltage characteristic with standard *si* also shows a saturation above the gap length of about 7.5 m. Breakdown characteristic for positive polarity *dc* would fall between the *ac* and positive polarity *li*.

Another interesting phenomenon observed with switching impulse voltages of longer time to crest ($T_{cr} > 250\ \mu s$), is that 50% breakdown voltage magnitudes are lower than as measured with standard *si* of time to crest of 250 μs. For a particular gap length and electrode configuration, a minimum breakdown voltage of a particular time to crest exists, known as "critical time to crest" T_{cr} (critical). For longer gap distances, this minimum breakdown voltage is measured with longer time to crest, as shown by the lowest curve in Figure 3.56.

Breakdown characteristics investigations for very long gap lengths have been made mostly with impulse voltages of various wave shapes. On applying an impulse voltage of sufficient magnitude to cause breakdown in a gap, the actual spark breakdown occurs after a time delay known as "time lag". The time lag primarily depends upon the shape and magnitude of the applied voltage and the electrode configuration. The phenomenon of time lag, which strongly affects the breakdown characteristics and the probability of occurrence of breakdown under impulse voltages, is explained below.

3.3.5.4 The Requirement of Time for the Formation of Spark Breakdown with Impulse Voltages As described in Section 3.2.2, for the initiation of a breakdown, presence of free electrons is essential in order to begin the ionization/avalanche process. With *dc* or with slowly changing *ac* voltages, the number of initiatory electrons produced by cosmic rays and natural ultraviolet illumination are sufficient to initiate a breakdown. However, with impulse voltages, especially with pulses of very short duration, for example *li*, it is possible that breakdown may not occur even when the peak voltage magnitude exceeds the lowest or the static breakdown voltage (U_S). Presence of sufficient number of initiatory electrons must therefore be ensured. This may cause the so-called "statistical time lag" t_s, or a time delay in beginning the discharge process with impulse voltages. The time t_s depends upon the presence of the amount of pre-ionization in the gap. The pre-ionization in turn depends upon the area of electrodes, the gap distance and the magnitude of UV irradiation to produce the primary electrons. The probability of the presence of such electrons is usually statistically distributed. The time t_s may extend from a few μs to over 100 µs. However, the time t_s can be reduced considerably by artificial irradiation of the electrode surface. It may also be reduced on applying higher magnitudes of voltages above U_s.

Once the discharge process has begun, it requires a certain finite "propagation time" t_p to reach the opposite electrode. The requirement of total propagation time in a gap depends upon the individual types of PB and their extent in the gap just before the breakdown. The extents of PB are affected by the field distribution, gap distance and the magnitude of the applied voltage. On an average, one may consider the propagation velocity for stable streamer corona to be 100–300 cm/µs and for stable leader corona about 10 cm/µs. Since the breakdown in very long air gaps is achieved when a stable leader is able to bridge the gap, the transit time required for propagation is mainly determined by the leader propagation velocity in these gaps.

Consider the general shape of an impulse voltage $u(t)$, shown in Figure 3.57 (a), having $\hat{U}$ its peak magnitude applied to a gap. Let U_s be the minimum magnitude of the voltage that leads to breakdown of this gap after a long time of its application, for example, a direct voltage. The times t_s, t_p and t_b represent the statistical time lag, the time required for propagation and the total time required for the formation of breakdown, respectively. These are related as,

$$t_b = t_s + t_p \tag{3.40}$$

The propagation time t_p has also been described as "formative time" in a number of references. The proportion of t_s to t_p in Figure 3.57 (a) has been chosen

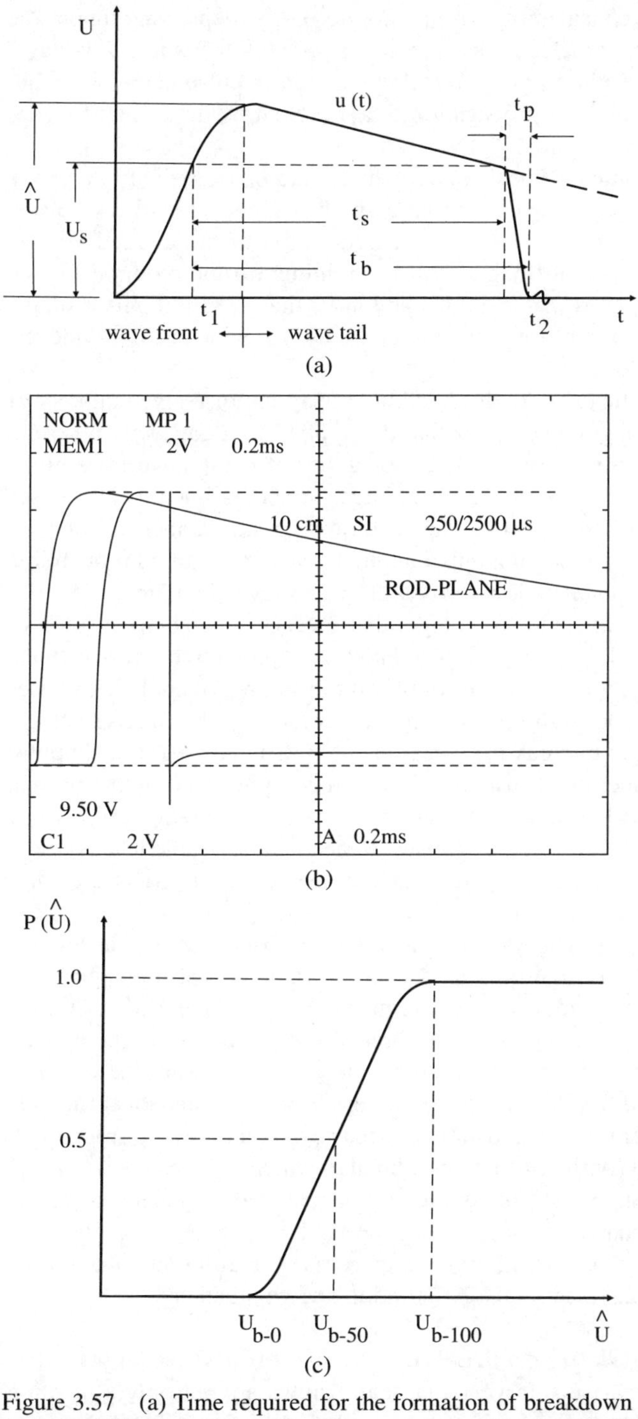

Figure 3.57 (a) Time required for the formation of breakdown with impulse voltages. (b) An Oscillogram showing breakdown with positive polarity li impulse voltage between rod-plane electrode. (c) Probability of breakdown with impulse voltages.

arbitrarily. It may vary considerably from case to case, depending upon the gap distance, surface area of electrodes and magnitude of the applied voltage. Figure 3.57 (b) shows an oscillogram of breakdown with *li* impulse in the wave tail. The statistical time lag t_s and the propagation time required by the unstable leader t_p can be clearly seen in this figure.

When an impulse voltage of peak value greater than U_s is applied to a gap, a certain probability but no certainty exists that a spark breakdown will take place. The occurrence of breakdown is conditional. The duration of the applied voltage magnitude above U_s must prevail longer than the minimum time required for the formation of breakdown, that is $(t_2 - t_1)$, must be longer than t_b. An impulse voltage having the same wave shape but lower magnitude may not be able to provide the required $(t_2 - t_1)$ time duration.

Because of the statistical nature of the "time lag" t_s, out of a given number of impulses of a particular magnitude higher than U_s applied to a gap, only a certain percentage may accomplish breakdown. The probability of breakdown for a given impulse voltage is estimated experimentally. A large number of identical impulses in shape and magnitude are applied on a gap. The ratio of the number of impulses accomplishing breakdown to the total number applied gives the required probability. The distribution function of breakdown probability is shown in Figure 3.57 (c). U_{b-100} represents the 100% breakdown voltage, that is, each voltage application of this magnitude leads to breakdown. This is also known as "protective level" voltage. U_{b-50} is the 50% breakdown voltage, that is, only half the number of applied impulses of the same magnitude accomplish breakdown. U_{b-0} is the highest impulse voltage magnitude that does not lead to breakdown at all. U_{b-0} is also known as the "impulse withstand level". In practice, the distribution function is represented by the normal (Gaussian) distribution with standard deviation. A particular voltage level may always be dubious whether it is U_{b-100} or greater. Hence it is advisable to work with a voltage magnitude lower than U_{b-100} for gaseous dielectrics. U_{b-50} is the most common level adopted in practice for investigations.

Consider impulse voltages of the same wave shape but of different peak voltage magnitudes applied to a gap, Figure 3.58. Let the wave 1 of peak magnitude $\hat{U}_1$, represent the 50% breakdown voltage $\hat{U}_{1b-50}$, requiring a breakdown time of t_{b1}. On raising the voltage, not only the probability of breakdown increases but the process of breakdown gradually shifts from somewhere on the wave tail toward the vertex of the applied voltage, t_{b2} and t_{b3}. On raising the voltage further the statistical time lag t_s may reduce so much that the breakdown may occur even on the wave front itself, t_{b4}. Thus the time required for the formation of breakdown t_b reduces on increasing the peak voltage magnitude or effectively the rate of rise of the applied voltage. A hyperbolic voltage-time characteristic is obtained, as shown in Figure 3.58. Each shape of the impulse voltage having different T_{cr} gives rise to a different voltage-time characteristic for a particular electrode configuration and gap distance.

3.3.5.5 *Effect of Wave Shape on Breakdown with Impulse Voltages*

The breakdown voltage characteristics in very long gap lengths are not only a function of the gap distance and electrode geometry but also the rate of rise, duration of the applied voltage and the wave shape. The time required for the development of breakdown t_b in relatively smaller gap distances mainly depends upon the statistical

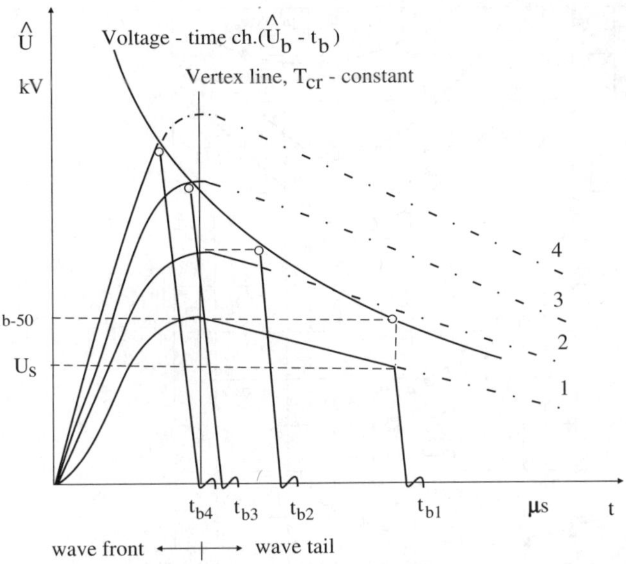

Figure 3.58 Breakdown voltage-time characteristics for impulse voltages of a particular wave shape but increasing rate of rise of the voltage.

time lag t_s, whereas for very long gap distances it mainly depends upon the time required by the propagation of leader t_p. Different breakdown voltage magnitudes are required for the same gap distance and electrode geometry on applying impulse voltages of either polarity. Additionally, the breakdown voltage is also found to depend upon the impulse voltage wave shapes having different time to crest T_{cr}.

Since the statistical time lag t_s with *li* is long, only applying higher voltage magnitudes increases the probability of breakdown. This is the reason for measuring high breakdown voltages with *li* compared to other types of voltages having a lower rate of rise, Figure 3.56. Because of very short duration of the *li* pulse, a stable leader corona is not able to take place in this case. Hence, a mean potential gradient of about 5 kV/cm (as for streamer corona) is required for breakdown in long air gap lengths with positive polarity *li* as shown in Figure 3.56.

The phenomenon of time required for the formation of breakdown with different shapes of *si* voltages is of immense practical importance. This was experimentally investigated in detail by Stekolnikov et al. [3.63] and Pigini et al. [3.64] for positive polarity voltages as shown in Figure 3.59. Similar measurements with negative polarity voltages were later conducted at Les Renardieres [3.51] as shown in Figure 3.60.

Stekolnikov et al. in Russia measured the curves for smaller gap distances, on rod-plane air gaps between 1 to 5.9 m. Pigini et al. recorded similar curves during their measurements in Italy for gap distances of 13 to 17 m with both rod and sphere-plane electrode configurations, shown in Figure 3.59. The negative polarity breakdown voltages for asymmetrical electrode configurations are much higher and therefore of less importance. However, the effect of time to crest (T_{cr}) of negative polarity voltages on thc breakdown voltages magnitude was later studied in France for gap lengths between 2 to 7 m, shown in Figure 3.60. As seen in these figures, a

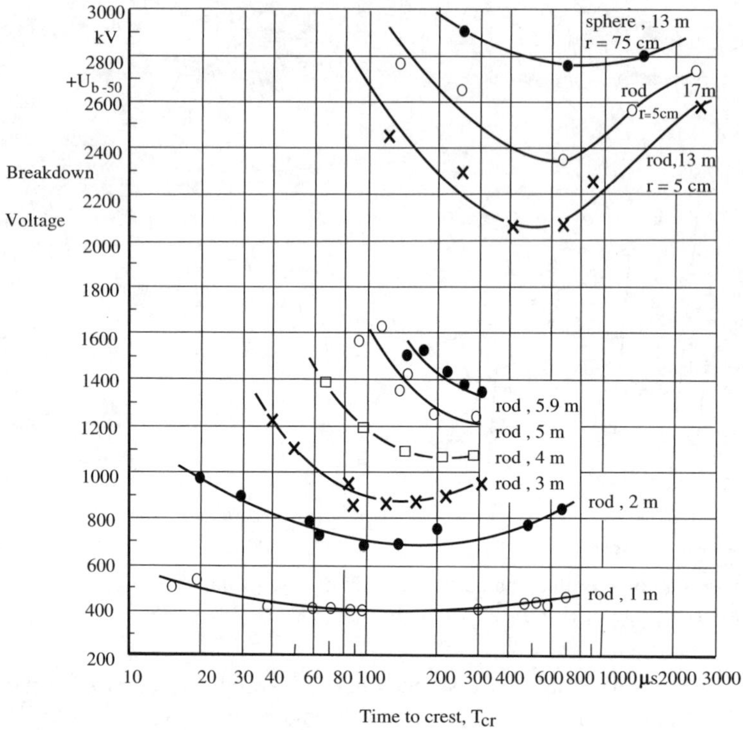

Figure 3.59 Positive polarity 50% breakdown voltage as a function of time to crest T_{cr} for rod and sphere-plane air gaps of different lengths, Stekolnikov and Pigini et al. [3.63, 3.64].

sort of a “U” characteristic is measured that is rather flat for small gap distances up to 2 m. For gap distances longer than 2 m, clear “U” characteristics predominantly affected by the gap length and electrode shape are measured, confirming the existence of a minimum breakdown voltage. This effect is clearly seen for 13 m rod-plane and sphere-plane characteristics, illustrated in Figure 3.59.

The critical time to crest T_{cr} (critical) is that rise time of the impulse wave for which the 50% breakdown voltage magnitude is minimal for a particular gap. The smaller the gap length, the shorter T_{cr} (critical) is. However, it increases for longer gap lengths as well as for larger and uniform shaped electrodes for the same gap length. Positive polarity 50% breakdown voltage, T_{cr} (critical) lies between 100 to 650 μs for gap lengths between 2 to 13 m of rod-plane electrodes, as seen in Figure 3.59. In cases of negative polarity, 50% breakdown voltage, T_{cr} (critical) varies within comparatively smaller range, that is, between 20 to 70 μs for gap lengths between 2 to 7 m, Figure 3.60. From these measurements in general, a rough estimation of the critical time to crest for gap distance d m in air can be made. For positive polarity, it works out about *50 x d* (μs, m) and for negative polarity *10 x d* (μs, m) in μs, Pigini et al. [3.64].

The 50% breakdown voltage characteristics for positive and negative polarity switching impulses of critical time to crest, measured by Pigini et al. [3.64] for very long air gap lengths between rod-plane and conductor-plane are shown in

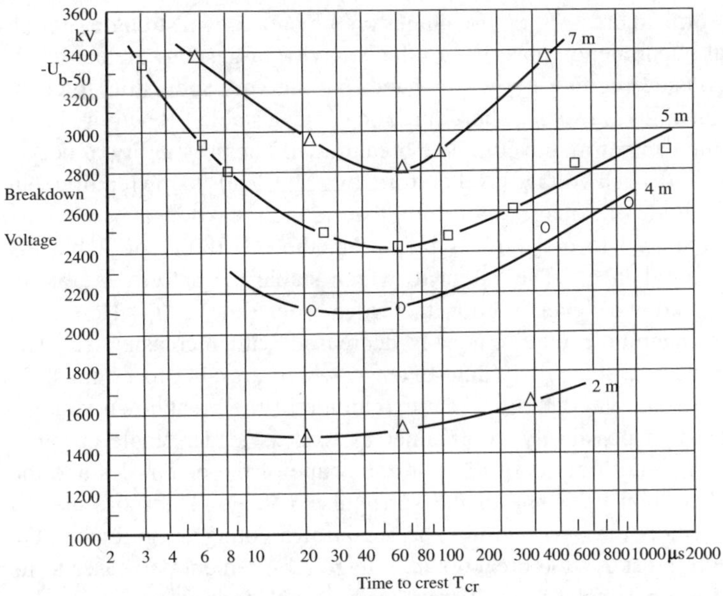

Figure 3.60 Negative polarity 50% breakdown voltage characteristics as a function of increasing time to crest T_{cr} for rod and sphere-plane air gaps of different lengths, Les Renardieres [3.51].

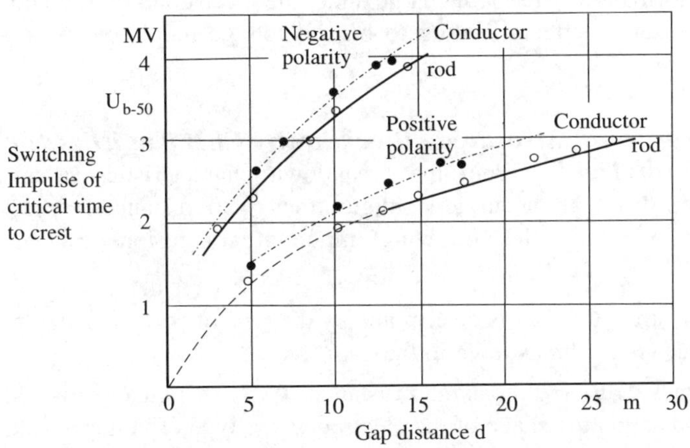

Figure 3.61 The 50% breakdown voltage characteristics of rod and conductor-plane very long gap lengths in air with positive and negative polarity switching impulses of critical time to crest, Pigini et al. [3.64].

Figure 3.61. These were measured up to a gap distance of 27 m with positive and a gap distance of 14 m with negative polarity voltages for rod-plane electrodes. For these experiments, the critical time to crest was evaluated with the help of estimation methods recommended above. The rods for their experimentation had a radius of 5 cm and the conductor was 50 m in length.

The effect of time to crest, T_{cr} of *si* voltages on the breakdown strength of long air gaps is a typical phenomenon observed in the behavior of gaseous dielectrics. It can be explained considering first the wave shapes having very short time to crest, say an *li* impulse. On applying such a short duration pulse, a stable PB activity is not able to develop in the gap before the complete breakdown. The probability of occurrence of breakdown with such voltages is therefore low. This leads to the requirement of quite a high magnitude of voltage for accomplishing the breakdown.

As the time required to crest, T_{cr} is increased, the possibility of stable PB activity in the gap is enhanced. The extent to which stable PB activity is able to grow before the breakdown depends upon the wave front time or T_{cr}. Hence the breakdown voltage magnitude requirement is decreased with increasing T_{cr}. The types of stable corona and their individual extents, which are able to establish in the gap before the breakdown, determine the magnitude of the breakdown voltage. In long air gaps exists a possibility of streamer as well as leader stable coronas. Their extents in the gap not only depend upon the shape of the electrodes and the gap distance but also upon the slope of the *si* impulse voltage. The more the PB activity is able to develop, the lower will be the breakdown voltage requirement. On increasing the time to crest T_{cr}, the breakdown voltage requirement decreases to its lowest magnitude for a particular time to crest known as T_{cr} (critical).

On increasing the T_{cr} beyond its critical value, stronger stable PB activity is expected to take place in the gap. The PB not only increases the loss of energy dissipation in the gap due to enhanced ionization activity but also gives rise to more effective space charge formation. The space charge under the given conditions could give rise to field equalization effects leading to higher voltage requirements for breakdown.

3.3.5.6 *Conclusions from Measured Breakdown Characteristics in Extremely Nonuniform Fields* From the breakdown characteristics in air described above under different electrode configurations, gap distances, types of voltages and their polarities, the following important conclusions can be drawn:

1. The breakdown strength of air is determined by the type of corona or stable PB, which precedes the breakdown in the gap.
2. The breakdown voltage U_b versus gap distance $d, (U_b - d)$ characteristics strongly depends upon the shape of the electrodes, the type of voltages and their polarities.
3. The asymmetric electrode configurations needle-plane or in case of long air gaps rod-plane represent the most unfavourable conditions. With positive polarity voltages, they have the lowest breakdown strengths.
4. The slopes of the line tangents drawn to the $(U_b - d)$ curves depend upon the extent of stable PB process/processes, which appear before the breakdown, representing the mean potential gradient across them.
5. A variation in the course of $(U_b - d)$ characteristics always represents a transition or change in the physical process of breakdown mechanism.

6. The $(U_b - d)$ characteristics with *dc* for rod-plane gaps up to 2.5 m show an average potential gradient requirement of about 5 kV/cm for positive and about 10 to 15 kV/cm for negative polarity. This verifies that breakdown with *dc* voltages are accomplished with the formation of only stable streamer (not leader) up to gap lengths of 2.5 m.
7. The $(U_b - d)$ characteristic with standard *li* for rod-plane electrodes and gap distances even above 5 m is linear. An average potential gradient requirement of about 5 kV/cm for positive polarity voltage verifies that the breakdown is accomplished with only stable streamer and not leader.
8. Development of stable leader corona is observed only above a certain gap distance, with *ac* power frequency, *dc* and *si* voltages. The inclination of $(U_b - d)$ curve is then determined by the mean potential gradient across the leader, which is quite low, of the order of 1 kV/cm for positive polarity and 2 to 3 kV/cm for negative polarity voltages and around 2 kV/cm peak for *ac* power frequency.
9. With positive polarity switching impulse voltages, the breakdown strength of a gap is measured lowest. It strongly depends upon the rate of rise of the applied voltage, that is, time to crest T_{cr}.
10. There exists a minimum switching impulse breakdown voltage of a particular time to crest, known as T_{cr} (critical), for each shape of electrode and gap lengths above 1 m.
11. The dielectric properties of atmospheric air for very long gap distances at higher voltages are poor. At higher voltages, more complicated new phenomenon occur, challenging mankind for their solution.

3.3.5.7 Estimation of Breakdown Voltage in Extremely Nonuniform Fields in Long Air Gaps Based upon the observed spatial structure of the stable PB processes, a Numerical Model for the estimation of breakdown voltage in long gap distances in air appears to be possible. A number of publications are made by Lemke [3.65] describing one such model. Other models, for example, Jones representing the discharge system by a uniform charge, Waters considering the leader current flow dependent upon the tip potential and Alexandrov's semi-empirical models are described by Meek and Craggs [3.13]. The model suggested by Lemke is described in the following paragraphs.

The photographs of stable PB phenomena and the $(U_b - d)$ breakdown characteristics suggest that the breakdown voltage can be estimated from the potential gradient requirement of individual PB process in the gap. Let us consider the already described worst case of positive rod-plane, applying an *si* voltage for which the air has its minimum breakdown strength. As shown in Figure 3.53, a gap distance of less than a meter length is bridged with stable streamer corona just before the spark breakdown (the final jump). The gap distances of more than a meter have in series both the streamer as well as leader stable coronas extended across the length just before the breakdown. From the schematic illustration in Figure 3.62, the breakdown voltage can be estimated as follows:

$$U_b = \begin{cases} U_s & for \quad d < 1\,m \\ U_s + U_l & for \quad d > 1\,m \end{cases} \tag{3.41}$$

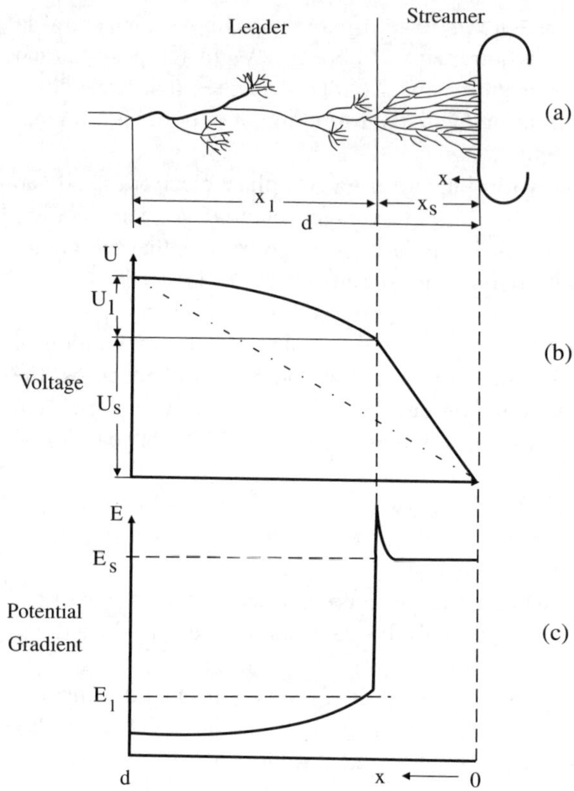

Figure 3.62 Last Stage of the development of breakdown in a long gap between rod-plane in air, (a) schematic of discharge, (b) distribution of voltage and (c) distribution of potential gradients in discharge channels.

where U_s and U_l are the potential drops across streamer and leader channels respectively.

The potential drop across the streamer corona channels for the PB activity that has just extended up to the opposite electrode is given by:

$$U_s = \int_{x=0}^{x_s} E_s \cdot dx \tag{3.42}$$

Similarly, for the leader channels the potential drop is given by:

$$U_l = \int_{x=x_s}^{d} E_l \cdot dx \tag{3.43}$$

where E_s and E_l are the average potential gradients across streamer and leader corona trajectories respectively; x_s and x_l are the range through which the two types of stable coronas extend themselves in the gap d, as shown in Figure 3.62.

Assuming the potential gradients E_s and E_l to be constant throughout the extents of respective streamer and leader channels in the gap, the following equations are derived form the above analysis:

$$U_b = \begin{cases} E_s \cdot d & for \quad d < 1\,m \\ E_s \cdot x_s + E_l(d - x_s) & for \quad d > 1\,m \end{cases} \tag{3.44}$$

The extent of streamer corona x_s, just before the final jump, is obtained from an empirical equation derived from experimental measurements on a given electrode system as a function of gap length d:

$$x_s = x_0\left(1 + \ln\frac{d}{x_0}\right) \tag{3.45}$$

where x_0 is the smallest gap length above which stable leader corona can take place in a given electrode system ($x_0 \geq 1\,m$).

On substituting equation (3.45) in (3.44), we get

$$U_b = \begin{cases} E_s \cdot d & for \quad d \leq x_0 \\ E_l \cdot d + (E_s - E_l)x_0\left(1 + \ln\dfrac{d}{x_0}\right) & for \quad d > x_0 \end{cases} \tag{3.46}$$

The average values of E_s and E_l for positive polarity voltage derived experimentally are 4.5 and 1 kV/cm respectively. These are substituted in equation (3.46) to get the breakdown voltage in kV. The values for d and x_0 must also be substituted in cm.

Lemke in [3.65] has given values for x_0 that were confirmed by practical measurements. For gap lengths above 2 m between rod-plane electrodes for *ac* power frequency, x_0 can be taken to be 1.2 m. For standard positive *si*, x_0 is given to be 1.4 m for toroidal ring-plane gap lengths of above 3 m. For long conductor-plane and bundle conductor-plane, it is recommended to be equal to 1.5 m.

If the value of x_0 is determined by equation (3.46) putting the measured values of breakdown voltages, this equation can be made applicable for other electrode configurations as well as types of voltages; for example, *ac* power frequency and *si* of different time to crest T_{cr} for which x_0 was not otherwise known.

3.3.6 Effects of Partial Breakdown or Corona in Atmospheric Air

Unlike in uniform and weakly nonuniform fields, in extremely nonuniform fields the inception of PB does not lead to a spark breakdown. PB to a certain extent is therefore acceptable in high voltage technology, especially in high voltage over head transmission lines and sub-stations in open air.

We have learned in this chapter about the three types of coronas, viz. star, streamer and leader coronas. High intensity of PB activity, which occurs with stable leader, is not permissible in practice. Under normal conditions leader erupts only when a spark breakdowns takes place. Luckily man-made spark breakdowns in atmospheric air are not so common in power systems. Lightning discharge, which occurs quite frequently, is a natural phenomenon and nature has always taken care of it. The most common types of man-made PB in air are star and streamer coronas. It is very important to know the physical and chemical implications of corona on environment, technology and human beings.

Stable PB or corona is a continuous process, which takes place on high voltage equipment and transmission lines in atmospheric air. While passing by a HV transmission line or nearing a substation, one may hear the "audible noise" (AN) produced by corona and often in dark the "visual corona" discharge can be observed. If the PB activity is intensive, one may even smell the chemically decomposed gaseous products (ozone generation) causing environmental pollution. The high frequency electromagnetic waves produced by corona interfere with the communication channels, known as "Radio Interference" (RI) and "Television Interference" (TVI), lately described as Electromagnetic Interference (EMI). The transmission lines also bear power loss due to corona. Besides the magnitude of electric stress, the intensity of corona activity also depends upon the weather conditions, that is, pressure, temperature, humidity, rain, dust, snow and even movement of air. Hence corona power loss W_c, AN, EMI and even corona luminosity are all affected by weather. The AN and EMI have their maximum intensity near the source, which reduces rapidly with the distance away from it.

So far amongst the most difficult problems caused by corona to engineers have been power loss in transmission lines under bad weather conditions and the electromagnetic interference, EMI. But with the rise in transmission voltages to 1150 kV and future plans up to 1500 kV, audible noise, AN, due to corona has also become a substantial problem. Tremendous experimental work has been carried out to study corona performance of transmission lines up to rated voltages of 1500 kV all over the world [3.66]. The electromagnetic interference due to corona has been studied in depth, particularly after the proliferation of electronic equipment and computers. Effects of corona on environment are described briefly in the following sections.

3.3.6.1 *Chemical Decomposition of Air by Corona* When corona excited gas molecules emit a short wave length light, it transforms the oxygen into ozone (O_3). Besides, due to oxidation of nitrogen, nitric oxides (N_2O, NO, N_2O_3, NO_2) and other decay components are formed. In the presence of water, formation of nitric and nitrous acids may take place.

$$2\,NO_2 + H_2O \Leftrightarrow HNO_3 + HNO_2$$

Molecules of CO_2, CO, OH and NO have also been detected in the electrical breakdown discharge channels [3.50, 3.51]. Appearance of these products in atmosphere may not cause much hazard except some air pollution. However, in enclosed chambers, especially in the presence of condensed water, these chemical components may damage the metallic and insulating materials by erosion. The life of fixtures and insulating materials is thereby affected. This problem is more predominant in the coastal regions, where the humidity is high and the salt deposition on equipment and conductors is a normal phenomenon.

Ryzko et al. [3.67] studied the effect of chemical products (ozone and nitric oxides) produced by corona in air on lightning impulse breakdown voltage (1/50 μs) at a gap distance of 5 mm in an enclosed chamber. Higher impulse breakdown voltages were measured in presence of ozone and nitric oxides that were produced outside the enclosed chamber in these experiments. The rate of rise of the breakdown voltage was found to be proportional to the rate of rise of concentration of ozone and nitric oxides in the chamber. This may be because of the presence of NO_2, which has a very large attachment energy coefficient, consequently a strong electron attachment property.

Transmission lines, when under active corona, generate ozone and other oxidants. It is important to assess the impact of this source of pollution on the environment, particularly in view of transmission at higher voltages. Research on ozone generation on transmission lines began in early 1970, [3.68–3.70]. Fair amounts of theoretical, laboratory and field work have been performed.

It is well established that ozone is a product of corona activity. The Westinghouse laboratory results indicated that ozone production rates for wetted conductors are approximately two times greater than for dry conductors. At the Apple Grove Project [3.69], field measurements of ozone were made on a 765 kV line using chemiluminescent monitors. Assuming the detection sensitivity of existing ozone instrumentation to be 1 ppb (parts per billion, 1 ppb $\approx 2\ \mu g/m^3$ air under 1 bar and 25°C), it was concluded that from a practical 765 kV line, fair weather ozone above ambient was impossible to detect at ground level. However, under worst case conditions during peak foul weather corona activity, intermittent ground ozone concentrations were detected to be as high as 6 ppb above ambient. The mean ozone concentrations at ground level due to the line, when averaged over all foul weather and wind conditions were not detectable above ambient. Thus, it was concluded that the normal 765 kV transmission lines contribute an insignificant amount of ozone to the environment.

The ozone contributions in the urban areas are comparatively very high, mainly from the automobile exhaust gases. A comparison of ozone concentration in

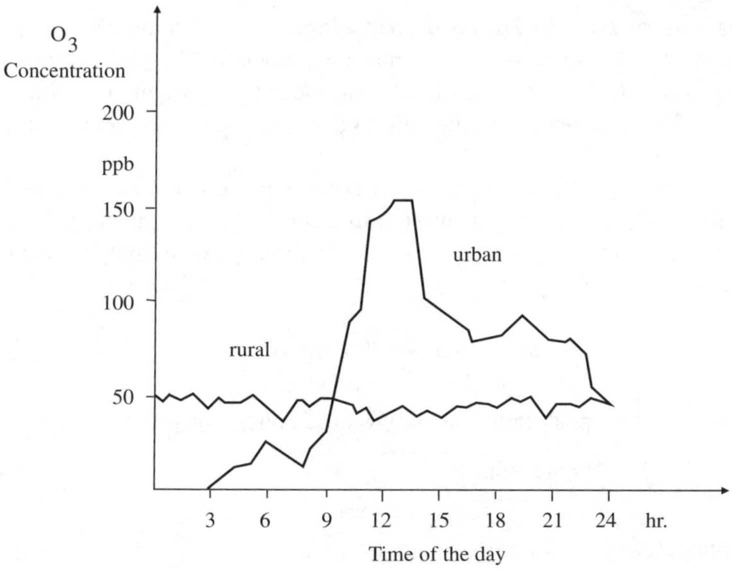

Figure 3.63 Comparison of rural atmospheric ozone concentration with the one in a city center over a day, Böhringer et al. [3.70].

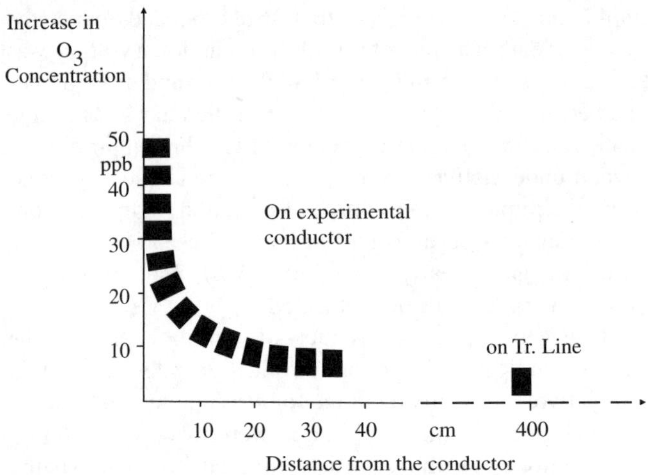

Figure 3.64 Increase in ozone concentration near an experimental line conductor for 400 kV, Böhringer et al. [3.70].

rural and urban areas is shown in Figure 3.63, given in [3.70]. As mentioned in this report, a concentration up to 100 ppb may not cause adverse effects on a healthy human being even up to five days, whereas 500,000 ppb may kill a person within a short time. Figure 3.64 shows the increase in ozone concentration in the near vicinity of an experimental line under dry weather conditions, which had a field intensity between 13.3 and 16.6 kV/cm on the conductor. The maximum field intensity on 400 kV line conductor also lies within this range.

3.3.6.2 Corona Power Loss in Transmission Lines Corona on high voltage transmission lines at 110 kV and above is a regular phenomenon. Corona is accompanied with power losses that are rather insignificant under fair weather conditions. As the intensity of corona increases during rain and snow, the corona power loss increases many fold.

The relation for the inception voltage, U_i of corona or the PB, also described as "critical disruptive voltage", for a transmission line conductor was given by Peek in 1910. This formula is still widely accepted. Although modified with time, it retains its basic feature as,

$$U_i = \frac{\hat{E}_i}{\sqrt{2}} m_0 \cdot \delta^{\frac{2}{3}} \cdot r \cdot \ln\frac{d}{r} \quad \text{kV (rms)} \tag{3.47}$$

where $\hat{E}_i$ is the PB inception peak field intensity on the conductor given by,

$$\hat{E}_i = 29.8(1 + 0.301/\sqrt{r \cdot d}) \quad \text{kV/cm} \tag{3.48}$$

where

r is the conductor radius in cm,

d is the distance between two phases in cm,

δ the air density factor taken to be equal to 1 for 20°C, and 1 bar of atmospheric pressure and

m_0 the roughness factor of the conductor.

The roughness factor m_0 is recommended to be considered as follows:

m_0 = 1.......... 0.93 for wires
= 0.72 for individual corona locations on a line
= 0.82 for corona along full length of a line

For estimating the corona losses in a 3-phase line, Peek later gave the following formula

$$W_c = \frac{241}{\delta}(f+25)\sqrt{\frac{r+6/d+0.04}{d}}(U_0-U_i)^2 \cdot 10^{-5} \quad \text{kW/km/phase} \qquad (3.49)$$

where U_0 is the rms, line to earth (phase) voltage in kV and f is the frequency in Hz.

A limitation of Peek's formulae is that these have been derived considering single conductor per phase. Perhaps the concept of bundle conductor did not exist when Peek lived. Although this is a widely accepted empirical formula for estimating corona losses, the effects of weather conditions are so strong that it may often provide misleading results. Hence measurement of losses under different weather conditions in the region of the proposed line is very important. Whenever a new project for the next higher voltage of transmission is taken up, measurements on actual experimental length of lines are conducted [3.69–3.77].

The results of corona power loss measurements on a 380 ± 10 kV actual experimental length line during different weather conditions in Rheinau, Germany are given in Figure 3.65 in [3.71]. Bundles with four conductors show a much lower corona loss over bundles with two conductors even in foul weather conditions. Rain increases the magnitude of corona losses tremendously. Fair weather losses in all the cases are insignificant. Table 3.6, given in [3.71] compares the corona losses

TABLE 3.6 Comparison of corona losses with (I^2R) losses for a 380 ± 10 kV line

Type of bundle/ conductor	I^2R losses at NTP kW/ km-3-phase	Corona losses					
		Dry weather kW/km		Rainy weather kW/km		Average kW/km	
		3-phase	% of I^2R	3-phase	% of I^2R	3-phase	% of I^2R
4 × 21/400 (S)	70	0.4	0.57	2.3	3.29	0.7	1.0
4 × 21/400 (D)	88	0.6	0.68	7.0	7.95	1.5	1.7
4 × 21.7/400	79	0.5	0.63	7.0	8.85	1.7	2.15
2.32/400	64	0.6	0.94	11.6	18.1	2.3	3.6
42 mmϕ Cu	49	0.6	1.23	18.8	38.4	3.3	6.7
42 mmϕ St-Al	31	0.6	1.94	23.2	74.8	4.5	14.5
2 × 26.6/400	59	0.5	0.85	34.8	59.0	5.9	10.0

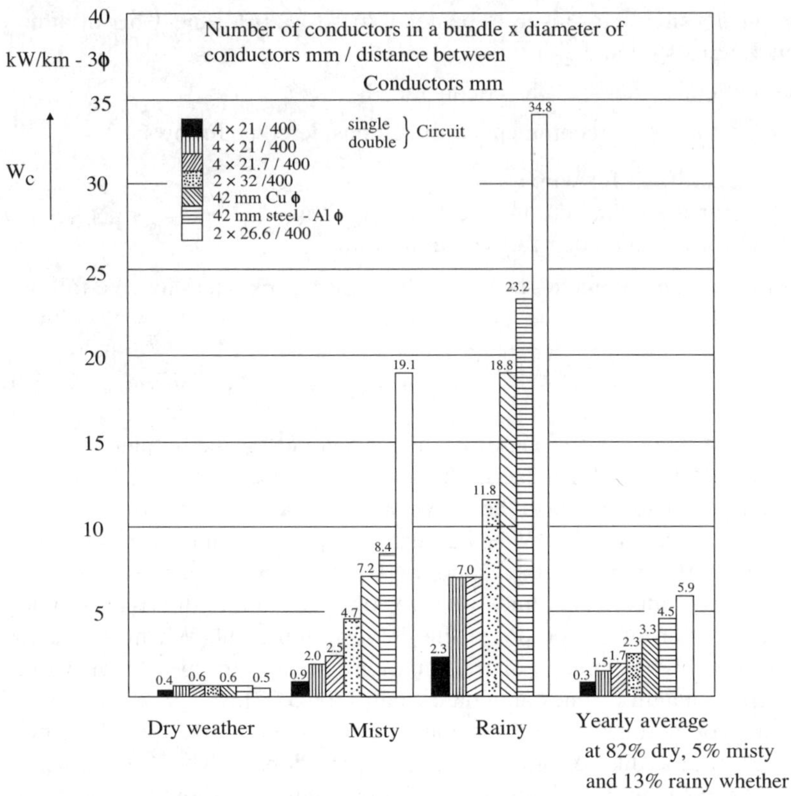

Figure 3.65 Mesurement results of corona losses on seven types of experimental conductors of 380 ± 10 kV line in different weather conditions [3.71].

with the copper losses (I^2R). As seen here, the corona losses are significant only during rainy weather conditions for single and twin conductor bundles.

Bundling of conductors for transmission voltages above 400 kV brings down the maximum field intensity at the conductor. The corona losses are, therefore, reduced considerably. The fair weather corona losses compared to copper loss become insignificant. Corona power losses as such in 800 kV and above levels of transmission are, therefore, less important a concern.

3.3.6.3 Electromagnetic Interference (EMI) and Audible Noise (AN) Produced by Power System Network EMI produced by three types of corona are described in Section 3.3.3.3. It was concluded that irrespective of their inception levels, the corona activity and hence the EMI increases with increase in applied voltage or field intensity. All types of PB are accompanied with current pulses. The rise time of such pulse current is a few nanoseconds and the 50% decay time may reach about 100 ns. Depending upon the weather conditions the frequency spectra of such electromagnetic interference (EMI) pulses may rise well above 1000 MHz on transmission lines and insulators. These cause interference to radio, TV and other communication channels. The transmission line conductors act like a huge antenna, radiating interference into the environment.

In transmission networks, there are two main sections as separate sources of PB, that is, the line and the substations, causing interference at different bandwidths. They are distinguished as follows:

1. Corona on transmission line conductors and fixtures give rise to their peak intensity disturbance in a relatively narrow frequency spectrum in the range of 0.1 to about 50 MHz. Since the communication frequencies begin at about 0.1 MHz (medium wave radio) and go above 300 MHz (Television Channels), the disturbance caused by corona on transmission line conductors affects a comparatively small range of the wide spectrum. This type of interference is known as Radio Interference (RI). As the weather, pollution and conductor surface conditions affect the intensity of corona, they also affect the frequency range and intensity of radio interference. RI has its maximum intensity nearest to the conductor (the source) and decreases rapidly with lateral distance away from it. At about 50 m away from the outer line conductor, it may be insignificant in fair weather conditions. The EMI in general decreases with conductor height, the effective diameter of the conductor and frequency [3.79]. Hence, the conductor size in transmission lines at 220 kV and above is determined by their corona performance rather than current carrying capacity. Bundle conductor configurations are therefore necessary to bring down the field intensity at higher voltages. To attain the desired corona performance, the number of sub-conductors is increased as the level of transmission voltage rises. Bundle conductors with 18 sub-conductors have been investigated for proposed 1500 kV lines.
2. The other types of sources of interference are corona at the surfaces of insulators, bushings and other solid dielectrics and between two electrodes (surface discharge and sparking gap types of sources); PB at weak points inside the solid dielectrics of cables, insulators, bushings, condensers, and so on (internal PB); and PB at poor electrical connections. The frequency range of interference produced by such sources is much higher than the one caused by corona on the line conductors since these involve surface discharge or tracking. Interference produced by these sources cause disturbance mainly to TV and higher frequency communication. EMI is produced more at substations than on the transmission lines and hence it is avoidable to some extent. Investigations made by Perry et al. [3.76] on 1200 kV line have revealed that the main EMI sources on a transmission line are at the towers. They also concluded that EMI is mostly a foul weather phenomenon.

The methods of EMI measurement recommended by different standards differ from each other. For example, the IEEE recommends standard measurement of RI at 1 MHz, whereas CISPR standard at 0.5 MHz resolution bandwidths. However, correction factors are worked out by which all these can be calculated for the same reference parameters. Table 3.7 gives comparison of measured AN, RI and TVI levels with predictions for 1200 kV level of transmission in the United States, Russia and Italy [3.76, 3.66]. The available measured values of RI on 400 kV level line in

TABLE 3.7 Comparison of measured and calculated AN, RI and TVI or EMI levels at different line voltages [3.66, 3.71, 3.74, 3.76, 3.80]

Rated line voltage/current →		400 kV, Kalwa-Bhusawal line, India single circuit $2 \times 31.77/450\,mm^2$	380 kV, Germany double circuit $4 \times 21/400\,mm^2$	765 kV, KEPCO, Korea double circuit $6 \times 30.4/480\,mm^2$	1200 kV, BPA Lyons test facility line, USA for 8×41 mm bundle (measured at 15 m from outer phase)			1150 kV line Russia (at 45 m from outer phase)	1050 kV CESI line, Italy (at 15 m from outer phase)	1500 kV proposed line, USA (at the edge of the right of way 125 m from outer phase)
% of the over all statistical distribution $L_{\%}$ (weather)		Measured according to IEEE at 15 m from outer phase	Measured at 12 m from outer phase	Measured at 15 m from outer phase	Measured according to IEEE standard	Prediction on BPA comparative (500 kV base case)	Calculated by semi analytical methods	Measured according to CISPR standard	Expected values after rain (wet) and in rains	Expected and required values
AN dB (A)	L_{50} (foul)	—	—	48.4	54.2	53.4	57.9	≤55.0	—	≤55.0
	L_{50} (fair)	—	—	40.9	38.0	—	—	—	—	—
	L_{80}	—	—	—	—	—	—	—	56–58 (wet) 58–60 (rains)	—
RI, db (above 1 μV/m) Quasi Peak	L_{50} (foul)	—	82.0	55.0–61.3	63.0	61.5	—	58.0 (at 0.5 MHz)	—	≤58.0 (at 1 MHz)
	L_{50} (fair weather)	50.0 (at 1 MHz)	64.0 (at 0.21 MHz)	44.0–48.6 (at 0.475 MHz)	46.0 (at 1 MHz)	42.0 (at 1 MHz)	47.0 (at 1 MHz)	—	—	—
	L_{80}	—	—	—	—	—	—	—	60.0	—
TVI, dB (above 1 μV/m) Quasi Peak	L_{50} (foul)	—	—	14.2–14.8 (75 MHz)	14.0 (at 75 MHz)	14.5 (at 75 MHz)	—	—	—	—

Germany and India, as well as expected values of proposed 1500 kV line in the United States are brought together in this table [3.71, 3.74, 3.76]. Results of long term corona interference studies on a double circuit 765 kV line in Korea by Dong-Il Lee et al. [3.80] are also given in this table. These investigations were first performed on a corona test cage and then on a full scale test line. A choice of bundle with six sub-conductors having 40 cm spacing could meet the criteria of specified requirements for AN and EMI. Measurements conducted by Lalli and Prabhakar [3.74] on 220 kV lines in India recorded maximum RI level of 56 dB during fair weather and 70 dB under rain conditions. These measurements were performed at 15 m away from the outer phase and at 1 MHz. These line designs are from the 1960s and their RI levels are even higher than the new 1200 kV level lines. This could be made possible with the use of bundle conductors. The field intensity at the bundle conductor is considerably lower; hence the reduced intensity of corona activity. One may conclude that earlier not enough emphasis was given to corona performance of transmission lines. Both high voltage technology and communication techniques have developed tremendously after 1955.

Audible Noise (AN) generated due to corona at the transmission line conductors, fixtures and spark gaps also becomes a cause of concern during foul weather. A forest on the way of a line is expected to attenuate the audible noise. As seen in Table 3.7, the mean AN level of a 1200 kV experimental line in fair weather is only 38 dB (A), but it is almost one and a half times under the foul weather. The United States' noise regulations limit the AN to a maximum of 50 dB (A) at the edge of the right-of-way in noise sensitive areas.

3.3.6.4 Other Effects of High Voltage Transmission Lines and Corona on Environment The effects of transmission line electrical phenomena that may have an impact on the growth of plants and behavior of animal communities are; electric field, audible noise and visible corona. Relatively few such studies have been conducted so far but there is a considerable public interest and concern on the environmental impacts of high voltage lines.

Rogers et al. [3.78] conducted environmental studies on a 1200 kV prototype line at Lyons, Oregon, USA. The effects on the growth of crop, plants, trees, shrubs and grass were investigated. The behavior of cattle, small mammals, birds and honeybees were also studied.

Damage due to corona at the leaf tips of some Douglas fir trees lying rather close (up to 18 m from the 1200 kV line) was observed. The leaf tip damage resulted from induced corona on the trees. This type of damage has also been documented earlier on trees near EHV transmission lines.

Barley plants grown near the 1200 kV line in the area having mean field intensity of about 7 kV/m were slightly taller and produced more straw compared to normal plants but grew smaller leaves. Earlier studies of crop growth under 765 kV lines had suggested that it was unlikely that electric field levels of 7 to 8 kV/m would affect crops. Further investigations are, therefore, required in this direction.

Cattle grazing under the line momentarily responded to audible noise when the 1200 kV line was energized. Later when they became used to it, they did not show any aversion to grazing or moving beneath the line. No definite effects of electric field or corona on the population size of small mammals and birds could be observed in their studies. Further, no statistically significant difference in the amount of honey and wax produced by honeybees in colonies located beneath and away from the 1200 kV line was recorded.

It is encouraging that no significant impact of 1200 kV line on the environment was observed. However, extensive experimental studies are yet to be conducted on a long term basis for gaining more information and experience in this field. There is an increasing concern about potential adverse effects of exposure to electric and magnetic fields at extremely low frequencies. Such exposures arise mainly from the transmission lines and use of electrical energy at power frequencies in buildings, homes and workplaces. Some investigators have reported that such exposure may suppress secretion of melatonin, a hormone connected with human's day-night rhythms. Melatonin is suggested to be protective against cancer, WHO publication, WHO/EHG/98.13, WHO Geneva, 2007.

An International Electromagnetic Field, (EMF) Project has been established by World Health Organisations (WHO) to work toward resolving the health issues raised by EMF exposure to ensure better health risk assessments.

3.4 ELECTRIC ARCS AND THEIR CHARACTERISTICS

The name "arc" was first used by Davy to describe the "arched column of horizontal electrical discharge in air between two charcoal electrodes". Since then, it might have more logical to confine the term "arc" to the column only. However, the term "arc" is commonly used in place of arched column.

As described earlier, when an electrically conductive channel acquires plasma state, it converts itself into an arc. The electrodes and their configurations do not only affect the characteristic of an arc but also the type of voltage, source impedance, the mediums and their conditions, and so on. It is difficult to classify the types of arcs. However, three types of electrical arc discharges, which commonly occur in electrical power system under different conditions, are distinguished as follows:

1. free burning arc column in air, a breakdown in very long air gaps, for example, "lightning discharge".
2. arc columns in air affected by electrode plasma jet phenomenon in relatively smaller electrode gap distances, for example, spark over on insulator strings, bushings and isolators in air.
3. arc columns produced due to high short circuit currents in power systems, for example, "high current arcs" in circuit breakers.

Atmospheric air is the most widely used insulation. Under transient conditions with lightning and switching impulses, air insulation systems are designed to have spark over between two electrodes known as protective gaps above a certain level of these voltages. In order to know the behavior of arcs, it is important to learn their characteristics under various conditions.

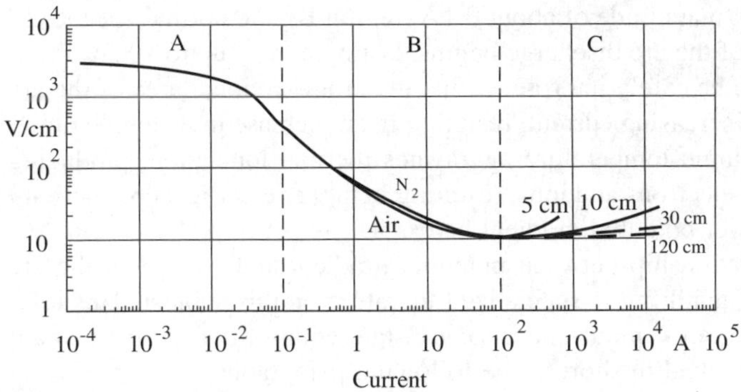

Figure 3.66 U-I characteristics of free burning arc in air and nitrogen, King [3.81].

3.4.1 Static Voltage-Current, U-I, Characteristics of Arcs in Air

Many physicists under various conditions and mediums have measured average arc voltage-current characteristic, with increasing electrode gap as a parameter, for free burning arc column in steady state. This is a very important characteristic and has received maximum attention in all the arc discharge experiments over a century. King in 1961 [3.81] came up with an interesting generalized U-I characteristic of arcs over a large current range extending from 10^{-4} to 10^{4} A, as shown in Figure 3.66.

Results of measurements made over a number of years in Electrical Research Association Laboratories, Leatherhead, Surrey, England have been collated under various conditions to give consistent characteristics. Similar curves measured in liquid nitrogen by Inaba in 1975 are given in [3.82].

As seen in Figure 3.66, no sudden transition occurs in the voltage gradient over the entire current range. This curve can be divided into the following three main regions on the basis of accompanied current in arc:

A: below 0.1 A

B: 0.1 to 100 A

C: above 100 A

Region A represents the electrical spark breakdown or the flashover with an unstable leader turning into an arc column. Breakdown with *li* voltage is a typical example for this region. Acceleration of charged particles under the influence of applied electric field continues to determine the main ionization process, the impact ionization in this region. The gas temperature acquired is, therefore, relatively low. Increasing power input, for example, with *ac*, *dc* or *si* voltages may gradually raise the gas temperature at particular locations in the stable PB or corona process, which precede the spark breakdown. Thermal ionization may supervene at stem and stable leader corona formation. The breakdown is ultimately accomplished with an unstable leader followed by an arc column. As seen in the figure, the potential gradient across the discharge column in this region are relatively high since it involves very low magnitudes of currents.

Above the current magnitude of about 0.1 A (region B), the normal "negative" or "fall" characteristic of the arc discharge begins, lasting almost up to 100 A. From this characteristic it can be safely interpreted that the cross sectional area of the arc column increases with increasing current, resulting in an increase in its temperature. The increase in arc column temperature aggravates thermal ionization, producing more charged particles, electrons and ions, resulting in higher electrical conductivity of the gas, and thus lower potential gradient across it.

An analytical relationship between potential gradient and current in the arc column is described in detail in the specialized literature on this subject. This relationship for free burning arcs in air, under steady state within the negative characteristic region B, is described in short in the following paragraphs.

Assuming a uniform distribution of the gas temperature and also the ion density over the total cross sectional area of an arc, and neglecting the current caused by the movement of ions, the following equation relating arc current density (J) and the potential gradient (E) is obtained,

$$J = \pi r^2 vNq = \pi r^2 k_e ENq \tag{3.50}$$

where

r—radius of arc column in cm

v—velocity of electrons in cm/s

N—number of electrons per cm^3

q—electron charge in C

and k_e—movability of electrons in $\left[\dfrac{cm/s}{V/cm}\right]$

Since N and k_e are dependent upon the gas temperature *(T)* of the arc column, the above equation can be written again as follows:

$$J = r^2 Ef(T) \tag{3.51}$$

Under steady state condition of an arc in the region B of Figure 3.66, often a generalized relationship between J and E is given as,

$$E = const \cdot J^{-n} = CJ^{-n} \tag{3.52}$$

Under the assumption that the radiation of heat from the arc is proportional to the column radius or the surface area, the theoretical value of n is calculated to be equal to 0.333. However, according to the investigations made by different research works, its practical value varies from 0.5 to 0.25, Sirotinski [3.83]. The value of n is taken to be equal to 0.5 for low arc currents up to 10 A. For higher currents, it may be taken between 0.35 and 0.25, which is close to the theoretical value. Equation (3.52) holds good also for the arcs in air fed by power frequency for the maximum current magnitudes, for example, in isolators. Value of the constant C depends upon the length of arc; thus, all the factors affecting the arc length, for example, wind, electrode shape and starting current of the arc, and so on, affect the value of this constant. The value of C equal to 100 cm/VA may be taken for the maximum length of an arc under given condition.

The current range above 100A in region *C*, Figure 3.66 invokes a number of new features. As the current is increased, the so called "plasma jets" and self generated gas blasts created by these jets are associated with the anode and cathode extending into the arc column for distances up to five to six mm. The plasma jets become prominent in region *C*, rendering it difficult to keep the arc free from electrode vapor. In order to obtain true characteristics of arc in this current range, experiments must be performed on longer gaps (a few *cm* to *m*) to be able to safely assume the plasma jet mechanism to have confined integrally within the electrodes. Because of the electrode effect, a definite positive characteristic results in the higher current region. If the gap length is increased, a contraction in the column in electrode region is affected. Thus a reasonably flat characteristic is obtained. A fairly constant voltage gradient is measured on a 30cm gap within a current range of 100 to 1000A as shown in Figure 3.66. This is the gap length region in which circuit breakers operate, breaking very high short circuit currents in kA range. The region of minimum voltage gradient of 10V/cm extends with increasing currents for long gap distances. This indicates that for very long arcs in air, the value of 10V/cm can be accepted from 100 to 10,000A current range. The longitudinal potential gradient in free burning arcs in air can be safely taken from this curve to vary between 10–50V/cm for arc currents beginning from 1A to the highest possible current in arcs.

3.4.2 Dynamic U-I Characteristics of Arcs

In cases in which the voltage applied is not steady state but changing with time, the magnitude of current and hence the plasma characteristics of an arc depend upon the instantaneous voltage. This characteristic of arc is accompanied with a time lag, known as "thermal time constant". An extremely rapid changing current (due to transient voltage) through the arc is not able to appreciably affect its temperature within a short time. Hence the arc can be considered to offer constant resistance during this time period. Figure 3.67 shows a schematic of U-I characteristics of arcs for different rate of change of current with time. Beginning with a current having no change $\frac{di}{dt} \to 0$ (with time static characteristic), to a current changing very rapidly $\frac{di}{dt} \to \infty$ (constant resistance characteristic), the predicted characteristics are shown in this figure.

Consider a power frequency source feeding an arc over a resistance R, Figure 3.68 (a). The instantaneous value of voltage variation in the arc is shown in Figure 3.68 (b). When the applied voltage passes through its natural zero, the voltage and, therefore, the current through the arc changes quite rapidly as the rate of change of applied voltage is very high in this region. Near the peak of the applied voltage, where the rate of its change is slow, the voltage and the current in the arc also do not change much with time. Accordingly, the area of cross section of the arc column changes in synch with the instantaneous values of the current. When the current approaches zero, the arc column becomes very thin and it even extinguishes completely for a very short duration (of the order of a few μs). It reignites as the voltage,

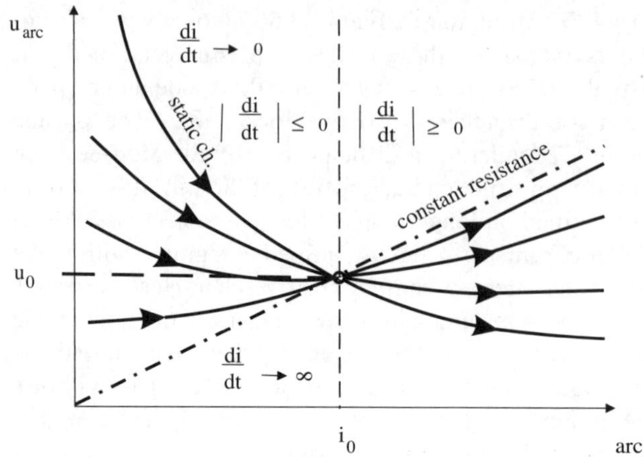

Figure 3.67 Effect of different time varying currents on U-I characteristics of arc.

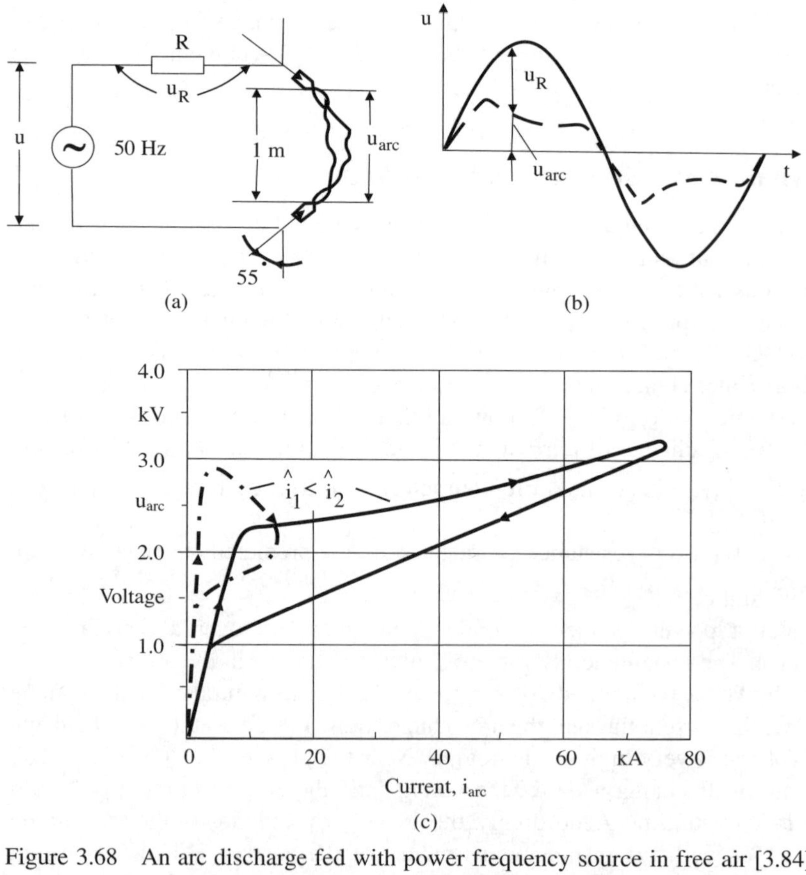

Figure 3.68 An arc discharge fed with power frequency source in free air [3.84]. (a) A power frequency source feeding an arc. (b) Variation of voltage across arc. (c) Hysteresis loop in U-I characteristic of arc, 100 ms after inception.

and hence the current increases. Because of the thermal time constant of arc column, a hysteresis loop is measured in such U-I characteristics as shown in Figure 3.68 (c). for 50 Hz current, a relatively large arc time constant, of the order of ms, develops [3.84]. Such arc discharge in free air may take place due to faults in the power system. These may also be caused due to the 50 Hz follow up current between two electrodes, for example, in isolators.

For a small peak arc current, $\hat{i}_1$, Figure 3.68, with increasing current a fall in voltage takes place and the area covered by hysteresis loop is small. On the contrary, for large current magnitude, $\hat{i}_2$, for increasing current, a rising characteristic is measured and the area covered by hysteresis loop is also large.

In Gas Insulated Systems (GIS) and in vacuum circuit breakers, up to a few tens of restrikes of current may take place in a single switching operation. This happens because the arc is not quenched and it restrikes several times before the final interruption of the current in these disconnector switches. Since the disconnector contacts in GIS and vacuum circuit breakers move slowly and the dielectric strength of the medium in these cases is quite high, a number of restrikes occur while opening operation of the switches. The time interval between the restrikes increases with increasing contact gap distance while opening. Each restrike of the arc generates "Very Fast Transient Overvoltage", (VFTO) having different levels of magnitude, of the order of up to 3.0 pu. The rise time of VFTO can be as small as 10 ns, Naidu, Thomas [3.85, 3.86]. Since the restrikes of arc are caused in turn by the rapidly changing transient over-voltages, the arcs thus formed can be considered to offer constant resistance during the operation period of the switch, as shown by the dynamic characteristic of arcs in Figure 3.67.

3.4.3 Extinction of Arcs

An arc in the free air is affected by the wind: its speed and direction, dispersion of heat and even the magnetic forces developed due to its own current. The wind conditions are unpredictable, whereas the dispersion of heat is mainly in an upward direction. The electromagnetic forces, developed due to the magnetic field of its own current, can be directed to act favorably and help the extinction of the arc. These forces are developed by design so that a minimum magnetic induction force is required to drift the arc away from its feeding point. This is an important aspect for consideration while designing the disconnector switch electrode system between which arcs may appear in the power system.

An arc in free air caused by a transient over voltage in the power system is extinguished when the over voltage passes away in the form of a travelling wave. But to protect the power system from such faults and power frequency follow up currents, the applied voltage must be tripped. This is taken care of mostly by the auto-reclosing schemes provided with the power system protection.

Rush cooling of arc column and high rate of rise of dielectric strength of the space between the electrodes in arc extinction chambers of circuit breakers is achieved by injecting fresh air, gas or oil. However, the techniques of quenching the arc caused by heavy short circuit current in circuit breakers is itself a specialized subject and it is not within the scope of this book.

3.5 PROPERTIES OF SULPHURHEXAFLUORIDE, SF_6 GAS AND ITS APPLICATION IN ELECTRICAL INSTALLATIONS

Although atmospheric air is the cheapest dielectric, it has very poor insulating properties. Hence, the insulation systems with atmospheric air have comparatively very large geometrical dimensions. The electrical field prevalent in these systems is an extremely nonuniform field. With increasing transmission voltages, the size of transmission towers, lines and substations have been increasing to meet the requirement of necessary clearances. These not only impair the aesthetics but also become an economic liability because of huge land as well as material requirements. The performance of electrical equipment and installations in open air is intensively affected by atmospheric conditions and environmental pollution. It was, therefore, a challenge for the engineers to produce compact, amicable and economically viable high voltage electrical installations that could provide more reliable service.

In order to tackle these problems, there existed a need to produce electrical installations of comparatively smaller dimensions. This is possible by shifting from extremely nonuniform fields in atmospheric air insulation systems to weakly nonuniform fields in gas insulated systems (GIS), also known as "Metalclad" systems. Better utilization of the dielectric properties is achieved in weakly nonuniform fields. Hence, it brings down the dimensions of the equipment at a given rated voltage.

The atmospheric air, even under high pressure, has relatively poorer dielectric strength. Gaseous dielectrics with better properties were desperately sought. A gaseous dielectric with high dielectric strength and other suitable properties could enable the reduction of the geometrical dimensions of installations more effectively. The most suitable gas being widely used since 1960s as an alternative to air and nitrogen is sulphurhexafluoride, SF_6. After seven years of the introduction of GIS in Japan, the first 550 kV full GIS system was installed there in 1976. By 1986 almost 40% of the newly built substations and replacements of existing substations had GIS in Japan [3.87]. Gas insulated switchgear with rated voltages up to 765 kV are in service for more than a quarter of a century. Toshiba, Japan first developed in the world complete SF_6 gas insulated substation (GIS) up to highest rated voltage of 1100 kV by the year 2000. Figure 3.69 shows an 1100 kV indoor, metal-clad or gas insulated substation (GIS) with its sectional schematic.

Since SF_6 gas was initially used for high voltage circuit breakers, the "Gas Insulated Switchgears" were suitably abbreviated as GIS. Later, when complete metal-clad substations with SF_6 gas were developed, the same abbreviation was adopted for "Gas Insulated Substations" by many countries in the world. Both these expressions are common but in different parts of the world. The SF_6 gas insulated coaxial cables are called "Compressed Gas Insulated Transmission Lines", (CGITL) by some. It is interesting to discover that the abbreviation GIS became more familiar than its full form. In order to streamline the terminology, a more general expression, "Gas Insulated Systems", for SF_6 gas insulated substations has been adopted by a number of authors lately in the literature.

There are a large number of other insulating gases, but they do not have the appropriate properties required for electrical insulation as good as SF_6 gas. Hence,

(a)

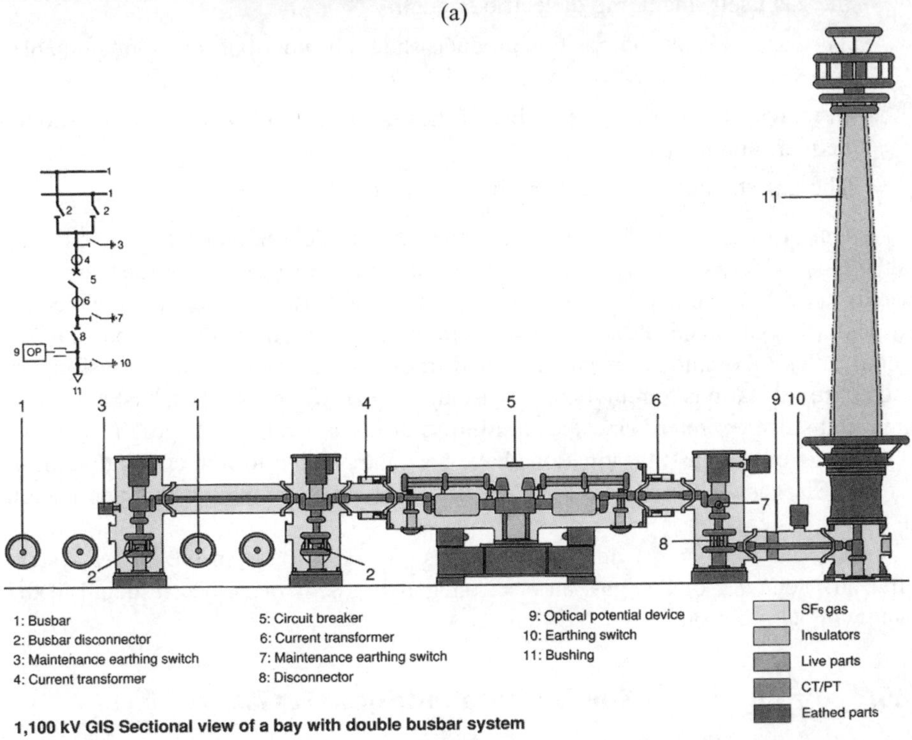

(b)

Figure 3.69 (a) An 1100 kV SF_6 gas insulated substation (GIS), courtesy Toshiba, Japan. (b) A Sectional view of the 1100 kV GIS bay with two busbar system.

gases other than SF_6 gas have found only a limited and specific application. The physical properties of different gases considered for insulation purposes have been brought together in Table 3.8 in a sequence according to their dielectric strength given in [3.88]. The values of electric strengths given in this table are for plane-plane electrode at a gap distance of 1 cm. The values for E_b, λ_e and W_l as given in the literature differ considerably from reference to reference.

Besides having a high dielectric strength, a gas proposed for electrical insulation system should also have the following properties for its suitable application.

- Over the working temperature range of −30° to +60°C and design pressure conditions, liquefaction of the gas should not take place. Liquefaction of the gas reduces the gas pressure in the enclosed volume, considerably affecting the dielectric strength.
- The gas and its decomposition products, produced due to electrical discharge in the gas, should not chemically react with other materials (metals, solid dielectrics, sealing materials, etc.). On the whole, the gas should be chemically stable as the decomposition products may have more inferior properties than the gas itself, including dielectric strength.
- The gas and its decomposition products should be nontoxic and nonflammable, which is safer for the personnel working on it.
- In view of the intricate assembly of the equipment, the gas preferably should be inert and have a high density.
- The gas should not have green house effect.

Out of the large number of gases enlisted in Table 3.8, other than air, SF_6 gas is the most widely used gas today. Compressed air or nitrogen circuit breakers were widely used but limited to a rated voltage of 400 kV due to their poor dielectric strength. SF_6 gas insulated switchgears (circuit breakers and isolators), coaxial gas filled cables (also known as gas-insulated transmission lines), power transformers and entire substations (circuit breakers, isolators, grounding switches, busbar, instrument transformers interconnected) are in use at highest rated voltages. To achieve economy, tremendous investigations have been carried out to find compatible mixtures of SF_6 gas with various other gases to determine their suitability of application in power systems.

SF_6 gas is the most important gaseous dielectric today in the power system after air. Behavior of SF_6 gas, an electronegative gas, is described in detail in the following sub-sections.

3.5.1 Properties of Sulphurhexafluoride, SF_6 Gas

French scientists Moissar and Lebeau first produced Sulphurhexafluoride in 1900 by direct fluoronisation of sulphur. In the beginning it was mainly used as a dielectric in atomic physics. During the late 1950s, it found application in high voltage circuit breakers. Ever since, its application in power systems has been continuously increasing.

TABLE 3.8 Physical properties of different gases in order of their Dielectric Strength, Mosch et al. [3.88]

Data for p = 0.1 MPa		Dielectric strength $\theta = 20°C$	Relative molecular mass	Density $\theta = 20°C$	Mean free path of electrons	Ionization energy	Normal condensation temperature
Gas or Vapour	Symbol	E_b kV/cm	M	ρ kg/m^3	λ_e μm	W_i eV	θ_c °C
Helium	He	3.7	4	0.17	1.10	24.6	–269
Neon	Ne	4.2	20	0.87	0.70	21.5	–246
Argon	Ar	6.5	40	1.73	0.39	15.7	–186
Krypton	Kr	8.8	84	3.59	0.30	14.0	–153
Hydrogen*	H_2	15	2	0.09	0.65	15.8	–243
Methane	CH_4	21	16	0.72	0.29	13.1	–162
Carbondioxide	CO_2	25	45	1.91	0.24	14.4	–29
Oxygen*	O_2	27	32	1.38	0.40	12.1	–183
Water vapour	H_2O	≈30	18	0.78**	—	13.0	+100
Air	—	32	(29)	1.25	0.37	—	–193
Nitrogen	N_2	33	28	1.21	0.35	15.7	–196
Tetra fluoromethane*	CF_4	36	88	3.81	—	15.4	—
Nitogen oxide	NO_2	40	44	1.94	0.24	12.9	–90
Carbonmonoxide	CO	42	28	1.24	0.36	14.1	–192
Chlorine*	Cl_2	52	71	3.22	0.16	11.8	–34.6
Sulphurdioxide*	SO_2	64	64	2.93	0.17	13.4	–10
Freon*	CCl_2F_2	80	121	5.33	0.30	≤18.0	–28
Sulphurhexafluoride*	SF_6	89	146	6.39	0.22	15.9 to 19.3	–63
Seleniumhexafluoride*	SeF_6	144	193	≈8.2**	—	—	+49
Tetrachloromethame*	CCl_4	180	154	6.65**	0.08	11.1	+77
Fluorine*	F_2	—	18	1.695	—	16.5	–188

* electronegative gases.
** $\theta = \theta_c$.
1 eV = 1.602×10^{-19} J.

3.5.1.1 Physical Properties In a SF_6 molecule, six fluorine atoms arrange themselves uniformly like an octahedron on a central sulphur atom. An excited sulphur atom can therefore form six stable covalence bonds with the strongly electronegative fluorine atoms by sharing the pair of electrons. Amongst halogens, the fluorine element and the sulphur atom both have very high coefficients of electronegativity, of the order of 4 and 2.5, respectively. This coefficient is a measure of the tendency to attract electrons of other atoms to form a dipole bond.

The rigid symmetrical structure, small binding distance and high binding energy between atoms of the SF_6 gas molecule provide a high stability, with the result that SF_6 gas properties are near that of rare gases at relatively low temperatures. Thermal dissociation in highly purified SF_6 gas begins at extremely high temperatures (above 1000 K). Such high temperatures in power systems occur only in electrical arcs. Even at continuous temperature up to about 500 K, neither thermal decomposition nor chemical reaction of SF_6 gas with other materials have been reported. SF_6 is a non-toxic, colourless, odourless, non-flammable, non-explosive gas besides being chemically inert and thermally stable. Thus SF_6 gas has many appropriate properties that has made it suitable for use in power system equipment. Some important physical properties of SF_6 gas, other than molecular properties, are brought together in Table 3.9.

The densities of gases increase with the relative molecular mass. As the molecular mass of SF_6 gas is quite high (146), it has a high density. Because of high density the charge carriers have a short mean free path. This property, along with the properties of electron attachment, that is, electro-negativity and high ionization energy, result in high dielectric strength of SF_6 gas, as shown in Table 3.8.

TABLE 3.9 Physical properties of SF_6 [3.88]

Property	Physical Conditions	Symbol	Unit	Value
Relative Permittivity	0.1 MPa, 25°C –51°C (liquid))	ε_r	—	1.002 1.81 ± 0.02
Dielectric loss tangent	0.1 MPa –51°C (liquid)	$\tan\delta$	—	$<5 \cdot 10^{-6}$ $<1 \cdot 10^{-3}$
Critical temperature	—	θ_{cr}	°C	45.5
Tripple point	p = 0.22 MPa	θ_T	°C	–50.8
Sublimation point		θ_s	°C	–63.8
Specific heat capacity	at 10°C and constant p = 0.1 MPa	C_P	$\frac{J}{molK}$	5.13
	at 10°C and const. volume	C_v	$\frac{J}{molK}$	4.06
Heat conductivity	30°C	—	$\frac{J}{cm\ sK}$	$0.82 \cdot 10^{-5}$
Heat transition No.	—	—	$\frac{J}{cm^2\ sK}$	$0.44 \cdot 10^{-5}$

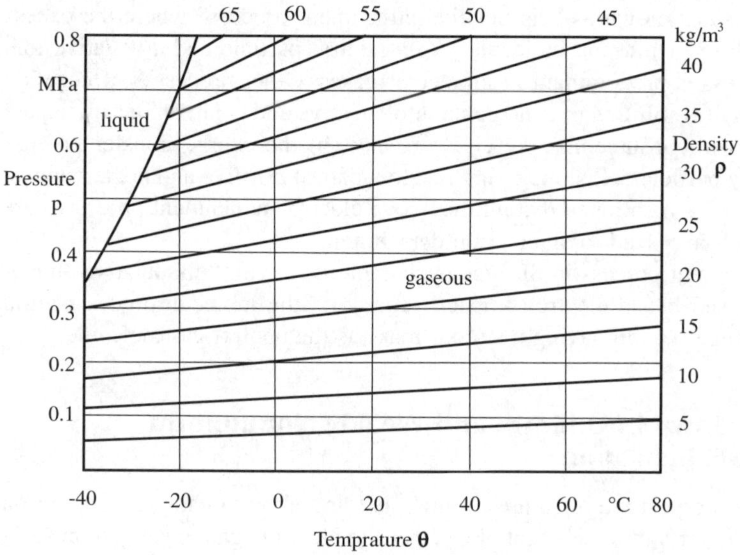

Figure 3.70 States of SF_6 gas for different constant gas densities and varying temperature and pressure within practical application range [3.88].

Density of a gas is actually a function of external pressure p and the temperature θ. In Figure 3.70 the two states of SF_6, liquid or gaseous, are shown at varying pressures and temperatures for different constant densities of the gas. This diagram is important from the point of view of practical application of SF_6 gas in GIS. Within the range of working pressure and temperature of the gas, it should not liquefy as its electrical insulation properties are affected.

3.5.1.2 Property of Electron Attachment Some atoms or molecules in their neutral gaseous state have an affinity to acquire a free electron and form a stable negative ion. This property of negative ion formation is known as "electron attachment" or "electron affinity". It has been proved that the atomic or molecular gases having electron affinity property lack one or two electrons in their outer shell and hence are known as electronegative gases. For example, the halogens (F, Cl, Br, I and At) have one electron missing in their outer shells, whereas O, S and Se have two electrons less in their outer shells.

For the attachment of an electron to a neutral molecule, the electron must have a certain minimum kinetic energy. This energy is of the order of 1 eV, very low compared to the energy required for impact ionization. For a negative ion to remain stable for some time, its total energy must be lower than that of a neutral molecule in the ground state. This difference in energy levels is known as "electron affinity" of the molecule. It is released as a quantum of kinetic energy on attachment.

There are several physical processes of negative ion formation; for example, "radiative attachment process" where excess energy is released on attachment in the form of a light quantum; the "collision attachment process" where the excess energy released on attachment is acquired as kinetic energy by a third body on collision.

The third process is known as "dissociative attachment process" where the excess energy is used to separate the molecule into a neutral particle and a negative ion. This latter process is predominant in molecular gases. Other processes of negative ion formation are (a) splitting of a molecule into negative and positive ions by impact of an electron without attaching the electron, and (b) the charge transfer process following heavy particle collision, giving rise to a pair of positive and negative ions. However, it must be mentioned that all the above electron attachment processes are reversible, which may lead to an electron detachment.

The attachment process in SF_6 gas is characterized with "dissociative attachment", It is caused by an electron impact, separating the molecule into a neutral particle and a negative ion having a similar mass as the neutral gas molecule.

3.5.2 Breakdown in Uniform and Weakly Nonuniform Fields with SF_6 Insulation

Negative ions, like positive ions, are too massive and slow to cause ionization on collision. The property of attachment of electrons in electronegative gases represents an effective way of absorbing electrons that otherwise would have contributed to the formation of more avalanches. Townsend's theory of development of avalanche in air, basically an electropositive gas, described in Section 3.2.2 has to be slightly modified to take care of the electron attachment property of the electronegative gas. Analogous to the ionization coefficient, α, the "attachment coefficient" η_a is introduced. η_a is defined as the number of attaching collisions made by one electron drifting 1 cm in the direction of the field. The ionization by electron collision is then represented by the "effective ionization coefficient", $(\bar{\alpha} = \alpha - \eta_a)$. Like $\alpha/p, \bar{\alpha}/p$ is also a function of E/p. Experimentally measured values of their dependency for SF_6 are shown in Figure 3.71. From these measured values, the following relationship for uniform fields can be derived:

$$\frac{\bar{\alpha}}{p} = K\left[\frac{E}{p} - \left(\frac{E_b}{p}\right)_i\right] \tag{3.53}$$

where K is a constant, and it is equal to $27\,\text{kV}^{-1}$ representing the slope of the line and $\left(\frac{E_b}{p}\right)_i = 890\,\text{kV/cm}\cdot\text{MPa}$, represents the intersection of the line with the abscissa at atmospheric pressure in Figure 3.71. The pressure p is taken to be normal at temperature $\theta = 20°\text{C}$.

In order for breakdown to take place, $\bar{\alpha}$ must be greater than zero, which in turn requires the fulfillment of the condition that E/p is greater than $\left(\frac{E_b}{p}\right)_i$. Accordingly, the magnitude $\left(\frac{E_b}{p}\right)_i$ represents the "inherent relative strength" of the gas. The field intensity E_{bi} given by the relation;

$$E_{bi} = \left(\frac{E_b}{p}\right)_i \cdot p = 890 \cdot p \quad \text{kV/cm} \tag{3.54}$$

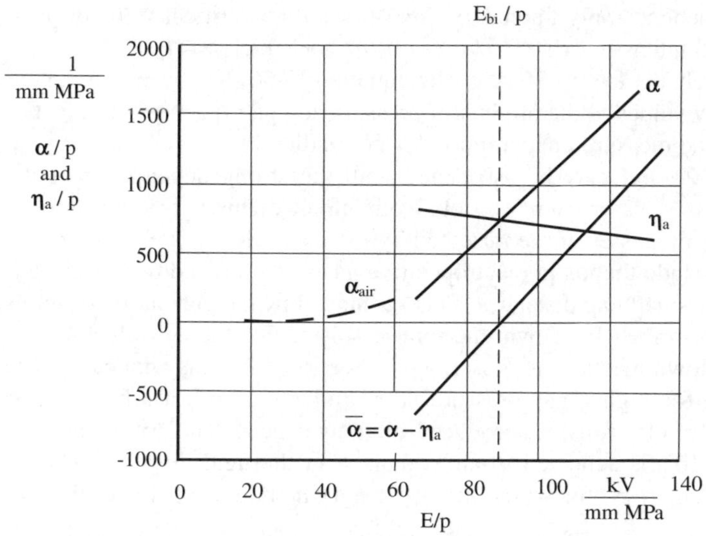

Figure 3.71 The ionization and attachment coefficients of SF_6 gas as a function of referred field intensity [3.88].

is known as the "intrinsic electric strength" of the gas. At atmospheric pressure, that is, at 0.1 MPa and 20°C it works out to be 89 kV/cm for SF_6 gas in uniform fields. This represents the minimum field intensity required for the inception of an independent discharge leading to breakdown.

Equation (3.21) describes the growth of average current under steady state conditions in a uniform field gap before the breakdown. In the presence of electron attachment process this equation can be rewritten as following relation:

$$I = \frac{I_0[\alpha/\bar{\alpha} \cdot e^{\bar{\alpha}d} - \eta_a/\bar{\alpha}]}{1 - \gamma\alpha/\bar{\alpha}[e^{\bar{\alpha}d} - 1]} \tag{3.55}$$

where α and γ are the Townsend's primary and secondary ionization coefficients defined earlier.

In a similar manner, as the original Townsend's criterion for spark breakdown, this criterion for self-maintained current independent of I_o in an electronegative gas is obtained when the denominator of equation (3.55) is made equal to zero. Then,

$$\gamma\alpha/\bar{\alpha}[e^{\bar{\alpha}d} - 1] = 1 \tag{3.56}$$

or

$$\bar{\alpha}d = \ln\left[\frac{\bar{\alpha}}{\alpha}\frac{1}{\gamma} + 1\right]$$

As long as $\alpha \geq \eta_a$, that is, on the right side of the line, E_{bi}/p in Figure 3.71, breakdown is always possible, irrespective of the values of α, η_a and γ, provided only that d is large enough. For low values of E/p, η_a is greater then α. With increasing value of d, equation (3.56) tends to be asymptotic; therefore, it can be written as,

$$-\gamma \cdot \alpha/(\alpha - \eta_a) = 1 \text{ or } \alpha = \eta_a/(1+\gamma)$$

a condition which depends only upon E/p. This shows that a critical value of pressure dependent field intensity exists, $(E/p) \rightarrow (E_b/p)_i$ for which $\alpha = \eta_a$.

To verify Paschen's law in SF_6 gas, the equation (3.56) has been experimentally verified for low values of the product of pressure and gap spacing, *pd* only. For high values of *pd*, the measured data reported is contradictory [3.56]. The disagreement as shown by Zaengl may be associated with the strong dependence of the coefficient γ upon the gas pressure and electrode effects. This is true because all secondary feedback processes are represented by the coefficient γ.

One may conclude that as in electropositive gases, the breakdown in SF_6 gas in uniform fields for small gap distances is also achieved by the generation of series of avalanches as described by Townsend's mechanism. For higher values of *pd*, the streamer breakdown mechanism, described in Section 3.2.4.2 is applicable. The breakdown criteria for higher gas pressures (or products of *pd*) are based upon the critical number of electrons, n_{cr} produced in an avalanche. The critical number of electrons (circa 10^8) is achieved in an avalanche in the field direction when it acquires a critical length x_c, for which the gap length d should be greater than x_c, refer Figure 3.17.

In practice, gap spacing in weakly nonuniform fields are of more interest. For such gap spacing, the breakdown criterion of streamer mechanisms can be extended, taking into account the attachment coefficient of electronegative gas.

Beginning with a single inception electron ($n_o = 1$), the number of charge carriers in an avalanche considering electron attachment is given by,

$$n = \exp \int_0^x \bar{\alpha} dx \tag{3.57}$$

and for critical amplification of an avalanche, which gives rise to initiation of a streamer,

$$n = n_{cr} = \exp \int_0^{d>x_c} \bar{\alpha} dx$$

or

$$\int_0^{d>x_c} \bar{\alpha} dx = \ln n_{cr} \approx 18.4 \tag{3.58}$$

From equation (3.53) and (3.58), breakdown voltage for SF_6 in weakly nonuniform field configurations can be theoretically estimated. Breakdown under these conditions takes place without any stable PB. A coaxial cylindrical electrode configuration has been chosen in the following example.

The field intensity distribution in a coaxial cylinder having inner radius r_i and outer radius r_o expressed in cylindrical coordinates as given by equation (2.34) is,

$$E_r = \frac{1}{r} \frac{U}{\ln(r_o/r_i)}$$

$$= \frac{r_i}{r} E_{\max} \tag{3.59}$$

When a breakdown in this electrode configuration occurs, the maximum field intensity in the geometric system E_{max} acquires the value of maximum breakdown field intensity of the gas $E_{b\max}$. Thus from equations (3.53, 3.54) and (3.59), the following expression can be derived,

$$\bar{\alpha} = K\left(\frac{E_{b\max} \cdot r_i}{r} - E_{bi}\right) \tag{3.60}$$

Substituting equation (3.60) in (3.58), and putting $r_c = (r_i + x_c)$, which is the radius of the depth of the dielectric, is just sufficient for critical amplification of avalanche. The relation satisfying the streamer criterion is then given as following,

$$K\int_{r_i}^{r_c}\left(\frac{E_{b\max} \cdot r_i}{r} - E_{bi}\right)dr = \ln n_{cr} \approx 18.4 \tag{3.61}$$

where r_c is determined from the following relation on substituting in equation (3.59),

$$E(r_c) = E_{bi} = E_{b\max}\frac{r_i}{r_c} \tag{3.62}$$

or

$$r_c = r_i\frac{E_{b\max}}{E_{bi}}$$

Introducing here a factor e_{max}, the "relative maximum breakdown field intensity" given by,

$$e_{max} = \frac{r_c}{r_i} = \frac{E_{b\max}}{E_{bi}} \geq 1 \tag{3.63}$$

Hence e_{max} is also described as "curvature factor". The expression (3.61) then reduces to

$$e_{max}(\ln e_{max} - 1) = \frac{\ln n_{cr}}{K \cdot r_i \cdot E_{bi}} - 1 \tag{3.64}$$

and x_c is given by,

$$x_c = (r_c - r_i) = r_i(e_{max} - 1) \tag{3.65}$$

Solution of equation (3.64) for e_{max} as an unknown is numerically possible. The calculated values of the relative maximum breakdown field intensity e_{max}, as a function of inner radius of coaxial cylinder r_i for $r_o \geq (r_1 + x_c) = e_{max} \cdot r_i$, are shown in Figures 3.72 and 3.73.

Similar results can also be derived for other electrode configurations in weakly nonuniform fields, for example, concentric spheres, sphere-sphere and sphere-plane, etc. The maximum field intensity in such electrode systems at breakdown, $E_{b\max}$ or the curvature factor e_{max} increases as the inner radius r_i of the electrode is reduced, as shown in Figure 3.72.

It is evident from Figures 3.72 and 3.73 that the factor e_{max} reduces with increasing pressure as well as inner electrode radius. Besides, the critical avalanche length, x_c increases the inner radius r_i and lowers the pressure of the gas.

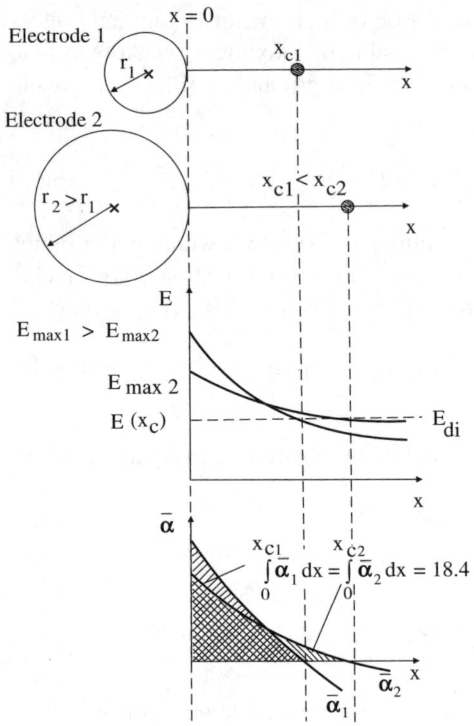

Figure 3.72 Effect of curvature/radius of electrode on the streamer criterion (critical amplification of avalanche), [3.88].

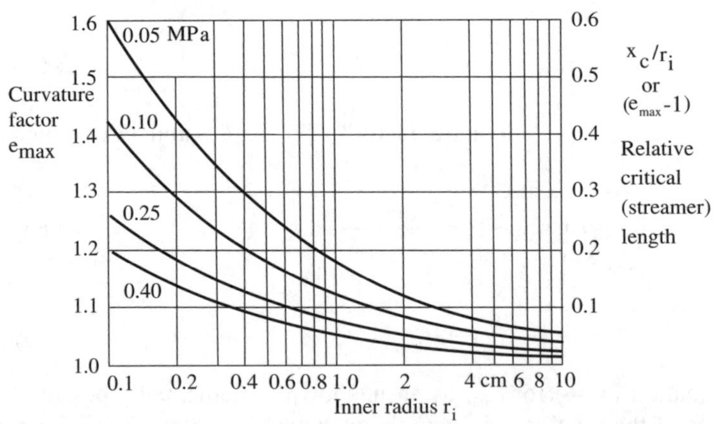

Figure 3.73 Relative maximum breakdown field intensity or curvature factor e_{max} as a function of inner radius of a coaxial cylinder for different gas pressures at 20°C, Mosch and Hauschild [3.88].

The assumption regarding critical number of charge carriers in an avalanche ($n_{cr} = 10^8$) according to streamer criterion for breakdown is slightly affected by the curvature factor e_{max} as shown in Figure 3.74 (a). The constant K and the inherent relative strength $\left(\frac{E_b}{p}\right)i$, required for the calculation of effective ionization

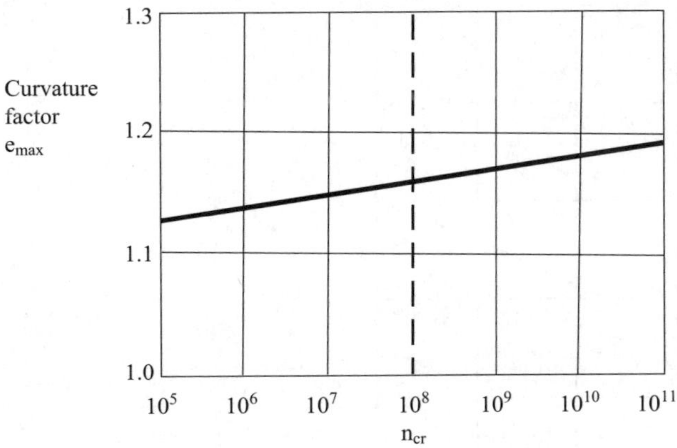

(a)

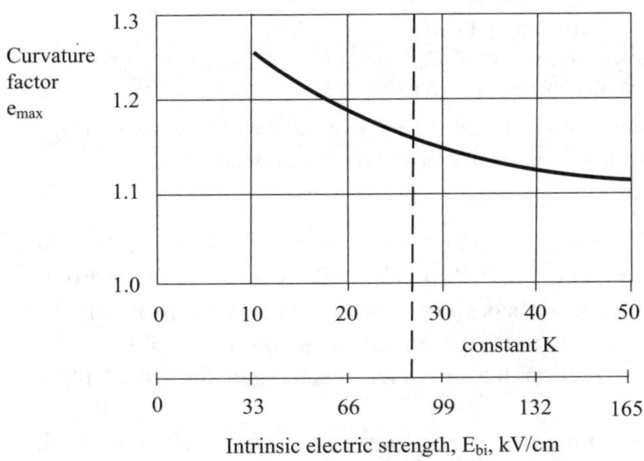

(b)

Figure 3.74 Effect of assumptions for streamer breakdown criterion on the curvature factor e_{max}. (a) Effect on critical number of charge carriers n_{cr}. (b) Influence of effective ionization coefficient $\bar{\alpha}$, [Equation 3.53], $E_{bi} = \left(\frac{E_b}{p}\right) i \cdot p$.

coefficient $\bar{\alpha}$, are also slightly affected by the curvature factor e_{max}, as illustrated in Figure 3.74 (b).

The value of $E_{b\max}$, the maximum breakdown field intensity of a gas for any configuration having weakly nonuniform field and given gas pressure can be estimated from equations (3.54, 3.63) and Figures 3.72, 3.73 and 3.74. Knowing $E_{b\max}$, the breakdown voltage U_b is estimated by the empirical relation given by Schwaiger,

$$\begin{aligned} U_b &\approx E_{b\max} \cdot d \cdot \eta \\ &\approx E_{bi} \cdot e_{\max} \cdot \eta \cdot d \end{aligned} \tag{3.66}$$

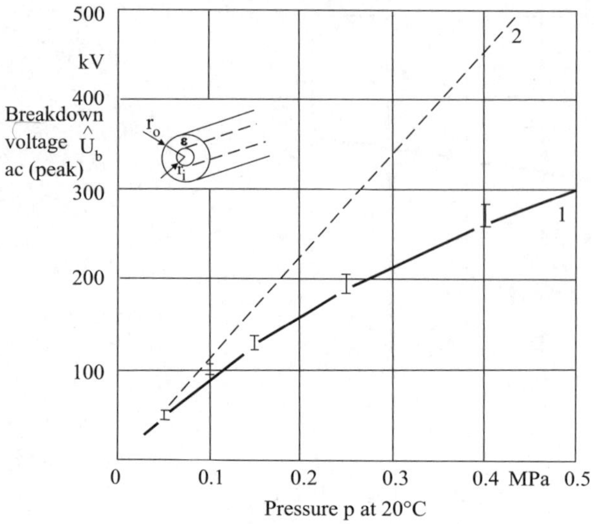

Coaxial cylinder : $r_i = 1.25$ and $r_o = 3.50$ cm
Curve 1 : "practical electric strength" E_{bt}
Curve 2 : "intrinsic electric strength" E_{bi}
Range I : experimentally measured values

Figure 3.75 Comparison of measured and theoretically estimated ac breakdown voltage characteristics for a coaxial cylinder filled with SF_6 at different pressures.

For a coaxial cylinder with $r_i = 1.25$ and $r_o = 3.5$ cm, that is, $d = 2.25$ cm and $\eta = 0.58$ (read from the Schwaiger curves), the peak breakdown voltages for different gas pressures are estimated as described above, shown by curve 2 in Figure 3.75. This curve represents the breakdown voltages under ideal conditions of the gas and electrodes, taking into consideration the streamer mechanism and the "intrinsic electric strength" E_{bi} of SF_6 gas.

The theoretical estimation made above is true for the ideal conditions of electrode surfaces and gas. It shows a good conformity with the measured values at lower pressures, around 0.1 MPa. As the pressure is increased, the two curves deviate more from each other. The probability range of measured values are shown in Figure 3.75, curve 1.

The breakdown phenomenon at higher gas pressures is strongly affected by certain unavoidable physical conditions, for example, roughness of the electrode surfaces, foreign particles in the gas, etc. These distort the field and act adversely, bringing down the breakdown voltages/PB inception voltages especially at higher gas pressures. In other words, the microgeometry of the electrodes prevents the full exploitation of the dielectric strength of the gas at higher pressure. With well polished and carefully produced gas insulated systems, where no foreign particles are allowed to be present, one may measure the breakdown voltages close to theoretically calculated values, the intrinsic strength.

For practical application purposes, it is necessary to determine experimentally the electric strength of the gas for different electrode configurations in weakly nonuniform fields and under given conditions. The so-called "practical electric strength"

TABLE 3.10 Experimentally determined practical breakdown strength of SF_6 and the factor z in weakly nonuniform fields

Type of voltage	Symbol	Polarity	$\left(\frac{E_b}{p}\right)t$ in $\frac{kV}{cm \cdot MPa}$	Factor z
Direct and Alternating	ac and dc	dc – ve	65	0.73
Power Frequency Voltages	dc	+ ve	70	0.76
Switching Impulse	si	+ ve	73	0.76
(250/2500 μs)		– ve	68	0.73
Lightning Impulse	Li	+ ve	80	0.80
(1.2/50 μs)		– ve	75	0.75
Theoretical value (inherent/ intrinsic strength)	$\left(\frac{E_b}{p}\right)i$	—	89	1.0

E_{bt} depends upon the type of applied voltage and satisfies the following nonlinear empirical relation:

$$E_{bt} = \left(\frac{E_b}{p}\right)t \cdot (10\,p)^z < E_{bi} \tag{3.67}$$

where E_{bt} is in kV/cm and p in MPa at 20°C. Equation (3.67) represents a straight line on a double logarithmic scale. The factor "z" represents the slope of the curve for different types of voltages. It is determined by lowest experimentally measured values of breakdown strength "E_{bt}" at different gas pressures. The factor z has a value less than unity. From such curves measured at normal temperature, the values of "practical relative strength", described as technical term, $\left(\frac{E_b}{p}\right)t$, and z for different types of voltages are brought together in Table 3.10.

For the practical breakdown characteristic, the required practical maximum breakdown field intensity "$E_{bt\max}$" can be estimated from the relation,

$$E_{bt\max} = e_{\max} \cdot E_{bt} \tag{3.68}$$

Once $E_{bt\max}$ is known, the breakdown voltage can be found out from the equation (3.66) given by Schwaiger. The practical breakdown voltage characteristic estimated in this way for the coaxial cylinder configuration mentioned above, conforms the experimentally measured values as shown in Figure 3.75, curve 1.

Using this method, estimation of breakdown voltages for other electrode configurations having weakly nonuniform fields based upon practical values also gives satisfactory results. This confirms that the method developed by Schwaiger for the estimation of breakdown voltage in weakly nonuniform fields closely agrees with the method based upon streamer mechanism. Thus, it can be accepted without reservation in practice with a reasonable approximation.

3.5.3 External Factors Affecting Breakdown Characteristics in Compressed Gases

At higher pressures (or densities of gas) where the breakdown field intensities are also high, deviation from Paschen's law in uniform field begins. As described in

Section 3.2.5, the breakdown voltage is more affected by the electrode materials, their size and surface finish (roughness), imperfections, dust and foreign particles, and so on, especially in cases of relatively smaller gap distances at higher pressures. These create a local enhancement of the field leading to PB initiated lower breakdown voltages. On the contrary, an increase in breakdown voltage is measured after the "conditioning" of the electrodes when the applied voltage is raised very slowly or in small steps, known as "stress conditioning". The reasons described for conditioning effect are first, the progressive destruction of high field intensity sites on the electrodes and second, the movement of free foreign particles to lower field regions where they cannot initiate a breakdown. Electrode surface roughness and particle-initiated breakdown in compressed gases with weakly nonuniform fields have attracted special attention from researchers. These have given rise to important practical problems in gas insulated systems [3.89–3.91 and 3.92–3.101].

3.5.3.1 *Effect of Electrode Materials and Their Surface Roughness on Breakdown* The quality of surface finish of a metal not only depends upon the metal itself but also upon the manufacturing and/or machining process. The surface roughness may often not be visible to the naked eyes, but it can be determined with the help of a microscope. The roughness profile of an electrode surface is depicted in Figure 3.76 (a). It is believed that one single protrusion of maximum height on an electrode may determine the breakdown voltage; however, it is difficult to measure each individual projection over a large surface area. Hence, a more practical method is necessary to define the roughness of a surface. The measured "mean roughness" $\overline{R}_d$ is μm is given by,

$$\overline{R}_d = \frac{1}{l}\int_o^l |R_{di}|dx \quad \mu\text{m} \tag{3.69}$$

where R_{di} is the depth of individual projections and $R_{d\max}$ the highest projection, l the length considered as shown in Figure 3.76 (a).

For theoretical estimation of the intrinsic electric strength "E_{bi}" under ideal conditions, the value of $\overline{R}_d$ is taken to be equal to zero. As the value of $\overline{R}_d$ increases, lower values of breakdown strength are measured approaching the "practical electric strength", E_{bt} given by equation (3.67).

By collating the empirical relations for ideal and practical conditions (equations 3.54 and 3.67) a generalized expression for "pressure dependent roughness factor", e_R can be derived,

$$e_R = \frac{E_{bt}}{E_{bi}} = \frac{(E_b/p)_t}{(E_b/p)_i} \cdot \frac{(10p)^z}{p} = f(p \cdot \overline{R}_d) \leq 1 \tag{3.70}$$

A plot of e_R for different constant gas pressures as a function of $\overline{R}_d$ is illustrated in Figure 3.76 (b). If the mean roughness $\overline{R}_d$ is known, the pressure dependent roughness factor e_R can be read from these curves. Knowing "intrinsic electric strength", E_{bi} of the gas, the practical breakdown strength E_{bt} for given conditions can be estimated. For an electrode configuration in weakly nonuniform field, the following relation therefore gives the practical breakdown voltage:

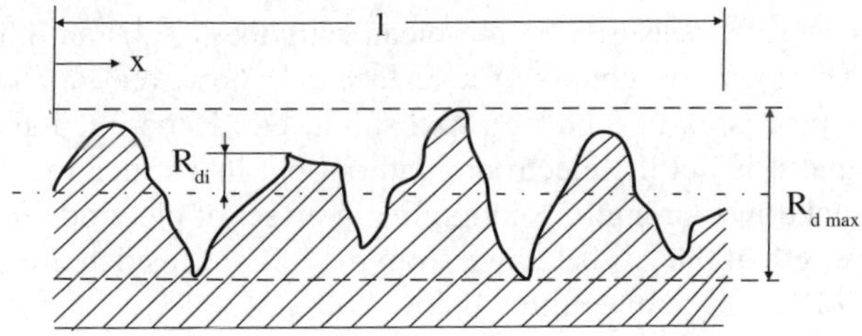

(a)

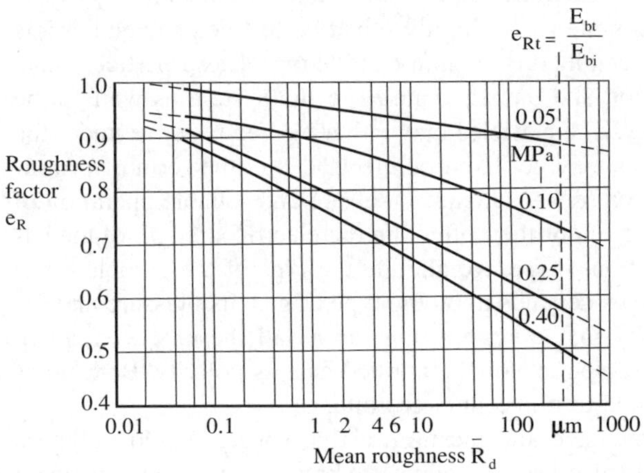

(b)

Figure 3.76 (a) Roughness profile of a metal surface. (b) Variation of pressure dependent roughness factor e_R with increasing mean roughness $\overline{R}_d$ at constant pressures and for uniformly rough surfaces [3.88].

$$U_{bt} = E_{bi} \cdot e_R \cdot e_{\max} \cdot \eta \cdot d \tag{3.71}$$

As seen in Figure 3.76 (b) a certain value of mean roughness $\overline{R}_d$ affects the roughness factor e_R more adversely at higher pressures, bringing down the breakdown strength/voltage of the gas for a given electrode configuration. Extra effort and care must be taken while preparing electrodes for use in high voltage gas insulated systems where SF_6 gas pressure of the order of 0.4 MPa is practiced.

As reported by Mosch and Hauschild [3.88], no effect of the type of voltage was measured on e_R. Therefore like $e_{\max}$, e_R can also be used for the estimation of breakdown strength/voltage for all types of voltage waveforms. According to their approximation, the value of $\overline{R}_d = 300\ \mu$m corresponds to give the "practical electric strength", E_{bt}, therefore,

$$e_{Rt}(at\ \overline{R}_d = 300\ \mu m) = \frac{E_{bt}}{E_{bi}}$$

as shown in Figure 3.76 (b).

It has been observed that, when different electrode materials are processed in the same way, different roughness profiles of their surface finish are obtained. One may infer that surface structure measured by the mean roughness $\overline{R}_d$ is an inherent property of a material. The mean roughness of a surface can, however, be changed by different mechanical processes like buffing, polishing, etc. Experimental measurements have shown that it is not the electrode material but its surface roughness factor that affects the breakdown strength. No significant effect of electrode materials on the breakdown strength of the gas is known from the SF_6 gas insulated systems in service at around 0.4 MPa and temperatures below 200°C, [3.88].

3.5.3.2 Effect of Particle Contaminants in Gas Insulated Systems (GIS) Compressed gas filled systems are highly sensitive to free or fixed particle contaminants, which may be conducting or non-conducting. These particles may result from mechanical abrasion and damage caused to the system as well as the enclosures during the assembly. The particles may either be free to move under the influence of the applied field or may get fixed on electrodes in the form of protrusions. Such particles constitute a serious threat to the safe and reliable operation of "gas insulated systems". The extent of their effect on the electric strength of the gas largely depends upon the location, shape, weight and the type of the particles. The damage caused by these particles depends upon the type of gas, its pressure and the designed electric field intensity in the system. On top of all these, the extent of damage also depends upon the type of voltage applied and its polarity. Both *ac* as well as *dc* GIS may be subjected to transient over voltages.

Depending upon the type, size and location, a free particle could initiate a local breakdown or PB. It could also lead to a complete breakdown by its movement in the gap or cause surface discharge (tracking) on the solid dielectric spacers.

3.5.3.2.1 Movement of Particles Consider a free conducting particle lying at the cathode in a coaxial cylinder system. On applying a voltage to the conductor (anode) the free particle gets charged to the same polarity as that of the coaxial electrode. The induced charge acquired by the particle increases with the increasing voltage and a lifting force is applied to the particle. At certain field intensity, say E_L, when the electrostatic force exceeds the gravitational force, the particle begins to get lifted towards the opposite electrode. On raising the voltage/field intensity further, the particle oscillates between the electrodes under the influence of the applied voltage and the gravitational force. The motion of the particle depends upon the type of applied voltage as illustrated in Figure 3.77, Mosch et al. [3.93]. These motions have been verified by the high-speed photographic techniques.

After getting lifted up, as the charged particle approaches the opposite electrode, it may lose its charge due to a gas discharge, a local breakdown or PB. It is assumed that such a discharge short-circuits the gap almost instantaneously, bringing down the particle to the electrode potential. The particle then acts as nothing but an extended protrusion on the electrode. Depending upon the size and the shape of a particle, as well as gap dimensions and gas pressure, electric stress at the tip of this particle may be sufficient for a critical avalanche (streamer) formation leading to complete breakdown, Rizk [3.94]. The reduction in breakdown voltage is more severe under *dc* applied voltages than for *ac* and impulse voltages [3.92]. Further, the effect on breakdown voltage is observed more in uniform fields than in nonuniform fields.

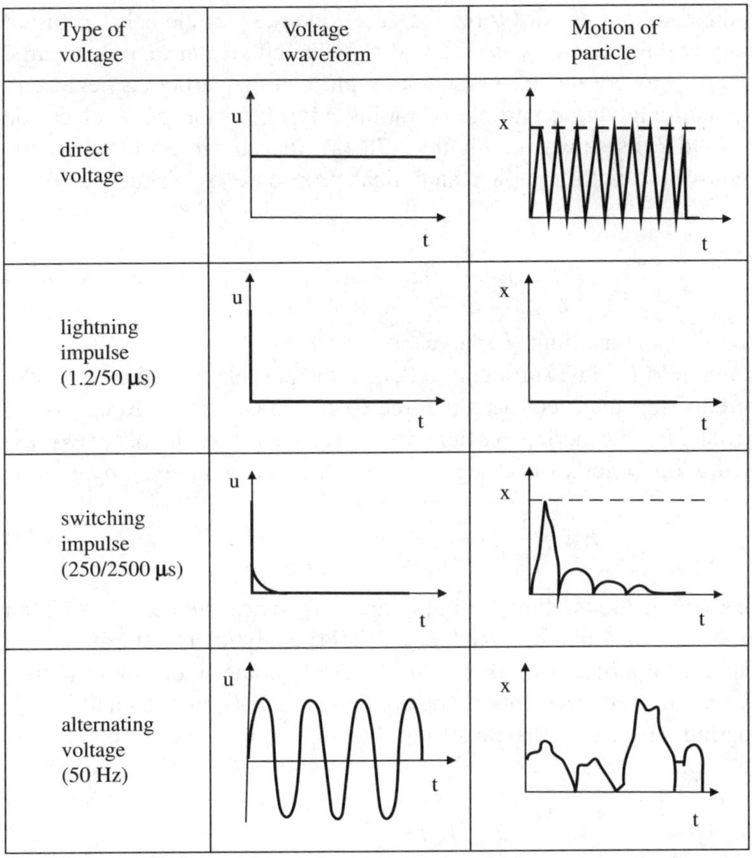

Figure 3.77 Motion of free particles in GIS under different types of applied voltages, Mosch et al. [3.93].

According to Wohlmuth [3.98], large PB signals are caused by particle impacts with the internal conductor, r_i and much smaller, but still clearly detectable signals on impacts with the enclosure, r_o. PB may also occur in mid-gap positions of the particles. Particle impacts with the internal conductor have high risk of an insulation failure; however, every impact may not lead to complete breakdown.

3.5.3.2.2 Estimation of Induced Charge and Lifting Field Intensity of Particles An analytical estimation of the charge induced on a freely conducting particle and the field intensity required to lift the particles is made by Rizk and Cookson and also given by Laghari [3.94–3.96 and 3.92]. They have chosen spherical and wire shaped particles for their analysis described below.

When a conducting particle is located freely on an electrode, it acquires a charge q. The magnitude of charge acquired depends upon the electrode field E besides shape, orientation and size of the particle. When the electrostatic force $q.E$ on the particle exceeds the gravitational force, the particle is elevated. Once the particle is lifted, the force of attraction on the particle due to the image charge decreases. Thus the resultant upward force increases and the particle is lifted faster. The charge density at the particle ρ_p is assumed to be much greater than the charge density of the gas ρ_g. Any corona discharge taking place at the particle is neglected.

For a spherical conducting particle of radius r lying on the plane electrode in a uniform *dc* field E as shown in Figure 3.78 (a), the charge acquired by the particle, as demonstrated with complex analytical procedure by Felici [3.101] is given by,

$$q_{pu} = \frac{2}{3}\pi^3 \varepsilon r^2 E \tag{3.72}$$

ε being permittivity of the medium (for a gas, $\varepsilon \approx \varepsilon_o$).

The *dc* lifting field E_L, just sufficient to detach the particle from the electrode, is calculated by equating the electrostatic force to the gravitational force due to weight of the particle. For a spherical particle of radius r in uniform field, the expression for E_{Lu} based upon Felici's work neglecting contact forces is given by,

$$E_{Lu} = \frac{1}{\pi}\sqrt{\frac{2.4\, gr(\rho_p - \rho_g)}{\varepsilon_o}} \tag{3.73}$$

The above two equations given by Felici are widely accepted and have been investigated further by many authors. Rizk et al. [3.94] analytically determined the charge q_p on a spherical particle of radius r in a coaxial geometry of inner radius r_i and outer radius r_o under a positive polarity dc applied voltage U, as shown in Figure 3.78 (b). Considering a practical coaxial system having r_o/r_i to be nearly equal to e, the charge q_p on the particle is given by,

$$q_p = \varepsilon_o r^2 E_o f\left(\frac{r}{r_i}\right) \tag{3.74}$$

where E_o is the field intensity at the outer cylinder.

The function $f(r/r_i)$ can be determined by computing q_p using charge simulation techniques (CSM) for different ratios of r/r_i. The relationship obtained between $q_p/(\varepsilon_o r^2 E_o)$ and r/r_i is shown in Figure 3.78 (c). It is seen from this figure that for very small particles compared to electrode dimensions, as r/r_i tends to zero, $q_p/(\varepsilon_o r^2 E_o)$ reduces to $2/3\pi^3$. The same result is also obtained from the analytical solution of particle in a uniform field, equation (3.72). For any arbitrary value of r/r_i and the same value of field, the charge acquired by a particle is lower in case of coaxial cylinders than in the uniform field. Further,

$$\frac{q_p}{\varepsilon_o r^2 E_o} = f\left(\frac{r}{r_i}\right) = K_c \tag{3.75}$$

where $K_c \le \frac{2}{3} r^3$, the exact value is obtained from the Figure 3.78 (c).

From equations (3.72) and (3.75), a general expression for the charge acquired by a particle can be given as:

$$q_p = K_c \varepsilon_o r^2 E = K_{nu} q_{pu} \tag{3.76}$$

where K_{nu} is the nonuniformity charge factor given by,

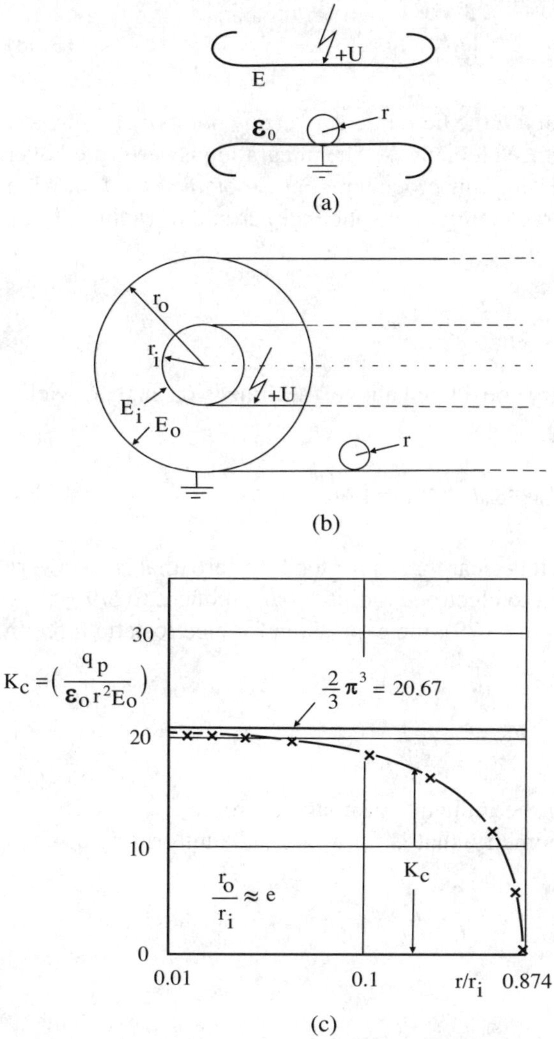

Figure 3.78 (a) Particle in uniform field, (b), Particle in coaxial cylindrical weakly nonuniform field, (c) Electrostatic charge acquired by a spherical particle lying on the outer conductor of a coaxial-cylinder system, Rizk et al. [3.94].

$$K_{mu} = \frac{K_c}{\frac{2}{3}\pi^3} \tag{3.77}$$

For a coaxial cylindrical field configuration, $E_o = \dfrac{U}{r_o \ln r_o/r_i}$, which reduces to $E_o = U/r_o$ for $r_o/r_i \approx e$.

As a consequence of the accumulated charge, the particle is subjected to an electrostatic force F_E, which tries to detach or lift the particle from the surface. F_E in the direction of a unit vector $\vec{p}$ is given by,

$$\vec{F}_E = \frac{1}{2}\int_s qE\vec{n}\cdot\vec{p}ds \tag{3.78}$$

where q is the surface charge density, E the field intensity at any point on the surface, $\vec{n}$ the unit vector normal to the surface element ds. The integration is performed over the total surface of the particle s. Using this procedure, Felici obtained the following expression for the electrostatic force acting on a spherical particle of radius r lying on the plane electrode in uniform field:

$$F_{Eu} = \frac{5}{9}\pi^3\varepsilon_o r^2 E^2 \tag{3.79}$$

For the coaxial cylindrical geometry considered above, the dimension analysis yields the following expression:

$$F_E = \varepsilon_o r^2 E_o^2 g\left(\frac{r}{r_i}\right) \tag{3.80}$$

From equations (3.79) and (3.80), it is clear that as r/r_i tends to zero, that is for a very small particle dimension compared to electrode radius, $g(r/r_i)$ reduces to 5/9 π^3.

Substituting equation (3.72) in (3.79), the expression for electrostatic force in uniform fields takes the form:

$$F_{Eu} = \frac{5}{6}q_{pu}E \tag{3.81}$$

where the factor 5/6 takes care of the influence of image charges.

For a coaxial cylindrical geometry, that is, in weakly nonuniform field,

$$F_E = \frac{5}{6}q_p E_o$$

or

$$= \frac{5}{6}K_{nu}q_{pu}E$$

or

$$= K_{nu}K_{Eu} \tag{3.82}$$

Accordingly, for a spherical particle of radius r in a coaxial cylinder, the lifting field intensity given by Rizk et al. works out to be:

$$E_L = \frac{1}{\pi}\sqrt{\frac{2.4\,gr(\rho_p - \rho_g)}{K_{nu}\varepsilon_o}} \tag{3.83}$$

Figure 3.79 (a) and (b) show a comparison between calculated and measured lifting fields in uniform and coaxial cylinder electrode geometries given by Rizk

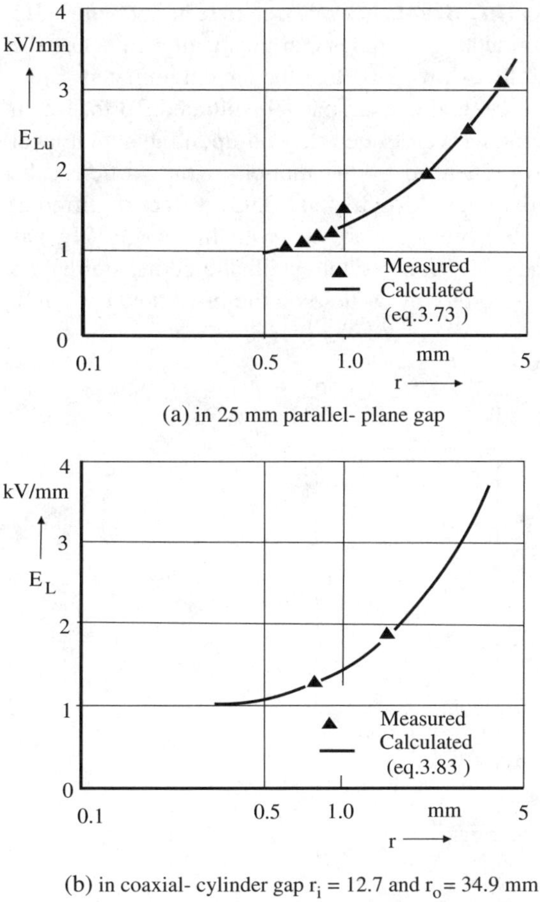

Figure 3.79 Calculated and measured values of lifting fields as a function of spherical steel particle radius, Rizk et al. [3.94].

et al. [3.94]. A satisfactory agreement between the measured and calculated values is seen in these figures confirming the derived expressions (3.73 and 3.83).

The equation of motion of the particle in a general form is given by:

$$m\vec{a} = \vec{E}_E + \vec{F}_D + \vec{F}_G \tag{3.84}$$

where m is the mass, $\vec{a}$—acceleration of the particle, $\vec{F}_E$—the electrostatic force, $\vec{F}_D$—the drag force and $\vec{F}_G$—the gravitational force.

A vast amount of analytical work has been taken up in this field considering several shapes of particles in different types of applied voltages. However, the practical approach to this problem has suggested interesting solutions.

Mosch et al. [3.93] concluded that from the practical point of view in SF_6 insulation system, the particles are dangerous only when their pressure dependent value $(l_p \cdot p)$ is greater than 7 μm·MPa. Long and light particles have been found to be more dangerous. Accordingly, lower design pressures and larger electrode dimensions are suggested for the construction of GIS.

3.5.3.3 Particle Initiated PB and Breakdown Measurements in GIS

Investigations have revealed that conducting particles can drastically influence the dielectric strength of SF_6, reducing it to as low as 10% of the uncontaminated value, as reported by Laghari and Qureshi [3.92]. However, particle initiated PB inception levels and reduction in breakdown voltage very much depend upon the actual local conditions involved. These are also influenced by the motion of the particles, the gas pressure and the type of applied voltage, Mosch et al. [3.93]. Effect of different types of voltages on "particle initiated breakdown", U_{bp} with increasing SF_6 gas pressure measured in a coaxial cylinder electrode system is illustrated in Figure 3.80 (a). It can be observed that there is no effect of particle on the breakdown strength

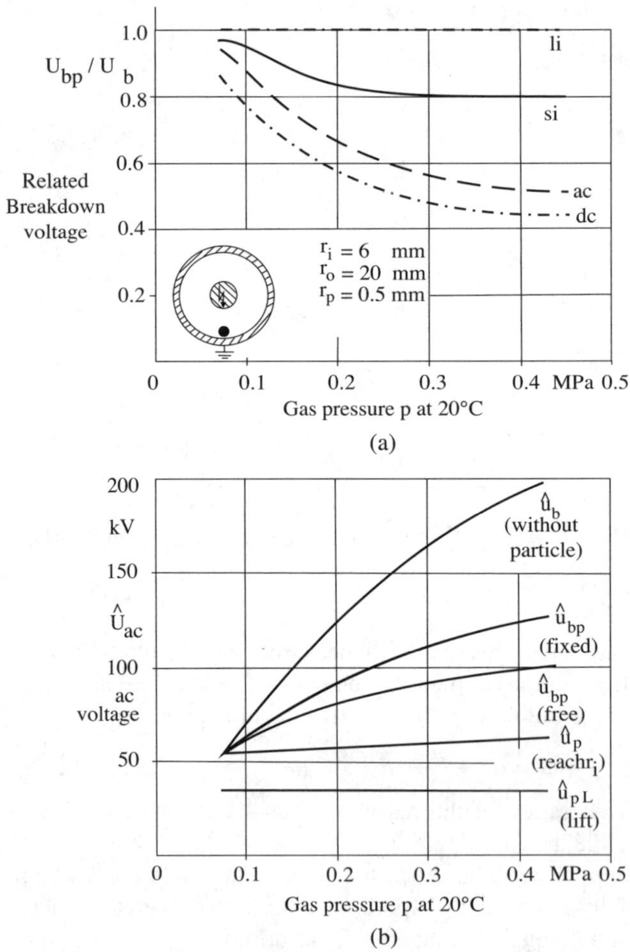

Figure 3.80 (a) Ratio of free moving particle initiated breakdown U_{bp} to breakdown without particle U_b with increasing SF_6 gas pressure for different types of voltages in coaxial electrodes for spherical conducting particle. (b) Voltage characteristics with increasing SF_6 gas pressure illustrating the particle effect in coaxial electrode system as in (a).

for *li* impulse voltage. As explained, this could be because the particle is unable to lift from its position due to short duration of *li* impulse, as shown in Figure 3.77. Breakdown voltages are measured considerably lower in presence of a particle with increasing gas pressure for all other types of voltages.

A free moving particle brings down the breakdown voltage more than a fixed particle in weakly nonuniform fields. The same electrode systems as shown in Figure 3.80 (a) has also been chosen to demonstrate this effect illustrated in Figure 3.80 (b). The alternating voltages at which a free particle gets lifted and rises to touch the inner electrode and finally lead to breakdown were measured with increasing SF_6 gas pressure. This figure also shows the breakdown voltage characteristics with increasing pressure for a fixed particle (protrusion) and without any particle, which are measured much higher, [3.88].

The insulating material particles are found to have less effect on the electrical performance of the gases compared to the conducting particles. However, a system becomes highly sensitive to contaminants at higher gas pressures, above 0.4 MPa [3.92].

Particle initiated PB inception voltages $\hat{U}_{ip}$ were calculated by Mosch et al. [3.93] in uniform field electrode configuration with SF_6, taking into account the streamer criterion. An ellipsoidal shaped particle was considered to be placed at the opposite polarity electrode. Figure 3.81 shows the variation in PB inception voltage related to the breakdown voltage in an undisturbed system, as a function of the product of particle length, l_p and the gas pressure, p ($l_p \cdot p$). These measurements were conducted at normal room temperature of 20°C for particles having different ratios between their length l_p and maximum diameter d_p of 1:1, 3:1 and 10:1. As seen from this figure, the undisturbed system is affected only when the product ($l_p \cdot p$) exceeds a value of 7 µm·MPa. Further, as the length of the particle is increased, the relative PB inception voltage reduces more with increasing gas pressure.

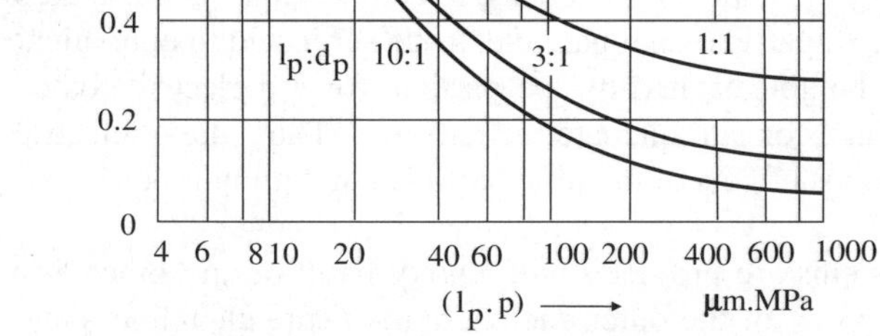

Figure 3.81 Effect of ellipsoidal shaped particles on PB inception in uniform field with SF_6 gas, Mosch et al. [3.93].

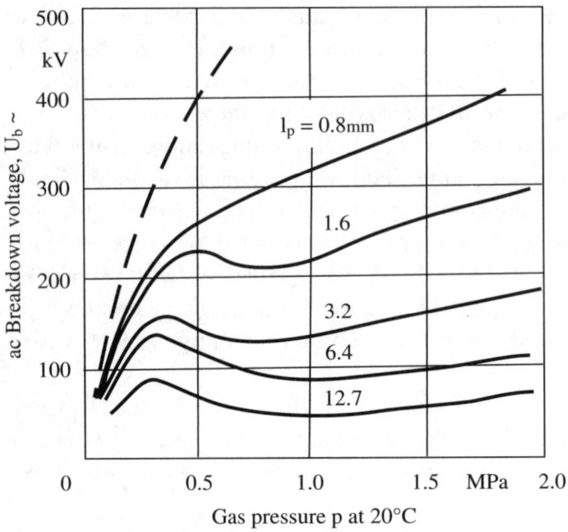

Coaxial cylinder system having, $r_i = 75$ mm, $r_0 = 125$ mm,
Copper wire diameter $d_p = 0.4$ mm of different lengths l_p

Figure 3.82 Effect of particle length on ac breakdown voltage in SF_6 gas, Cookson and Farish [3.102].

Effect of the length of particle on *ac* breakdown voltage with increasing gas pressure in coaxial cylindrical electrode system was studied by Cookson and Farish, shown in Figure 3.82, given in Mosch et al. [3.88]. It is evident from this figure that as the length of the particle is increased, the breakdown voltage reduces considerably. The dotted curve in this figure shows breakdown voltage characteristic without any particle.

3.5.3.4 Preventive Measures for the Effect of Particles in GIS In order to control and minimize the effect of particle contamination, one of the techniques adopted is to apply a dielectric coating to the inner surface of the outer GIS enclosure. Dielectric coatings help to improve the insulation performance in several ways. For example, it is known that the dielectric coated electrodes in compressed gas give higher breakdown strength and reduce the PB conduction current through the gas. Such coatings also provide a smooth and clean surface.

A metal particle resting on a dielectric coated electrode may acquire less induced charge as compared to a particle on a bare electrode subjected to equivalent electric field. The maximum height reached by the particle for a dielectric coated enclosure is much lower than for an uncoated enclosure. Thus the dielectric coating minimizes the deleterious effects of metallic particle contamination, Morcos et al. [3.103].

Some gas insulated systems are provided with epoxy paint or any other thin dielectric coating on the inner side of the outer enclosure to ensure cleanliness and deactivate particle movement, Cookson [3.104]. Adhesive coatings have also been proposed along with other types of particle traps.

Often GIS are provided with "particle traps" to deactivate the particles either by physically moving them to zero or low field regions or by preventing their movement altogether. The compressed gas insulated cables are provided with such particle traps. These are made of perforated/slotted grounded metal cylinder, or flat sheets inside the enclosure, or a trough at the bottom of the enclosure. The traps are designed so that particles elevated by the electric field move towards the traps and fall through the openings into the zero or low field regions. Different types of electrostatic trap designs are described by Cookson in [3.104].

3.5.4 Breakdown in Extremely Nonuniform and Distorted Weakly Nonuniform Fields with Stable PB in SF_6 Gas Insulation

Experimental investigations have revealed that in compressed gas insulated extremely nonuniform field electrode arrangements, the inception of PB not only depends upon the electrode geometry but it is also strongly affected by the gas pressure. From a practical point of view, it is therefore important to know the extent of applicability of the Schwaiger relation for estimating the PB inception and breakdown voltages. The pressure dependent limiting uniformity factor, η_{lim} for SF_6 gas is shown in Figure 3.83. In the region $\eta > \eta_{lim}$, no stable PB occur since the field is weakly nonuniform. Whereas in the region $\eta < \eta_{lim}$, being an extremely nonuniform field stable PB are observed before the breakdown. It can be observed that as the pressure is increased, the value of η_{lim} reduces. This means that more nonuniformity in the field is permissible at higher gas pressures while remaining within the limits of weakly nonuniform field.

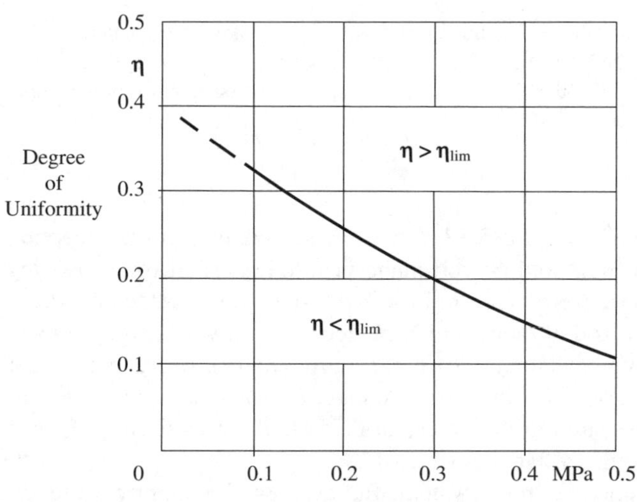

Figure 3.83 Pressure dependent variation in limiting degree of uniformity η_{lim} for electrode configurations in SF_6.

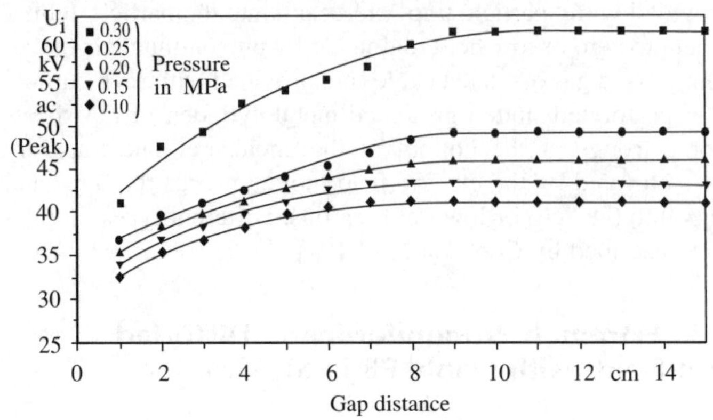

(a)

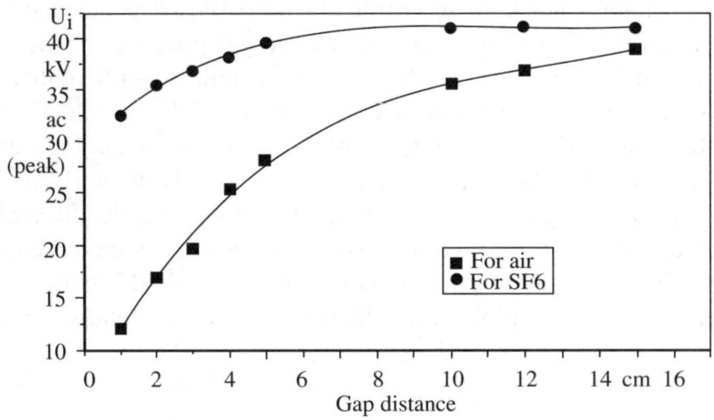

(b)

Figure 3.84 (a) PB inception voltage, U_i characteristics with increasing gap distance between 6 mm diameter rod-plane for increasing SF_6 gas pressures. (b) PB inception voltage, U_i characteristics for air and SF_6 gas at one atmospheric pressure with increasing gap distance, Arora and Haldar [3.105].

The effect explained in Figure 3.83 can be observed in the PB inception voltage, U_i characteristics measured on rod-plane electrode system for increasing constant gas pressures. These were measured for increasing gap distance between a smooth hemispherical brass rod of diameter 6 mm electroplated with chromium and stainless steel inverted bowl of diameter 70 mm, Figure 3.84 (a), Arora and Haldar [3.105]. A significant increase in PB inception voltage, U_i was measured at SF_6 gas pressure of 0.3 MPa as compared to the values at 0.25 MPa. On increasing the gap distances, these curves acquire a flat characteristic.

Measured on the same electrode system, Figure 3.84 (b) compares the PB inception voltage characteristics for air and SF_6 gas. It can be observed that for smaller gap distances, the U_i in SF_6 gas is measured to be much higher, almost 2.5 times. Field equalization effect of negative space charge due to electro negativity of

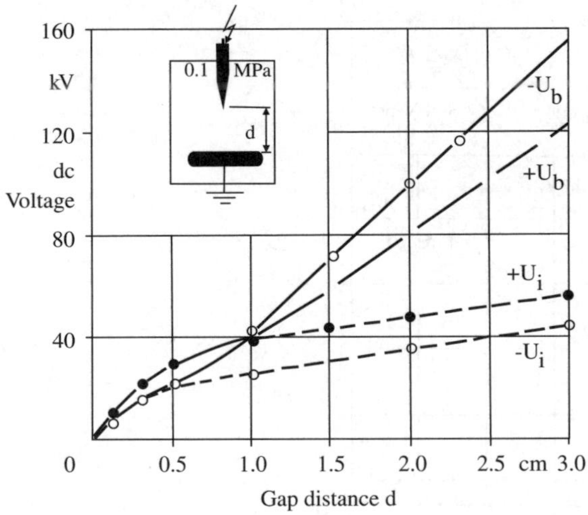

Figure 3.85 PB inception and breakdown with dc voltages on a needle-plane electrode in SF_6 gas with increasing gap distance.

SF_6 gas can be attributed to this measurement. As the nonuniformity in the field is increased on increasing the gap distance, this difference reduces gradually since the PB inception in air is measured at higher voltages. It is measured to be only marginally higher for the gap distances of 15 cm and above. At longer gap distance the negative space charge in air due to above critical amplification of avalanche also plays its role more effectively to equalize the field, resulting in higher PB inception voltages.

PB inception and breakdown voltage measurements were made on a needle-plane electrode system with increasing gap distance in SF_6 gas at atmospheric pressure, as shown in Figure 3.85. On applying a direct voltage, lower PB inception levels were measured for negative polarity needle electrode than for positive polarity. As in the case of atmospheric air, in SF_6 gas too, a compact star corona is observed at the sharp electrode in extremely nonuniform fields. The nature of the PB pulse current is the same as in the case of air. Stronger PB current pulses are measured with negative polarity needle electrode than with positive polarity. Since the field equalizing effect of space charge is stronger with negative polarity, higher voltages are required for accomplishing spark breakdown, Figure 3.85 ($-U_b > +U_b$). These polarity effects on PB inception and breakdown voltages are valid for all types of voltages in SF_6 gas.

Similar curves showing the PB inception and breakdown voltages of a distorted weakly nonuniform field electrode configuration are shown in Figure 3.86 with standard *li*, *si* and *ac* voltages. The breakdown voltages were measured for varying gap distances and a fixed length of needle projection (0.5 cm) on a sphere-plane electrode system having constant SF_6 gas pressure of one atmospheric.

In distorted uniform or weakly nonuniform field electrode systems at low SF_6 gas pressures (less than atmospheric), stronger space charges are able to build up due to PB, resulting in higher field equalizing effects. This results in a significant

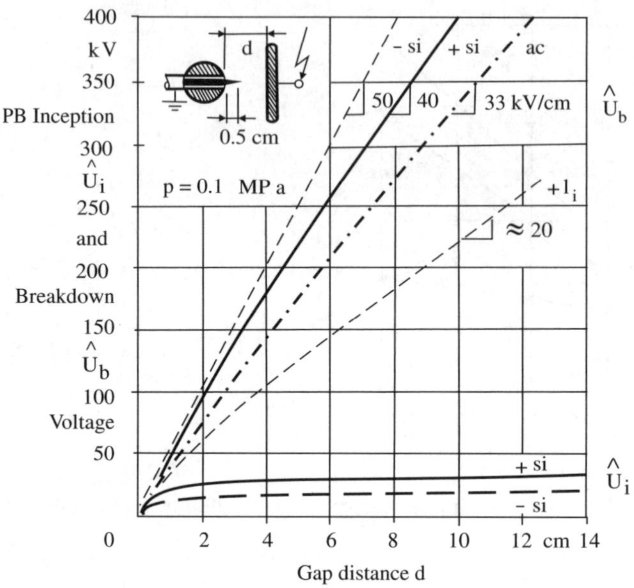

Figure 3.86 PB inception and breakdown voltages in a distorted weakly nonuniform field with SF_6 gas.

difference between PB inception and breakdown voltages at low pressures. On increasing the gas pressure above atmospheric, the intensity of PB is weakened, hence the space charge effect is also. With this result, the breakdown voltage characteristic experiences a maximum at relatively low pressure itself, whereas the PB inception characteristic increases uniformly, as shown in Figure 3.87. At higher gas pressures, the positive polarity inception and breakdown voltage characteristics both meet. It is well known that the breakdown with *ac* occurs at positive polarity half cycle. This so-called typical "N shape" characteristic was measured with alternating voltage and for the given dimensions of electrodes by Inuishi [3.106].

On applying *ac* voltage, the PB inception first takes place at the negative polarity and then at the positive polarity half cycle on increasing the voltage. In Figure 3.87 as reported by the author, the breakdown mostly occurred during the negative half cycle for pressures below 1.9 kg/cm^2 (representing nearly the peak of the U_b characteristic). At higher gas pressures, the breakdown occurred during the positive half cycle. This suggests that the space charge field-equalizing phenomenon is more effective at negative polarity and at low gas pressures. Further investigations in this direction are needed to be able to provide a more satisfying explanation of such measured characteristics.

3.5.5 Electrical Strength of Mixtures of SF_6 with Other Gases

Sulphurhexafluoride, SF_6 is a superior dielectric gas for nearly all high voltage applications. It exhibits exceptional insulation and arc interruption properties and

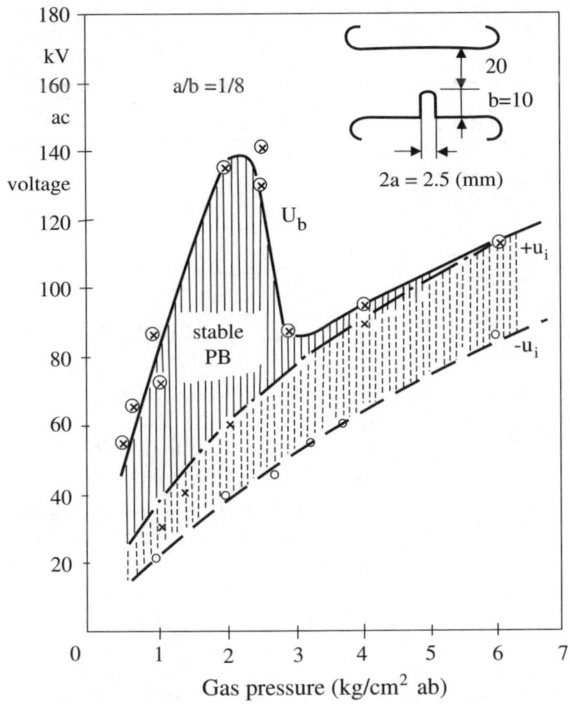

Figure 3.87 PB inception and breakdown voltages for a parallel plane gap in SF_6 with spheroidal protrusion of a fixed dimension and for varying gas pressure, Inuishi [3.106].

has proven its performance for over four decades of ever increasing use. However, the extremely high global warming potential of SF_6 mandates that its release into the environment is minimized. One way to achieve this is the use of other gases or mixtures of SF_6 in place of pure SF_6 gas, Christophorou et al. [3.107].

In search of improved dielectric properties, low cost to achieve economy, reduced sensitivity to surface roughness and conducting particles, low toxicity of decomposed by-products, compatibility with solid dielectrics and other desired properties, a considerable amount of work has been performed to investigate the behavior of mixtures of other gases with SF_6. The other main gases considered so far are air, nitrogen, carbondioxide and helium. Gases with very strong attachment properties, usually halogenated hydrocarbons, have also been considered to mix with SF_6 in order to obtain superior dielectric properties to those of pure SF_6.

Various gas mixtures show considerable promise for use in new equipment. However, the equipment should be designed specifically for use with a gas mixture [3.107]. The breakdown strength characteristics of a mixture of SF_6 with N_2 in uniform field for a fixed gap distance of 1.5 cm are shown in Figure 3.88. These peak *ac* breakdown voltages were calculated as well as measured for increasing volume percentage of SF_6 gas in the mixture at different constant gas pressures by Wieland [3.108]. As seen from these curves, at about only 25% of SF_6 in the mixture, more than 75% of the electric strength properties of pure SF_6 are achieved. Similar

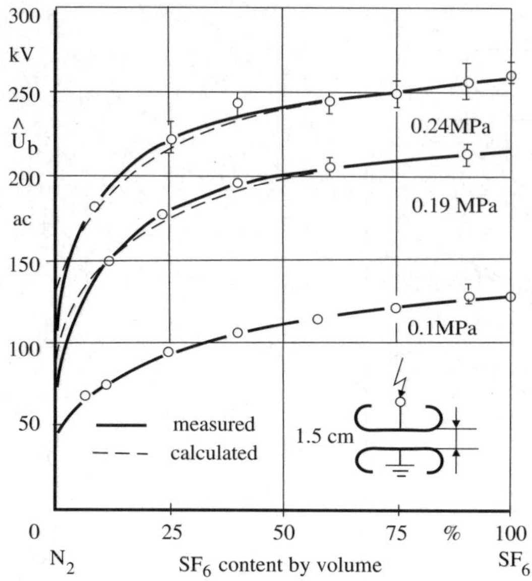

Figure 3.88 Alternating voltage breakdown characteristics in uniform field for mixture of SF_6 and N_2 at constant pressures, Wieland [3.108].

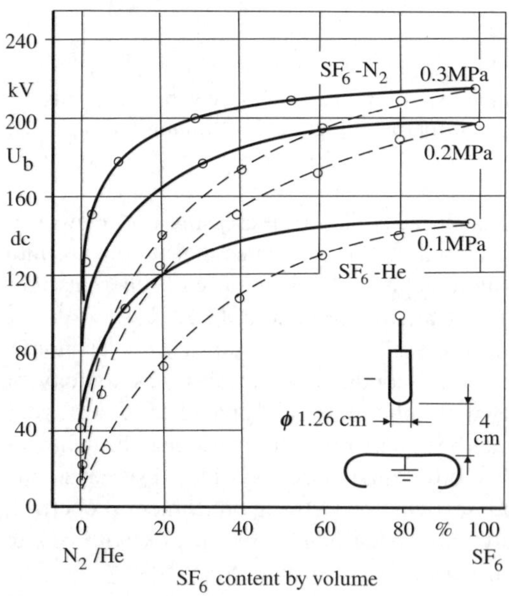

Figure 3.89 Direct voltage breakdown characteristics for negative rod-plane electrode system with SF_6-N_2 and SF_6-He mixtures, Malik et al. [3.110].

breakdown characteristic curves were also measured by Pelletier et al. [3.109] for the mixtures of SF_6 with N_2 and CO_2 in uniform field with alternating voltage.

Malik and Qureshi [3.110] measured breakdown characteristics for the mixtures of SF_6 with He and N_2 on a negative rod-plane gap applying direct voltage. A fixed gap distance of 4 cm was maintained between the stainless steel electrodes, the rod having a diameter of 12.6 mm. These characteristics for different gas pressures

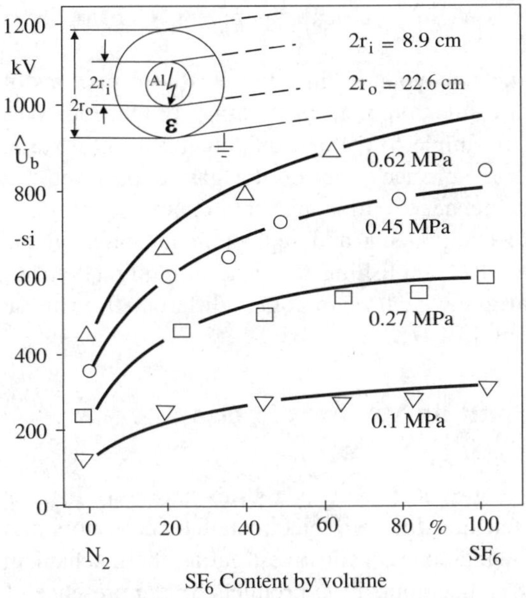

Figure 3.90 Average negative switching impulse breakdown voltage characteristics for a coaxial cylinder electrode geometry with SF_6-N_2 mixtures, Cookson et al. [3.111].

are shown in Figure 3.89. As seen from this figure, the electric strength of SF_6-N_2 mixture is superior to that of SF_6-He.

Cookson and Pedersen [3.111] have reported similar characteristics for SF_6-N_2 mixtures. Their measurements were performed on a coaxial cylinder electrode geometry of $2r_i = 8.9$ and $2r_o = 22.6\,\text{cm}$ having a length of 1 m, with negative *li*, *si* and alternating voltages. Characteristics measured with average negative *si* (250/3000 μs) are shown in Figure 3.90. The coaxial Al electrode system for these measurements represents the actual 145 kV CGIT system. The conductors were given a standard bus finish and there were no insulators in the test length. An internal fan was used to ensure thorough diffusion of the gases.

In practice, mixtures of SF_6 with N_2 have so far found application in gas insulated power cables, also known as "compressed gas insulated transmission lines" (CGIT). Since the amount of gas required for CGITs is very large, any savings in the quantity of costly SF_6 would effectively economize the total investment.

Mixtures of nearly equal amounts of SF_6 and N_2 exhibit dielectric properties that suggest that they could be used as a "universal application" gas for both electrical insulation and arc/current interruption purposes. In this connection, standard procedures for mixtures handling, use and recovery would need to be further developed. Mixtures of low concentrations (<15%) of SF_6 in N_2 show excellent potential for use in gas insulated transmission lines. A mixture of SF_6 and helium has shown promise when used in gas insulated circuit breakers and should be further investigated. It is clear that a significant amount of research must be performed for any new gas or gas mixture to be used in electrical equipment, Christophorou et al. [3.107].

Although there appears to be no fundamental limit to the use of mixtures of SF_6 gas, application of mixture of gases in equipment in a definite proportion gives rise to newer practical problems. It is simple to fill up equipment with a mixture, but for some applications routine maintenance requires the gas to be removed, recycled, refilled or disposed. The maintenance routine may entail pumping, purging, filtering and separating the mixture components and replacing the gases in the desired ratio and volume percentage. Accomplishing these tasks requires various chemical management practices and associated technologies different from those used for pure SF_6, Christophorou et al. [3.107].

3.5.6 Decomposition of SF_6 and Its Mixtures in Gas Insulated Equipment

The increased application of SF_6 in gas-insulated switchgears/sub-stations (GIS), gas insulated cables (CGIT), gas insulated transformers, electrostatic accelerators and x-ray equipment etc. have led to growing concern for investigating the mechanism of SF_6 decomposition and the effects of decomposition products. In the presence of corona (PB), spark breakdown and electric power arc, SF_6 decomposes into lower oxy-fluorides of sulphur. These may react with the electrodes or gas impurities or other solid dielectrics to form a number of chemically active products. Although SF_6 gas is chemically inert and nontoxic, the decomposition products of SF_6 are known to be toxic and corrosive, Chu [3.112]. The accumulation of decomposition products in the equipment has caused concern regarding personnel safety and material compatibility, hence the life of the equipment.

The decomposition of SF_6 gas is greatly influenced by gaseous impurities. In industrial grade SF_6 the typical impurities are CF_4, N_2, O_2 (air) and H_2O. The gaseous impurities are generally introduced during filling and partly due to the desorption of moisture into the dry SF_6 after filling. A survey of major North American utilities revealed that the average air concentration in SF_6 compartments is 5000 ppmv and the average moisture content is about 500 ppmv, Chu [3.112]. In practical GIS environments the presence of such impurities is unavoidable.

An excellent review of the research work performed on the problems of SF_6 decomposition is made by Chu [3.112] giving a large number (204) of reference literatures, the majority of which were published in 1980s, but are still valid. The subject of SF_6 decomposition mechanisms is vast and complex. It is described briefly in the following paragraphs.

Three different fundamental processes can cause the decomposition of SF6: electronic, thermal and optical. In high voltage gas insulated equipment, the electronic and thermal processes are dominant, resulting in the formation of stable products. Both electronic and thermal processes can occur in discharges such as spark breakdown and power (heavy current) arcs. The decomposition of SF_6 due to PB is mainly under nonthermodynamic equilibrium conditions, where the electron temperature is higher than the gas temperature.

At room temperature and in the absence of an external applied electric field, the only charged particles, in otherwise neutral SF_6 gas, are generated from natural ionization as a result of cosmic rays or natural radioactivity. The electrons, generated from the natural ionization, quickly attach themselves to the SF_6, forming SF_6 negative ions as represented below:

$$SF_6 + e \rightarrow (SF_6^-)^* \rightarrow SF_6^- \tag{3.85}$$

$$\rightarrow SF_5^- + F \tag{3.86}$$

The metastable association complex $(SF_6^-)^*$ can be stabilized to SF_6^- by collision. Alternatively, the complex can decay to form SF_5^- and F. The life time of the $(SF_6^-)^*$ against auto detachment is short, about 25 μs. It is reported in the literature that the formation of SF_6^- and SF_5^- ions occur at electron energy of less than 2 eV by resonance capture process. The formation of these ions at low electron energies of less than 0.4 eV by direct dissociative electron attachment processes has also been reported.

In electric fields the electron energy increases as the ratio E/n between the applied electric field and the number density of the SF_6 molecules is increased. As the electron energy increases, the production of SF_5^- is also enhanced. At higher electric field levels, that is, higher E/n, the collision between the SF_6^- ion and the SF_6 (neutral) to form SF_5^- is explained as follows:

$$SF_6^- + SF_6 \rightarrow SF_5^- + F + SF_6 \tag{3.87}$$

at even higher E/n, F^- appears as shown by,

$$SF_5^- + SF_6^- \rightarrow F^- + SF_4 + SF_6 \tag{3.88}$$

or

$$SF_5^- + SF_6 \rightarrow F^- + SF_5 + SF_5$$

The formation of SF_5^- and subsequently SF_5 and F is the first sign of decomposition of SF_6. Even at low E/n and moderate temperature, SF_6 may dissociate into SF_5^- which may convert into other long lived species if impurities are present in the gaseous form. In the presence of oxygen and water vapours, reaction occurs between the lower fluorides of sulphur and the contaminants to form sulphur oxyfluoride HF and metallic fluorides. Some examples of these reactions are expressed as follows:

$$SF_5 + OH \rightarrow SOF_4 + HF \tag{3.89}$$

$$SF_5 + O \rightarrow SOF_4 + F \tag{3.90}$$

$$SF_4 + O \rightarrow SOF_2 + 2F \tag{3.91}$$

$$SF_3 + O + OH \rightarrow SO_2F_2 + HF \tag{3.92}$$

In the presence of electrical arc, flashover and PB or corona, a portion of the SF_6 decomposes into lower fluorides of sulphur, which can react to form a number of chemically active byproducts. The possible formation of SF_4, SF_2, S_2F_{10}, SO_2, SOF_2, SOF_4, SO_2F_2, SOF_{10}, $S_2O_2F_{10}$, HF and H_2S during degradation of SF_6 is well documented and widely accepted, Constantine [3.115]. The compounds formed include gaseous sulphur fluorides and sulphur oxyfluorides besides metal fluorides

TABLE 3.11 Typical concentration of by-products found in a test sample of sparked SF_6 [3.113 and 3.114] also in [3.115]

By-Product	Concentration in ppm	Percentage (%) relative to SF_6^*
SOF_2 (SF_4)	3870	1.10
SOF_4	720	0.22
S_iF_4	78	0.03
SO_2F_2	140	0.015
SO_2	—	0.002
Total	—	1.37

For the experimental conditions, E-the total discharge energy = 36 kJ, p-pressure of SF_6 = 0.133 MPa, t-time following discharge = 24 hr.

formed by the reactions with electrode materials and spacers. Experiments have revealed that production of some of the contaminants e.g. S_2F_{10} may be affected by the presence of water, oxygen or even surface reactions originated in the presence of organic insulating materials under corona, flashover and arcs, [3.115].

SF_4, a byproduct produced due to arc, is more stable than SF_5. The presence of SOF_4 and the subsequent conversion of SOF_4 to SO_2 F_2 may be responsible for the presence of a significant amount of these two species due to corona or in low energy sparks, Table 3.11 as given by Sauers [3.113 and 3.114]. The formation of other ions and neutral species such as SOF_3^+, $SO_2F_3^+$, SF_5O_2, SF_5O has also been observed but they may act as only intermediates in the production of observed oxyfluorides.

Outside the PB region, slower reactions between the long lived lower sulphur fluorides, such as SF_4 and stable oxyfluoride, with contaminants further lead to the formation of other types of compounds such as SO_2, SO_2F_2 and SOF_2.

The decomposition products of SF_6 due to arcs are calculated as a function of temperature from known thermo chemical data of SF_6 and its decomposition products. As the temperature increases beyond 1500 K, there is a sharp drop in SF_6 concentration with corresponding increases in SF_4 and F densities. The equilibrium concentration of SF_6 and the decomposition products at temperatures between 1000 and 3500 K have been verified by the so-called shock tube studies and the decomposition rates have been obtained [3.112]. At 4000 K, the molecules are almost completely dissociated while most of the electrons are still bound due to the formation of negative ions. At temperatures beyond 15000 K, the particles are mostly positive ions and electrons. At the center of a high current arc column, the temperature has been measured to be about 20,000 K.

As the arc cools, recombination of Sulphur and Fluorine to form SF_6 occurs rapidly. In the presence of oxygen, H_2O and metal vapor as a result of electrode heating, the recombination process is altered, which leads to the formation of various arc by-products having the main constituents as SF_2, SF_4 and F. As the most stable SF_4 diffuses out from the arc zone, reaction with H_2O dominates to form the lower oxyfluoride.

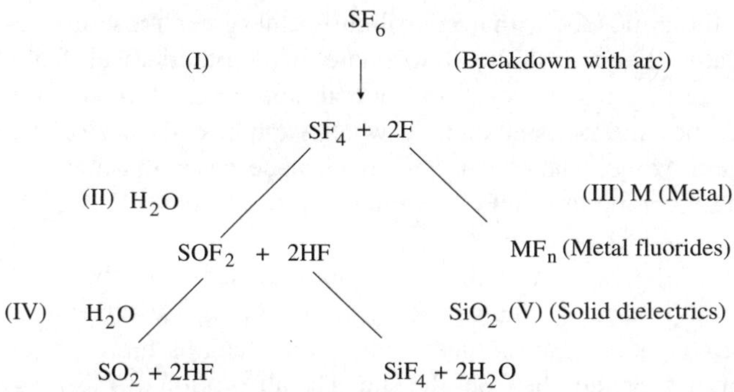

Figure 3.91 Reaction scheme for arced SF_6 leading to the formation of the long lived decomposition products, Sauers et al. [3.114].

A reaction scheme of entire family of decomposition by products derived from the formation of SF_4 as a primary decomposition product is illustrated in Figure 3.91, as given by Sauers et al. [3.114].

At the electrode regions, much higher field intensities, current densities and temperature gradients are present than in the arc column. Heating and fluid dynamic effects in the electrode regions give rise to "plasma jet effect". Molten vapor from the electrode surface is forced into the arc column by the pressure gradients near the electrode region. The metal vapor reacts with the free fluorine atoms to form the metal fluoride compounds as given by the following reactions:

$$Al + 3F \rightarrow AlF_3 \tag{3.93}$$

$$Cu + 2F \rightarrow CuF_2 \tag{3.94}$$

$$3SOF_2 + Al_2O_3 \rightarrow 2AlF_3 + 3SO_2 \tag{3.95}$$

The decomposition mechanism of SF_6 due to spark breakdown is very close to that of a heavy current arc discharge as the spark breakdown is also accompanied with a short duration arc.

Some studies on the toxicities of the SF_6 by-products have revealed that there is a cytotoxicity effect of individual arc by-products; SO_2, HF, SOF_2, SiF_4, SF_4. The arced SF_6 exhibits cytotoxic actions with increasing exposure time [3.112]. However, as the author concludes, there are still some unanswered questions regarding health and safety from the SF_6 decomposition products. The same conclusion is also drawn regarding the solid insulating material compatibility with the decomposition by-products.

3.5.7 SF_6 Gas and Environment

An important desirable property of gaseous insulations is that they should be environment friendly. Unfortunately, the potency of SF_6 as a "greenhouse gas" is extremely high. Greenhouse gases in the atmosphere are partially absorbing the infrared radiation emitted by the earth and radiating it back to earth. Greenhouse gases are components of the atmosphere that contribute to the greenhouse effect. Some greenhouse gases naturally occur in the atmosphere, while others result from human activities such as burning fossil fuels, coal, wood, and so on. Greenhouse gases present naturally in the environment include water vapour, carbon dioxide, methane, nitrous oxides and ozone. The greenhouse man made gases that may be released are fully fluorinated compounds (FFC), combustion products such as CO_2, nitrogen and sulphur oxides and SF_6.

When sunlight reaches the surface of the Earth, some of it is absorbed and warms the surface. The Earth's surface radiates energy at much longer wavelengths than does the sun. The atmosphere absorbs these longer wavelengths more effectively than it does the shorter wavelengths from the sun. The absorption of this long wave radiant energy warms the atmosphere. Greenhouse gases emit long wave radiation both upward to space and downward to the surface. The downward part of this long wave radiation emitted by the atmosphere is the "greenhouse effect".

Potent greenhouse gases have strong infrared absorption in the wave length range from ~7 to 14 μm. The effective trapping of long wavelength infrared radiation from the earth by the naturally present greenhouse gases in the atmosphere and its re-radiation back to earth results in an increase in the average temperature of the earth's surface. An imbalance in the earth's normal greenhouse effect occurs when the man-made or anthropogenic emission of greenhouse gases contribute to an enhanced greenhouse effect by disturbing the balance between the incoming and outgoing radiation [3.107].

Sulphurhexafluoride is an efficient absorber of infrared radiation, particularly at wavelengths near 10.5 μm, Edelson and McAffe [3.117]. Unlike most other naturally present greenhouse gases, SF_6 is immune to chemical and photolytic degradation over a long period of time. Hence its contribution to global warming is expected to be cumulative and virtually permanent.

Although the potency of SF_6 as a greenhouse gas is extremely high, the amount of SF_6 in the atmosphere, compared to the concentrations of the naturally occurring and other man-made greenhouse gases, is extremely low. Estimates of relative contribution of SF_6 to non-natural global warming, using 1993 estimated SF_6 concentration levels, range from 0.01% to 0.07%, [3.118, 3.119]. It is estimated that with the present growth in use of SF_6, this figure could rise to as high as 0.2% in 100 years. However, due to its long lifetime in the atmosphere, it is feared that even small volume emissions of SF_6 gas may have significant cumulative effects on the environment. This has lead to growing concern over the possible long-term environmental impact of SF_6. The electric power industry has responded in ways for better control of the use of SF_6.

There has been an increasing demand for SF_6 in the last four decades (since 1965) because of its increasing application in electrical equipments. The estimated world production of SF_6 has increased steadily to approximately 7000 metric tons per year in 1993 [3.118, 3.119]. The best response to the concerns over the possible impact of SF_6 on global warming would be to prevent its release into the environment. The use of the quantity of SF_6 gas could be reduced by using gas mixtures. That, however, introduces newer problems. These include difficulties in handling, mixing, maintaining constant mixture composition, reclamation of individual constituents of the mixtures, possible inferior performance with regard to thermal, insulation and arc interruption properties and associated equipment design changes. The use of mixtures of SF_6 with other gases requires investigation of decomposition of the new mixture and the effects of newer byproducts on the electrical equipment.

REFERENCES

[3.1] Saha, M. H., Philosophical Magazine, 40 (1920), p. 427.

[3.2] Rieder, W., "Plasma and Arcs", Vieweg, Braunschweig. FRG, (1967) (in German).

[3.3] Townsend, J. S., Nature, 62 (1990), p. 340.

[3.4] Townsend, J. S., Philosophical Magazine, 1 (1900), p. 198.

[3.5] Engelmann, Eberhard, "A contribution to the discharge behavior of large electrodes having field distortions in air with positive switching surge", Dissertation, TU Dresden (1981) (in German).

[3.6] Raether, H., "Entwicklung der Elektronen in den Funkenkanal", Zeitschrift für Angewandte Physik, 112 (1939) pp. 464–489 (in German).

[3.7] Raether, H., "Electron Avalanches and Breakdown in Gases", Butterworths, London (1964).

[3.8] Wagner, K.H., "Investigation of the development of electron avalanche in plasma channel by highgain image intensifier and image converter streak shutter technique", Zeitschrift für Physik, 189 (1966) pp. 465–515 (in German).

[3.9] Raether, H., "Development of Kanal discharge", Archiev fur Elektrotechnik, 34 (1940)1, pp. 49–56 (in German).

[3.10] Loeb, L.B., Meek, J.M., "The mechanism of spark discharge in air at atmospheric pressure", Journal of Applied Physics, 11 (1940) pp. 438–447.

[3.11] Davies, A.J., Davies, C.J. and Evans, C.J., "Computer simulation of rapidly developing gaseous discharges", Proceedings IEE, 118 (1971), 6, pp. 816–823.

[3.12] Chalmers, I.D., Duffy, H. and Tedford, D.J., "The mechanism of spark breakdown in nitrogen, oxygen and sulphurhexafluoride", Proceedings of the Royal Society of London, A, 329 (1972), pp. 171–191.

[3.13] Meek, J.M and Craggs, J.D., "Electrical Breakdown in Gases", John Wiley& Sons Ltd., (1978).

[3.14] Alston, L.L., "High Voltage Technology", Oxford University Press (1968).

[3.15] Townsend, J.S., "Electricity in Gases", Oxford (1914).

[3.16] Rees, D. B., Ph.D. Thesis, University of Wales, U.K. (1963).

[3.17] Kluckow, R., "On the variation of current with time for a gas discharge in hydrogen", Zeitschrift für Physik, 161 (1961) pp. 353–369 (in German).

[3.18] Bandel, H. W., "Measurement of the current during the formative time lag of sparks in Uniform fields in Air", Physical Review, 95 (1954) 5, pp. 1117–1125.

[3.19] Hoger, H., "Dielectrics", 1 (1963) p. 94.

[3.20] Mielke, Heinz, "Increase of current with time during static breakdown voltage in air, hydrogen and oxygen", Zeitschrift für Angewandte Physik, 11 (1959) 11, pp. 409–414 (in German).

[3.21] Schröder. G. A., "Measurement of static breakdown field strength in air in uniform field gap distances from 2 to 9 cm", Zeitschrift für Angewandte Physik, 13 (1961) 6, pp. 296–303 (in German).

[3.22] Pedersen, A., "On the Electrical Breakdown of Gaseous Dielectrics, An Engineering Approach", IEEE Transactions on Electrical Insulation, Vol. 24, No. 5, October 1989.

[3.23] Loeb, L.B. and Meek, J.M., "The Mechanism of the Electric Spark", Stanford University Press, Stanford, 1941.

[3.24] Dawson, G. A. and Winn, W.P., "A model for streamer propagation", Zeitschrift für Physik, 183 (1965) pp. 159–171.

[3.25] Meek, J.M. and Craggs, J.d., "Electrical breakdown in gases", Clarendon Press, Oxford (1953).

[3.26] Loeb, L.B., "Basic Processes of gaseous electronics", University of California Press Berkeley, USA (1965).

[3.27] Cueilleron, C. R., Academy of Science, Paris 226 (1948), p. 400.

[3.28] Banwari, "Experimental Investigation of Electrical Properties of Ambient Plasma", M.Tech. Thesis, Department of Electrical Engineering, Indian Institute of Technology, Kanpur, April 2000.

[3.29] Schumann, W. O., Archiev für Elektrotechnik, 12 (1923), p. 593.

[3.30] Sohst, H., "Breakdown voltage of air (760 Torr) in uniform fields (Gap distance 7 to 30 cm)", Zeitschrift für Angewandte Physik, 14 (1962) 10, pp. 620–627 (in German).

[3.31] Dakin, T. W., Luxa G., Oppermann, G., Vigreux, J., Wind G. and Winkelnkemper, H., "Breakdown in gases in uniform fields, Paschen's curves for Nitrogen, air and Sulphur Hexafluoride", Electra, 32 (1974), pp. 61–82.

[3.32] Trump, J.G. Cloud, R.W., Mann J.G. and Hausen, E.P., "Influence of electrodes on dc breakdown in gases at high pressure", Transaction on American IEE, 69 (1950), pp. 961–965.

[3.33] Cookson, A.H., "Electrical breakdown for uniform fields in compressed gases", Proceedings IEE, 117 (1970) 1, pp. 269–280.

[3.34] Dokopoulus, P., "Wachstumsgesetze der Durchschlagswahrscheinlichkeit von Hochspannungsisolierung", Ph.D. Thesis, TH Braunschweig (1967) (in German).

[3.35] Alston, L.L., "High temperature effects on flashover in air", Proceedings IEE, 105 (1958) pp. 549–53.

[3.36] Trichel, G. W., "The mechanism of the positive point-to-plane corona in air at atmospheric pressure", Physical Review, 55 (1939), pp. 382–390.

[3.37] Woboditsch, W., "The characteristics of technical discharge gaps having extremely nonuniform field", Wissenchaftliche Zeitschrift der T. H. Dresden, 8 (1958/59) H. 4, p. 870 (in German).

[3.38] Lemke, Eberhardt, "Breakdown mechanism and characteristics of nonuniform field electrode configurations in air for switching surges", Dissertation, TU Dresden (1967) (in German).

[3.39] Lemke, Eberhardt, "The breakdown mechanism in air with switching surges", Wissenschaftliche Zeitschrift der Technischen Universität, Dresden. 17 (1968) H. 1, A. 105 (in German).

[3.40] Nasser, Essam, "Discharge development at positive point in nonuniform field in atmospheric air", Dissertation, T U Berlin (1958) also in Archiev für Elektrotechinik, 44 (1959) 3, pp. 157–176, (in German).

[3.41] Gänger, Berthold, "The electrical breakdown of gases", Springer Verlag, Germany (1953) (in German).

[3.42] Petropoulos, G.M., "Constriction of a spark discharge during impulse breakdown", British Journal of Applied Physics, Vol. 15 (1964) pp. 169–176.

[3.43] Petropoulos, G.M., "Growth of Positive Channels in Non-Uniform Fields during Impulse Breakdown", Nature, Vol. 197 (1963), pp. 174–176.

[3.44] Menemenlis, C., "Growth of positive channels in nonuniform fields during impulse breakdown", Nature 196 (1962), pp. 564–565.

[3.45] Allibone, T.E., and Schonland, B. F J., Nature, 134 (1934) p.735.

[3.46] Allibone, T.E., Meek, J.M., "The development of the spark discharge", Proceedings Royal Society A 166 (1938) p. 97.

[3.47] Waters, R.T., Jones, R.E., "High Voltage Impulse Breakdown of Rod-Plane in Air", Proceedings of the Fifth International Conference on "Ionization Phenomena in Gases", Munich (1961) Vol. I, pp. 992–1001 Edited by Maecker, H., North-Holland Publishing Co., Amsterdam.

[3.48] Dyson, J., Hemmings, R.F., Waters, R.T., "An f/1 Streak Camera for Spark Studies in Both Ultraviolet and Visible", Journal of the Society of Motion Picture and Television Engineers, Vol. 70 (1961) Part II, pp. 725–730.

[3.49] Les Renardieres Group, "Research on Long Air Gaps Discharges at Les Renardieres", 1973 Results, Electra, CIGRE, No. 35 (1974).

[3.50] Les Renardieres Group, "Physics of Discharges, Positive Discharges in Long Air Gaps at Les Renardieres", 1975 Results and Conclusions, Electra, CIGRE, No. 53 (1977).

[3.51] Les Renardieres Group, "Physics of Discharges, Negative Discharges in Long Air Gaps at Les Renardieres", 1978 Results and conclusions, Electra, CIGRE, No. 74 (1981).

[3.52] Srivastava, R.K., "An Investigation of Electromagnetic Interference (EMI) caused by Different Types of Coronas in Air", M.Tech. Thesis, Department of Electrical Engineering, Indian Institute of Technology, Kanpur, May 1991.

[3.53] Arora, Ravindra, Kalra, Prem K. and Srivastava, Rakesh K., "An Investigation of Electromagnetic Interference caused by Different Types of Coronas in Air", IEEE International Symposium on Electrical Insulation, Baltimore, MD, USA, June 1992, pp. 287–290.

[3.54] Terren, Peter, "Step Away from the Vehicle", The big picture, IEEE Spectrum, May 2007.

[3.55] Uhlmann, Erich, "Electrical breakdown in air between concentric cylinders", Archiev für Elektrotechnik, 23 (1930) 3, pp. 323–350 (in German).

[3.56] Kuffel, E., Zaengl, W.S., "High Voltage Engineering Fundamentals", Pergamon Press (1984).

[3.57] Arora, Ravindra and Choudhary, Jagdish, "Investigation of the Effect of Polarity of Switching and Lightning Impulse Voltages on the Performance of Air for Short Gap Distances in Extremely Nonuniform Field", Proceedings, Gaseous Dielectrics IX, NIST, MD, USA, May 2001, Kluwar Academic/Plenum Publishers.

[3.58] Arora, Ravindra and Prem, Sunil, "Experimental Estimation of Schwaiger Factor Limit (η_{lim}) in Atmospheric Air", Proceedings of the "Eighth International Symposium on Gaseous Dielectrics", by NIST and IEEE, Virginia Beach, USA, June 1998.

[3.59] Prem, Sunil, "An investigation of the Schwaiger Factor Limit in Air", M.Tech. Thesis, Department of Electrical Engineering, IIT Kanpur, India, April, 1994.

[3.60] Vaibhav, Kishore, "Study of the effect of Degree of Uniformity of Field on Breakdown Strength of Air in Weakly Nonuniform field Configuration with Lightning and Switching Impulse Voltages", M.Tech. Thesis, Department of Electrical Engineering, IIT Kanpur, India, July, 2002.

[3.61] Kind, Dieter and Kärner, Hermann, "High Voltage Insulation Engineering", Friedrich Viewag & Sohn. FRG (1982) (in German).

[3.62] Alexandrov, G.N., Evanov, W.L., Kiesewetter, W.E., "Electrical breakdown strength of high voltage insulation", Energia, Leningrad (1969) (in Russian).

[3.63] Stekolnikov, I.S., Brago, E.N., Baselyan, E.M., "Decrease in breakdown voltages of long air gaps with switching impulses", Journal of Technical Physics Vol. 32 (1962) No. 8, pp. 993 (in Russian).

[3.64] Pigini, A., Rizzi, G., Brambilla, R. and Garbagnati, E., "Switching impulse strength of very large air gaps", 3rd ISH, Milan (1979), No. 52.15.

[3.65] Lemke, E., "Contributions to the estimation of breakdown voltage for long air gaps", Zeitschrift für elektrotechnische Information and Energietechnik, Leipzig (GDR), 3 (1973) 4, pp. 186–192 (in German).

[3.66] Study Committee 31 report (Power System Planning), "Electric power transmission at voltages of 1000 kV and above, plans for future ac and dc transmission", Electra, CIGRE, No. 91 (1983).

[3.67] Ryzko, H., Muller-Hillebrand, D., "The influence of chemical products left by spark in the air on the impulse breakdown voltage of enclosed air gap", Proceedings, Fifth International conference on Ionization Phenomena in gases, Munich, (1961), pp. 1065–1074.

[3.68] Scherer, Jr., Harold N., Brendan, J. Ware, Shih, C.H., "Gaseous Effluents due to EHV Transmission", IEEE PAS-92 (1973) pp. 1043–1049.

[3.69] Roach, J.F., Charter, V.L., Dietrich, F.M., Nowak, H.J., "Ozone concentration measurements on the c-line at the Apple Grove 750 kV Project and theoretical estimates of ozone concentration near 765 kV lines of normal design", IEEE-PAS-97 (1978) No. 4 pp. 1392–1401.

[3.70] Böhringer, Angelika, Rheinbaben, H.V., Stößer, Bernd, Röhsler, Heinz, "Ozone development on High Voltage Transmission lines", Elektrizitätswirtschaft, 87 (1988) H. 21, pp. 1017–1022 (in German).

[3.71] 400 kV Research Committee, "The 400 kV Research Project Rheinau", Vol. 1–III, Heidelberg (1955), West Germany (in German).

[3.72] Edison Electric Institute, "EHV Transmission line reference book", USA (1960).

[3.73] IEEE Radio Noise Sub-Committee, "Comparison of radio noise prediction methods with CIGRE/ IEEE survey results", IEEE, PAS-92(1973), pp 1029–1042.

[3.74] Lalli, M. S., Prabhakar, B.R., "Radio Interference field measurements on HV and EHV Transmission lines of India", 5th ISH. Braunschweig (1987), 83.13.

[3.75] Electric Power Research Institute, USA, "Transmission line reference book 345 kV and above", Palo Alto, CA, USA (1975) and (1982).

[3.76] Perry, D.E., Chartier, V.L., Rainer, G.L., "BPA 1100 kV Transmission System Development, Corona and electric field studies", IEEE, Vol. PAS,-98 (1979) No. 5, pp. 1728–1738.

[3.77] Gens, Ralph S., Gehrig, Eward H., Eastvedt., Robert B., "BPA 1100 kV Transmission System Development—Planning, Program and Objectives", IEEE, Vol. PAS-98 (1979) No. 6, pp. 1916–1921.

[3.78] Rogers, L.E., Gilbert, R.O., Lee, J.H. and Bracken. T.D., "BPA 1100 kV Transmission System Development-Environmental Studies", IEEE, Vol. PAS-98 (1979) No. 6, pp. 1958–1967.

[3.79] Nayak S.K. and Thomas M. Joy, "Computation of EMI Fields Generated Due to Corona on High Voltage Overhead Power Transmission Lines", Proceedings of INCEMIC 2001–2002, pp. 15–19.

[3.80] Dong-Il Lee, Koo-Young Shin, Jeong-Boo Kim, "Characteristic of Corona Interference from a three phase double circuit 765 kV test line", (Korea Electric Power Research Institute), Fourth Workshop and Conference on EHV Technology, IISc. Bangalore, India, July 1998.

[3.81] King, L.A., "The voltage gradient of the free burning arc in air or Nitrogen", Proceedings of the 5th International Conference on Ionization Phenomena in Gases, Münschen (1961), Volume I, North-Holland, pp. 871–877.

[3.82] Inaba, T., "Voltage gradients of free burning arc column in liquid Nitrogen", Proceedings of the 12th International Conference on Phenomena in lonized Gases' Eindhoven (1975), part I, North-Holland, p. 141.

[3.83] Sirotinski, L.I., "High Voltage Technology", Vol. I, Part I, Discharge in Gases VEB Verlag Techink, Berlin (1955) (In German).

[3.84] Rutsher, Alfred et al., "Wissenspeicher Plasmatechnik", VEB Fachbuchverlag, Leipzig (1983), (in German).

[3.85] Naidu, M.S., "Very fast transient over voltages in GIS", Workshop on GIS, CPRI, Bangalore, India, Dec. 1998.

[3.86] Kumar K.V., Thomas, Joy M., "Estimation of VFTO in a 420 kV GIS", Fourth Workshop and Conference on EHV Technology, IISc. Bangalore, India, July 1998.

[3.87] Hayashi, K, Yamauchi, T., Marutani, T., and Sasamari K, "Gas insulated Switchgear and Interrupter", IEEE-Trans. Vol. EI-21, No.6, Dec. 1986, pp. 975–979.

[3.88] Mosch, Wolfgang and Hauschild, Wolfgang, "High Voltage Insulation System with SF_6", Dr. Alfred Huthig Verlag, also VEB Verlag Technik, Berlin Germany (1979) (in German).

[3.89] Pederson. A, "The effect of surface roughness on breakdown in SF_6", IEEE Trans. PAS-94 (1975). p. 1769.

[3.90] Mc Allister, I.W., "The influence of electrode microscopic curvature upon surface roughness in compressed SF_6", Archiev für Electrotechnik, 62 (1980), p. 43.

[3.91] Berger, S., "Onset of breakdown voltage reduction by electrode surface roughness in air and SF_6", IEEE Trans, PAS-95(1976), pp. 1073.

[3.92] Laghari, J. R. and Qureshi, A.H., "A review of particle-contaminated gas breakdown", IEEE Trans, EI-16 (1981), p. 388.

[3.93] Mosch, W., Hauschild, W., Speck, J. and Schierig, S., "Phenomena in SF_6 insulation with particles and their technical valuation", 3rd ISH, Milan (1979) 32.01.

[3.94] Rizk, F.A.M. Masetti, C., Comsa, R.P., "Particle initiated breakdown in SF_6 insulated systems under high direct voltage", IEEE Trans. PAS-98 (1979) 3, pp. 825–836.

[3.95] Cookson, Alan H., Farish, O., Sommerman, G.M.L., "Effect of conducting particles on AC corona and breakdown in compressed SF_6", IEEE Trans. PAS-91 (1972) pp. 1329–1388 also PAS-92 (1973) p. 871.

[3.96] Cookson, Alan H., et al, "Recent research in the United States on the effect of particle contamination reducing the breakdown voltage in compressed gas insultated system", CIGRE (1976), 15.09.

[3.97] Srivastava, K.D. and Anis, H., "Particle initiated breakdown in compressed gas insulation under time varying conditions", IEEE Trans. PAS-100 (1981), pp. 3694–3702.

[3.98] Wohlmuth, Michael, "Partial Discharges and Breakdown in GIS Initiated by Free Moving Particles", Paper No. 72, Proceedings of Seventh International Symposium on Gaseous Dielectrics, by IEEE PES & DEIS, Knoxville, Tennessee, USA, April 1994.

[3.99] Ooishi, T., Yoshimura, M., Hama, H., Fujii, H. and Nakanishi, K., "Charging Mechanisms of a Conducting Particle on Dielectric Coated Electrode at ac and dc Electric Fields", Paper No. 73,

Proceedings of Seventh International Symposium on Gaseous Dielectrics, by IEEE PES & DEIS, Knoxville, Tennessee, USA, April 1994.

[3.100] Hasegawa, T., "Improvement of Withstand Voltage at Particle Contamination in dc GIS due to Dielectric Coating on Conductor", Paper 86, Proceedings of Eighth International Symposium on Gaseous Dielectrics, by IEEE PES & DEIS, Virginia Beach, Virginia, USA, June 1998.

[3.101] Felici, N.J., "Forces et charges de petits objects en contact avec une electrode affectee dun champ electrique", Rev. Gen. Elect., Oct. 1996, pp. 1145–1160.

[3.102] Cookson, A.H., Farish, O., "Particle Initiated Breakdown in Coaxial Systems in Compressed SF_6 Gas", IEEE Trans., PAS 92 (1973), page 871–876, also at ISH München (1972), pp. 343–349.

[3.103] Morcos, M.M., Srivastava K.D., Holmberg, M. and Gubanski, S., "Dielectric Coatings and Particle Movement in GIS/GITL Systems", Paper No. 29-I, Proceedings of the Eighth International Symposium on Gaseous Dielectrics, Virginia Beach, USA, 1998, by NIST and Old Domino Univ., IEEE DEIS.

[3.104] Cookson, Alan H., "Gas insulated Cables", IEEE Trans., El-20 (1985) 5, pp. 859–890.

[3.105] Arora, Ravindra and Haldar, Partha, "Investigation of the Inception of Streamer Corona in SF_6 Gas", 12th International Symposium on High Voltage Engineering, IISc. Bangalore, India, August 2001, also M.Tech. Thesis, Department of Electrical Engineering, IIT Kanpur, India, December 2000.

[3.106] Inuishi, Y., "Electrical discharges of SF_6 gas in nonuniform fields", Electra, CIGRE, Oct. 1982, pp 45–52.

[3.107] Christophorou, L.G., Olthoff, J.K. and Green D.S., "Gases for Electrical Insulation and Arc Interruption: Possible Present and Future Alternatives to Pure SF_6", NIST Technical Note 1425, United States Department of Commerce Technology Administration, National Institute of Standards and Technology, Gaithersburg, USA, November 1997.

[3.108] Wieland, A., "Breakdown behavior of SF_6 and SF_6-gas mixtures at high gas pressures", Dissertation, TH Darmstadt (1978), Germany, published in Electrotechnische Zeitschrift (ETZ)-A 94 (1973) 7, pp. 370–372.

[3.109] Pelletier, T.M., Gervais, Y., Mukhedkar, D., "Dielectric strength of N_2-He mixtures and comparison with N_2-SF_6 and CO_2-SF_6 mixtures", 3rd ISH, Milan, Aug., 1979, Paper 31.16.

[3.110] Malik, N.H. and Qureshi, A.H., "Nonunifrom field breakdown in SF_6 and its mixtures with Helium and Nitrogen", 3rd ISH, Milan, Aug. 1979, Paper 31.03.

[3.111] Cookson, A.H. and Pederson, B.O., "Analysis of the high voltage breakdown results of mixtures of SF_6 with CO_2, N_2 and air", 3rd ISH, Milan, Aug. 1979, Paper 31.10.

[3.112] Chu. F.Y., "SF_6 Decomposition in Gas- Insulated Equipment", IEEE Trans. EI-21 (1986) 5, pp. 693–718.

[3.113] Sauers, I., "Sensitive detection of By- Products formed in electrically discharged SF_6", IEEE Trans. EI-21 (1986) 2, pp. 105–110.

[3.114] Sauers, I., Ellis, H.W. and Christophorou, L.G., "Neutral decomposition products in spark breakdown of SF_6", IEEE Trans.EI-21 (1986) 2, pp. 111–120.

[3.115] Constantine, T. Dervos and Vassilion, Panayota, "Byproducts in the Insulating Gaseous Matrix of a GIS", Gaseous Dielectrics IX, Kluwer Academic/Plenum Publishers, New York 2001, pp. 403–412.

[3.116] Diaz, Joseph; Casanovas, Anne-Marie; Godard, Christine and Casanovas, Joseph, "Chemical Decomposition of High Pressure SF_6/N_2 (5:95) Mixtures under Negative dc Corona Discharges", Gaseous Dielectrics IX, Kluwer Academic/Plenum Publishers, New York 2001, pp. 543–547.

[3.117] Edelson, D. and McAffe, K.B., "Notes on the Infrared Spectrum of SF_6", J. Chem. Phys., Vol. 19, 1951, pp. 1311–1312.

[3.118] Mauthe G., Niemeyer L., Pryor B.M., Probst R., Bräutigam H., O'Connell P.A., Pettersson K., Morrison D., Poblotzki J., and Koenig D., Task Force 01 of Working Group 23.10, "SF_6 and the Global Atmosphere", Electra No. 164, February, 1996, pp. 121–131.

[3.119] Mauthe G., Pettersson K., Probst, R., Poblotzki, J., Koenig D., Niemeyer L. and Pryor B.M., Members of WG 23.10 Task Force 01, Draft document on "SF_6 and the Global Atmosphere", March 1995.

CHAPTER 4

LIGHTNING AND BALL LIGHTNING, DEVELOPMENT MECHANISMS, DELETERIOUS EFFECTS, PROTECTION

The phenomenon of lightning existed on earth perhaps even before life began. In the year 2007, NASA reported occurrence of lightning on Venus.

Lightning has always inspired people with awe and reverence. The Greeks, for instance, attributed to their supreme god Zeus the lightning bolt as the ultimate weapon of control. At the same time, the fury of the lightning strike always scares the human being. Some still consider the lightning to be "God's anger or curse", a "super natural power" or even "extra terrestrial force" [4.1].

Distribution of lightning activity on our globe is not uniform because it strongly depends upon differing weather and climatic conditions in various regions. The current lightning activity at any time on earth can be viewed on the site wwlln.net. Figure 4.1 shows the pattern of lightning activity over the world for 31 day average in the month of July 2009, courtesy Prof. Robert Holzworth, University of Washington, USA. Almost ten million lightning strokes are shown in this figure. WWLLN has shown to detect nearly 100% of all lightning generating storms, but only about 10% of total currently active lightning activity. Lightning activity is much higher near the equator region. A comparison between the months of July and December on another such plot reveals the shift in lightning activity from northern to southern hemisphere.

This information is continuously mapped by installing less than 50 sensors on the ground all over the world. These are located several thousand kilometers away from each other. The other method is to obtain this information with the help of satellites. Each technique has its advantages and limitations. These techniques need more development in order to capture 100% lightning activity and to distinguish completely between cloud-to-cloud and cloud-to-ground lightning strikes.

High Voltage and Electrical Insulation Engineering, First Edition. Ravindra Arora, Wolfgang Mosch.
© 2011 Institute of Electrical and Electronics Engineers. Published 2011 by John Wiley & Sons, Inc.

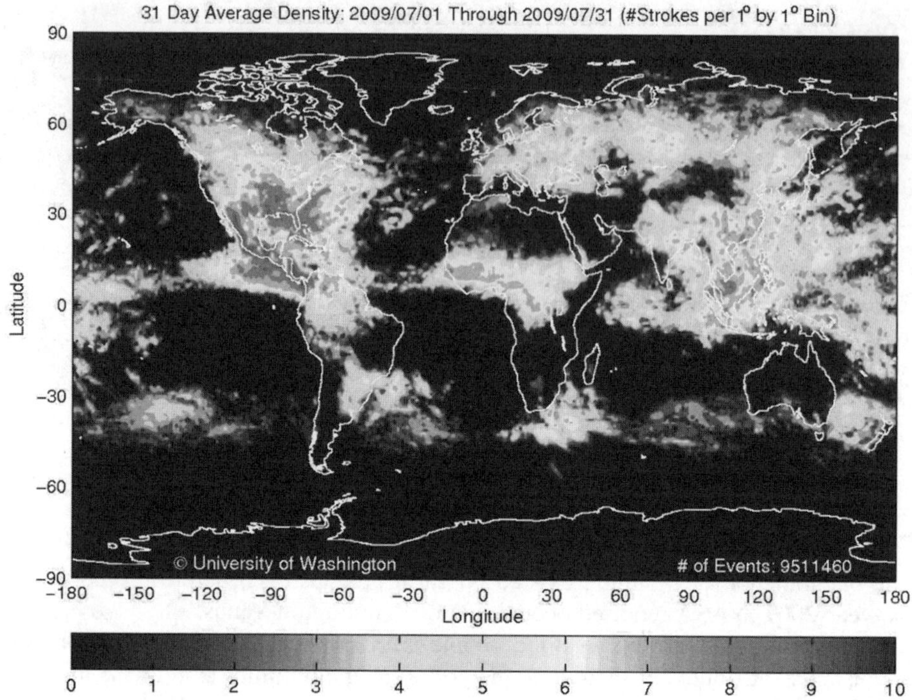

Figure 4.1 Distribution of lightning activity in different regions on the earth for July 2009.

The unpredictable behavior of lightning and the deleterious effects caused by lightning strike continue to terrify the living beings. Each incident of lightning strike is typical in its characteristic. In spite of so much of research, scientists and researchers are still making efforts to explain the mechanism of lightning strike and the peculiar types of damage caused by lightning. Continuous efforts are being made to find more effective measures of protection from lightning. It is important for people to be aware of the huge damage to life and property caused by lightning and to inform them of simple methods of protection [4.2].

This beautiful and unique phenomenon of nature cannot be exploited or tamed. However, a better understanding of lightning has been acquired with the knowledge gained through research and experimental investigations in the high voltage and high current laboratories as well as measurements at sites known for frequent, actual lightning strikes.

4.1 THE GLOBE, A CAPACITOR

Our earth can be considered to be a huge natural capacitor. The electric charge resting on the clusters of clouds, aerosols, and so on, gain potential and form the high voltage electrode arm of this capacitor, whereas earth forms the ground

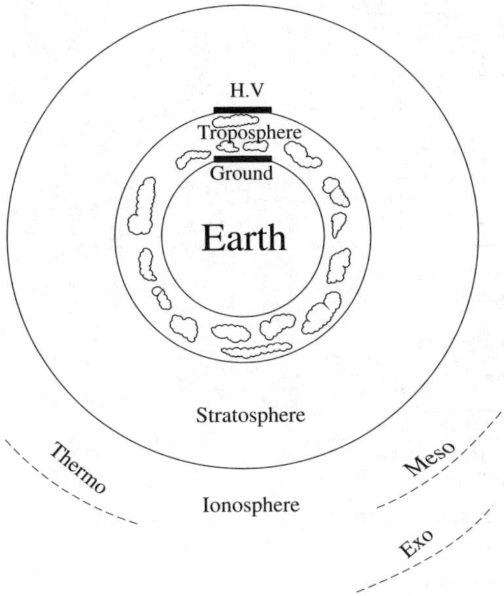

Figure 4.2 The earth and the atmospheric layers forming a capacitor.

electrode. In between the two electrodes, atmospheric air—the cheapest and the most widely used dielectric—provides the insulation, Figure 4.2.

Every dielectric has its own conductivity. The conductivity of gaseous dielectrics, although minimum at normal pressure, changes with pressure. The dielectrics are subjected to partial or complete breakdown when stressed with high electric field intensity. An electrical breakdown or failure of insulation results in equalization of potential between the two electrodes. The earth has the unique property of absorbing all kinds of waste created by humans, including the electrical charge. Vast amounts of electrical charge in the form of current is injected into the ground but the ground still remains at zero or ground potential. Each lightning strike on the earth also injects large amount of charge in a few milleseconds of time by an impulse form of current. The potential of the ground may rise temporarily around the location of lightning strike due to this charge injection but on finding a passage to go inside the ground, the potential is restored back to the ground or the zero potential level of its own. The magnitude of this impulse current varies in a wide range of 1 to 300 kA depending upon the atmospheric conditions. Figure 4.3 shows the probability of lightning strike current magnitudes in the European region.

4.1.1 The Earth's Atmosphere and the Clouds

Layers of gases surround the earth, each with slightly different composition but mainly comprised of nitrogen and oxygen. These layers of gases are held in place by gravitational attraction of the earth. More than three quarters of the gases are squashed into the troposphere, the lowest of the five layers of atmosphere. The other

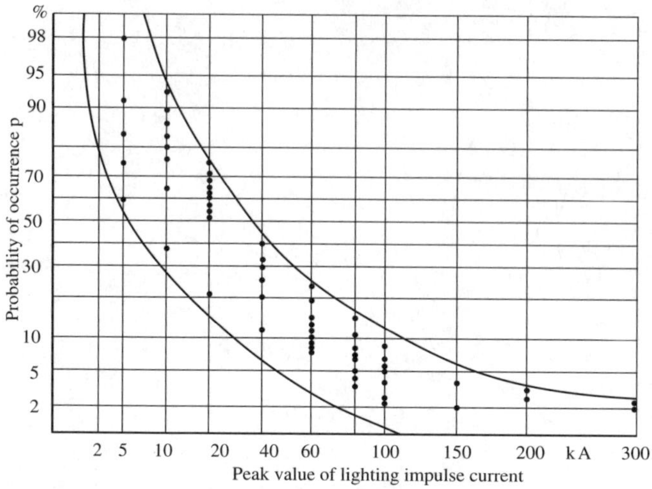

Figure 4.3 Probability of lightning strike impulse current magnitudes in European region.

layers described with increasing height above the ground are stratosphere, mesosphere, thermosphere, and exosphere. The density of gas decreases rapidly as the altitude increases. In the exosphere are hardly any gases at all. In a broad sense, the earth's atmosphere can be described by the following three layers.

4.1.1.1 The Troposphere International Scientific Vocabulary (ISV) (ca.1909) describes troposphere as the portion of atmosphere, that extends outward to about 7 to 10 miles or 11 to 16 km from the earth's surface. The temperature in this layer generally decreases rapidly with altitude and the presence of various forms of clouds. It decreases by around 2°C for every 1000 ft gain in altitude. Most weather phenomena occur within this zone. The convection of air in this layer is active. Cloud build-up is confined within this layer. It is however, rare to find clouds above the altitude of 11 km. The "thermal", the ascending current of air caused by heat, is confined within the troposphere. Hence, the troposphere's atmospheric conditions play most the important role in determining the lightning activity. Most of the long distance aircrafts fly in the upper region of this layer.

4.1.1.2 The Stratosphere The atmospheric layer above the troposphere is known as stratosphere. Depending upon the earth's latitude, season and weather it is the portion of the atmosphere above approximately 15 km and extends up to about 50 km. Clouds and water vapor are very rare in this layer. Unlike in the troposphere, the temperature varies very little with changing altitude in this region. The much talked about "ozone layer" lies in the lower part of stratosphere.

4.1.1.3 The Ionosphere This is the part of the earth's atmosphere above the stratosphere, beginning at an altitude of about 50 km and extending outward to

500 km or more. This region contains free electrically-charged particles and overlapping several regions within which more layers occur, such as meso, thermo and exo-spheres. These layers vary in height and free charge particle level with the time of the day, season, the solar cycle and are strongly affected by disturbances on the sun. Above 80 km and up to an altitudes of 200 km, the atmosphere is mainly composed of nitrogen molecules. Above this height, oxygen is the dominant constituent and the temperature rises with altitude to 1650°C in this region. This is a comparable region with respect to the presence of charged particles surrounding celestial bodies such as Venus and Mars.

4.1.2 Clouds and Their Important Role

Clouds play an important role not only in determining the climate but also lightning activity. Study of clouds is one of the basic factors in predicting weather conditions and lightning.

The World Meteorological Organization (WMO) defines the clouds as a visible aggregate of minute particles of water or ice, or both, in the free air. This aggregate may include larger particles of water or ice and also particles such as those of fumes, smoke or dust (aerosol). The moisturized air is lighter compared to dry air as the molecular weight of water vapor is lower. Hence the clouds are mostly at heights above the ground.

4.1.2.1 Classification of Clouds Luke Howard (1772–1864), an English manufacturing chemist and amateur meteorologist, developed the basis for the classification of clouds on their appearance such as, wispy, layered, heaped and puffy. The commonly-used word "cumulus", meaning a heap or accumulation, is the appropriately given name to the massy cloud form having a flat base and rounded outlines, often piled up like a mountain. The other common word, "stratus" is a cloud form with greater horizontal extension. The word cirrus, meaning a slender flexible animal, is also used often for wispy white cloud comprising of minute ice crystals found at high altitudes. These are described in the following sections with respect to their altitudes, Sharma [4.3]. The forms of clouds as sketched in Webster are shown in Figure 4.4.

4.1.2.1.1 Low Altitude Clouds Clouds below the height of 2 km are of either the cumulus or stratus family. Cumulus family clouds are puffy and heaped. Clouds in the stratus family form layers or sheets that cover broad expanses of the sky and these are found at comparatively lower altitudes than cumulostratus or cirrostratus. Low clouds are dark. These produce rain and are called "nimbus", the Latin word for rain. Nimbostratus clouds cover the whole sky with broad sheets and produce steady rain. Cumulonimbus clouds have dark bases and puffy tops. They tend to produce heavy precipitation.

4.1.2.1.2 Middle Altitude Clouds Clouds in the altitude of above 2 km and up to 6 km are designated as "alto". If in layers, they are called altostratus and if heaped and puffy, they are known as altocumulus.

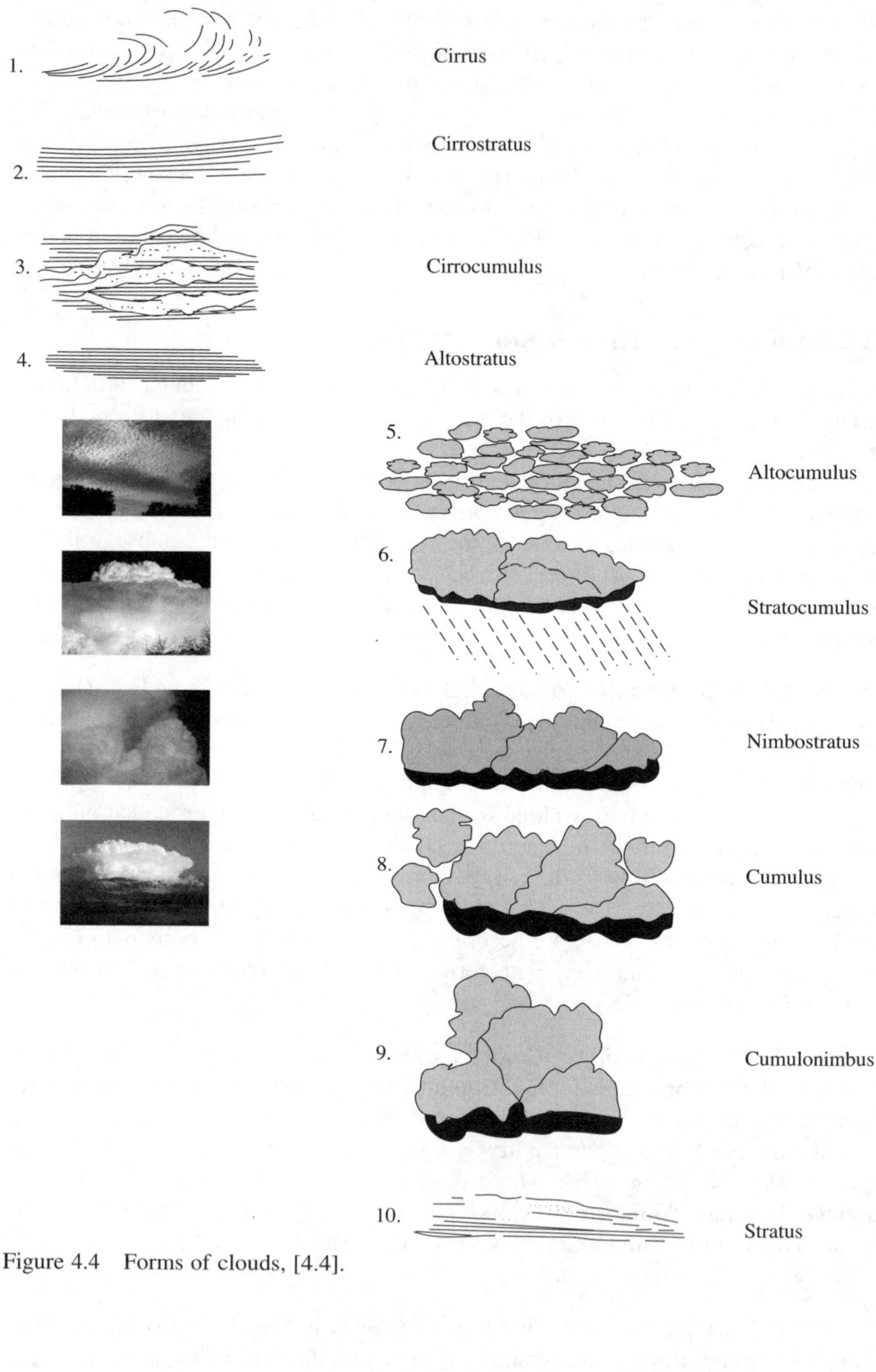

Figure 4.4 Forms of clouds, [4.4].

4.1.2.1.3 High Altitude Clouds The wispy clouds that are in the form of thin strip and white found at altitudes of above 6 km in the sky are described as "cirrus". These wispy white clouds are usually comprised of minute ice crystals. When cirrus clouds contain waves or puffs, they are called cirrocumulus, a cloud form of small white rounded masses found at high altitudes usually in regular groupings forming a mackerel sky. When they form fairly uniform continuous darker layers that seem to cover the sky, they are described as cirrostratus.

4.1.3 Static Electric Charge in the Atmosphere

Earth's atmosphere is full of free electric charge: electrons, ions, metastables etc. Whereas the sun is the prime external source of electric charge, friction due to high velocity wind currents within the atmosphere rapidly generates electric charge.

The electric charge is restful in the dielectrics but it is restless in the conductors. On the surfaces of the dielectrics the charge is turbulent due to much lower surface resistivity as compared to the volume resistivity. Thus the earth's atmosphere, basically air, a dielectric, stores vast amounts of static charge. The charge rests on the minute particles in the clouds and on the aerosols. The property of polarisability segregates the positive and the negative polarity charges within the dielectrics giving rise to the like polarity charge clusters on the clouds. The sources of electric charge in the earth's atmosphere are described in the following sections.

4.1.3.1 External Source of Electric Charge Solar radiation and the cosmic rays are a significant source of electric charge to the earth's atmosphere. Solar activity embraces a variety of conditions, including localized changes in the sun's temperature, the strength or direction of its magnetic fields, and the eruption of gases and plasmas from its surface. All these together create a wind of electrons, helium and hydrogen atoms and ions, and other charged and uncharged particles, some of which percolate from within the sun. This "wind" of atomic particles is thrown from the outer atmosphere of the sun and moves across the solar system. The solar storms on the earth are driven by the coupling of the earth's own magnetic field with this "solar wind". Their intensity and output can vary with the 11-year cycle in which sunspots—dark areas on the sun's surface—wax and wane.

As shown in Figure 4.5, the earth interacts with space at higher altitudes, for example, in the "plasma sphere", a torus of plasma encircling the earth and rotating with it. The principal connection of the interaction is through the aurora ovals, located around the poles, where electrons move up and down in parallel with the earth's magnetic field giving rise to the field aligned currents. The energetic electrons, emitted by the sun, penetrate deep enough to hit the earth's upper atmosphere and generate colorful aurora displays, Figure 4.6. The polar winds are outward flow of atmospheric gases and the electro jet currents around the poles and equator, Hekkila [4.5]. The region beyond is known as "magnetosphere", defined by the earth's magnetic field and the solar wind, where the radiation and the plasmas are trapped. "Solar wind" or the "Cosmic rays" are the high-energy electrons and atomic nuclei plasma particles continuously originating from the sun or outside the solar

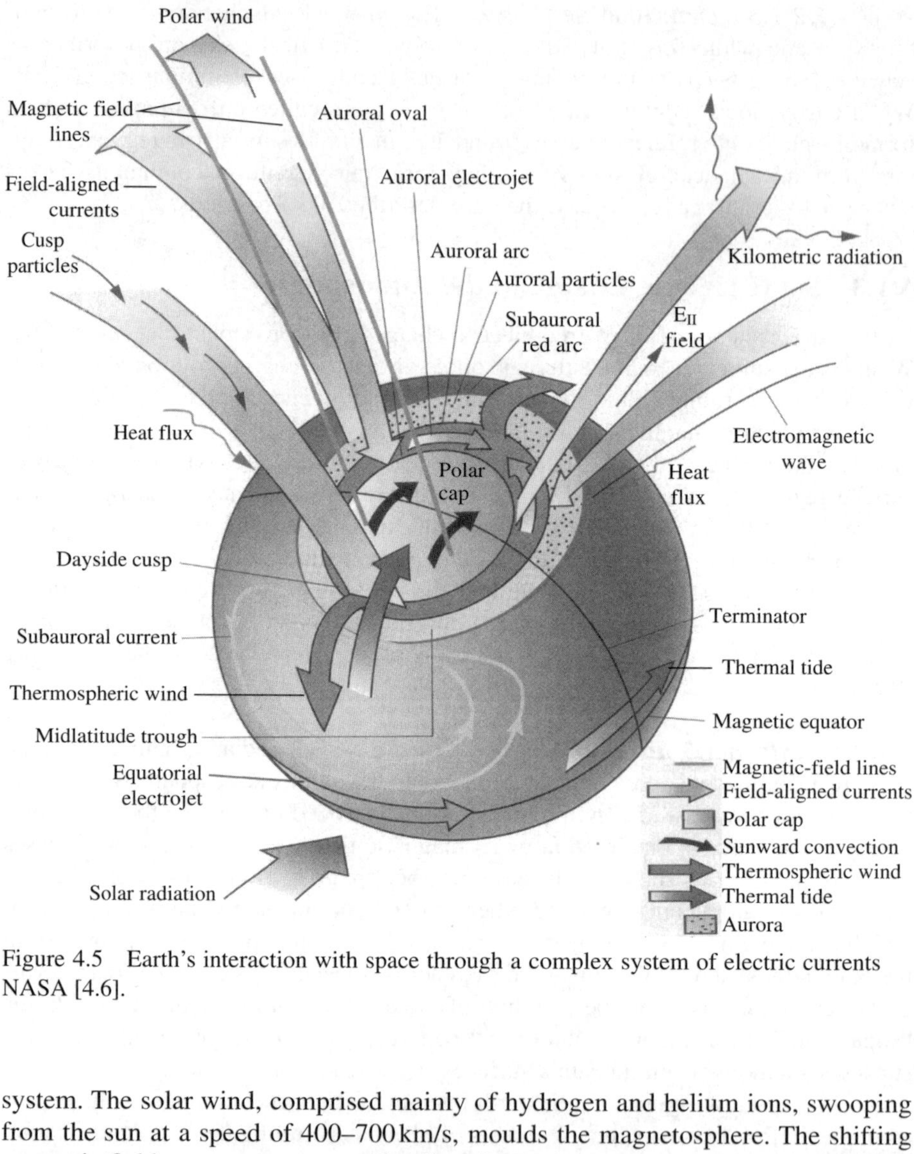

Figure 4.5 Earth's interaction with space through a complex system of electric currents NASA [4.6].

system. The solar wind, comprised mainly of hydrogen and helium ions, swooping from the sun at a speed of 400–700 km/s, moulds the magnetosphere. The shifting magnetic field can open areas and/or directions where the solar wind finds escape routes. The sun's rotation brings those areas toward earth, increasing auroras. At local noon time, it is compressed as near the earth as 66,000 km and drawing it out over two million kilometers on the night side of the earth [4.5].

4.1.3.2 Charges Due to Ionization within the Atmospheric Air Besides injecting electric charges into the earth's atmosphere from the solar system, the charge is also generated within the atmosphere due to ionization of air. Ionization of air molecules produces normally positive ions and electrons. However, oxygen, being a weak electro-negative gas, has a tendency to absorb free electrons and form

Figure 4.6 A photograph of Aurora.

negative ions. There are several processes by which the air is ionized. Some major processes are described in the following paragraphs.

4.1.3.2.1 Radiation from the Sun The sun emits a wide variety of ionizing and non-ionizing radiations. Infrared, visible light and ultraviolet are the non-ionizing radiations, whereas X-rays and γ-rays are forms of ionizing radiations. The dividing line between these two types of radiations in the electromagnetic spectrum falls in the ultraviolet portion. Radiation at the high-energy end of the UV spectrum can be as dangerous as X-rays or γ-rays. The UV wavelengths that reach the Earth's surface are in the range between 290 to 400 nm. Ultraviolet radiation is basically a form of non-ionizing radiation since it does not have enough energy to ionize atoms. However, it can cause molecules to vibrate and rotate, which means these would start heating up. Hence, UV irradiation on electrode surfaces in atmospheric air is used in high voltage laboratories to ensure the presence of initiatory electrons required for breakdown with impulse voltages.

The region of electromagnetic spectrum just below that of visible light is called "near ultraviolet" region. This type of radiation is strongly absorbed by most solid substances and even appreciably absorbed by air. Ozone layer in the upper atmosphere absorbs most of the harmful ultraviolet radiation from the sun but the higher frequencies can pass through the layer and produce harmful effects. The shorter UV wavelengths can reach the ionization energy for certain types of molecules. Termed as "far ultraviolet" region, it may trigger similar ionizing radiations [4.7].

4.1.3.2.2. Friction and Air Currents It is well known that friction within insulating bodies produces charge by ionization. Fast moving hot and cold air currents, having suspended aerosols, are active sources of charge particles in the atmosphere. The aircrafts and rockets/missiles moving at a speed of ~250 m/s and above also contribute electric charges to the atmosphere. Every human being insulated from the ground carries a lot of charge on his body produced by friction between the clothes. Even clapping is a process by which electric charge is produced.

Negative ions are created in nature, as air molecules break apart due to sunlight, radiations, and moving air, water, and ice crystals during thunderstorms. Negative ions of oxygen are found in invigorating environments such as near waterfalls, pine forests, or on the seashores where waves are breaking on the rocks.

4.1.3.3 Charging Mechanisms and Thunderstorms The free charge rests on the aggregate of minute particles of water or ice (the cloud) suspended in the air. A rapid charge generation takes place in the atmosphere during thunderstorms. Several charge generation processes may take place during thunderstorm electrification. A number of scientists have tried to explain the charge generation and segregation (or separation of charge) in thunderclouds. Some theories from the last 100 years are: Convection theory by B. Vonnegut; Wilson's Selective Ion Capture theory (1929); Freezing Potential theory of Workman and Reynolds (1950); Breaking Drop theory by Simpson (1909); Inductive and Non-inductive mechanisms by Elester and Geital (1913) and Reynolds, Brooke and Gaurly (1957) respectively [4.8]. However, there appears to be controversy over mechanisms taking place in various regions of the world having entirely different climatic conditions and the atmosphere. The occurrence of electrified warm thunderclouds, observed in the tropical regions, appear to involve different mechanisms as compared to the theories that describe the charge formation and separation process with ice crystals or hailstones. Based on experimental observations, occurrence of the following conditions is required together for the development of thunderclouds.

- Presence of strong upward draught is essential for rapid charge formation (ionization).
- Presence of humid, cold as well as dry winds.
- In a typical cloud, the negative charge center may exist near the isothermal level of −10°C present at an altitude of 3 to 4 km where ice crystal formation may take place.
- It is considered that for strong ionization leading to lightning, ice crystals are not enough; super cooled water drops or precipitation should coexist with ice.
- The positive charge centers may be present above the negative charge center isotherm and also locally at lower level of the cloud.

During a thunder storm, water droplets within the cloud are propelled upward by the strong internal winds. They turn into ice crystals and hailstones on arriving at isothermal levels of −10°C at an altitude of 3 to 4 km. As the ice formation picks up at higher altitudes, the ice crystal pieces become larger and heavier. These fall back through the clouds and collide with the rising water droplets. During these impacts, subatomic negatively charged (electrons) are transferred from the rising droplets to the falling ice crystals/hailstones. It gives rise to the negatively charged bottom of the stormy clouds.

According to the Freezing Potential theory based on laboratory experiments by Workman and Reynolds, when the ice crystals or hail having generally negative charge fall, a potential difference arises across the freezing front, the region in which the ice turns into water. Gravitational separation of shading water from ice crystals

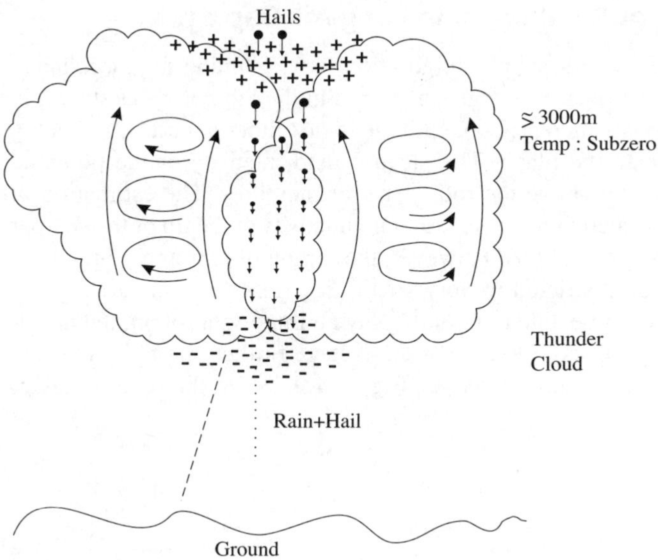

Figure 4.7 Falling hailstone, shading water and the freezing front.

or falling hailstones takes place resulting in an upward movement of shaded water drops carrying positive charges. The downward moving heavier hailstones with negative charge result in the concentration of negative charges at the cloud bottom and positive charges at the top, as explained with a schematic in Figure 4.7.

4.2 MECHANISMS OF LIGHTNING STRIKE

The physics of the phenomenon of lightning strike is, unfortunately, a matter of controversy among scientists even after the research work over centuries. Some believe that it is just a "discharge", the sudden movement of electric charge. Yes, it is a sudden movement of electric charge causing flow of impulse current. However, it is important to learn what makes this high magnitude of current able to flow? Inspired by the knowledge of the mechanism of breakdown of insulating properties of atmospheric air in very long gap distances developed in the high voltage laboratories, the phenomenon of lightning can be explained suitably. Breakdown mechanism of long gap distances in atmospheric air has been studied by engineers in the high voltage laboratories spread over a number of countries in the world [3.49–3.51], [3.62–3.65]. Breakdown characteristics of atmospherics air have been measured with different types of voltages, *ac*, *dc* and impulse (lightning and switching) of either polarity for different electrode systems up to a gap distance of 25 meters. On the basis of the knowledge acquired from these investigations about the development and propagation of breakdown phenomenon in atmospheric air, an appropriate inference can be drawn for the mechanism of lightning strike.

4.2.1 Mechanisms of Breakdown in Long Air Gap

The most appropriate electrode system configuration for explaining the mechanism of lightning strike is a large plane and a grounded rod in the laboratory as shown in Figure 4.8. Let a continuously increasing magnitude of *dc* (either polarity), switching impulse or *ac* be applied to the plane. The electric field intensity or the potential gradient below the plane and above the rod in the air increases. The estimation of field intensity by FEM revealed highest electric field intensity at the tip of the sharper electrode, the rod, in this system [4.9]. However, the region of extremely high field intensity at the rod will be restricted to very small area (mm) as compared to the large plane. Hence it may not be able to initially play an important role in the development of the breakdown process. On raising the voltage further, the field intensity at the large plane also substantially increases, Figure 4.9. When the required mag-

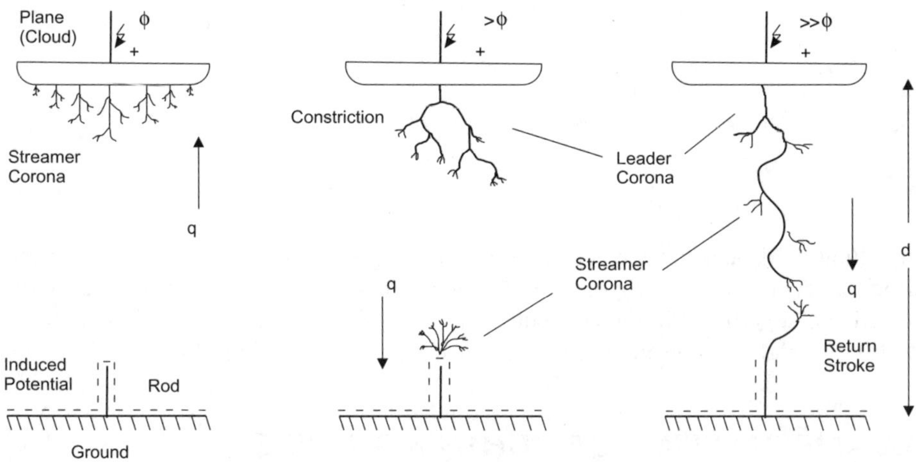

Figure 4.8 Schematic of the development of breakdown in air between a large plane and a small rod electrode system.

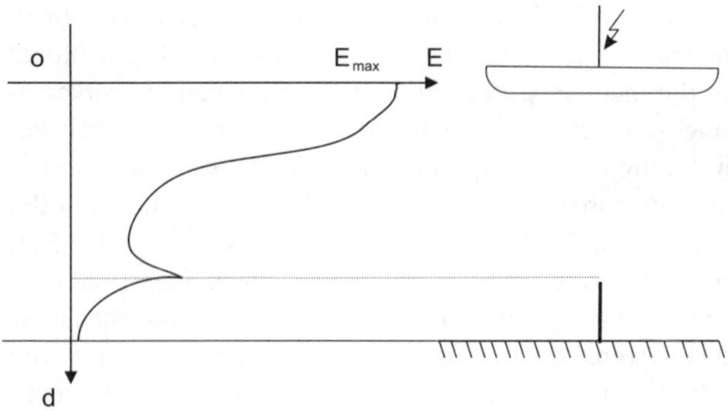

Figure 4.9 A Schematic of electric field intensity in the gap between HV plane and grounded rod electrode.

nitude of field intensity is acquired, local breakdown or PB begins at the locations of high field intensities. As the plane is a large electrode body, high field intensity is present underneath in a wide spread area. It can be observed from the Figure 4.9 that in almost 25% of the gap length, fairly high field intensity prevails.

Because of the smooth shapes of both the electrodes, the PB development in this case is with above critical amplification of avalanches. This gives rise to "streamer corona" at both these electrodes. Unlike the rod, wide spread streamer corona activity takes place at the plane covering the whole surface beneath.

PB generates free charge. The free charge or the current always follows the least resistance path for its movement. Since the streamer trajectories offer much higher conductivity as compared to the healthy part of air, the pulse form of the current generated by streamer corona at the two locations flows upward towards the plane, and downward towards the grounded rod, respectively. On further raising the voltage, the constriction phenomenon takes place and the stem and leader corona begin at the electrodes as explained in Section 3.3.5.3.

The leader corona, a very strong PB activity from the plane electrode, propagates and extends itself downward towards the ground electrode or the cathode. A much weaker streamer PB activity developed at the rod propagates upward in the air covering a relatively much shorter gap length. The propagation of leader is mostly observed in steps because it develops in the pockets of air having weaker dielectric conditions. The leader channels offer much higher conductivity as compared to the streamer trajectories. At each tip of the stepped leader, streamer corona takes place giving rise to pulse current that finds its least resistance path through the leader channels towards the plane electrode. When the cathode directed leader and the anode directed streamer come closer and the voltage is continued to increase, the gap length is bridged with unstable leader followed by an arc, letting the breakdown current from the plane to flow to the grounded rod.

4.2.2 Mechanisms of Lightning Strike on the Ground

The mechanism of lightning strike on an object on the ground is quite similar to the one in long air gaps described above in Section 4.2.1. The large plane in the laboratory simulates the clouds carrying static potential. In place of the grounded rod, it could be any metallic object above the ground. The gap distance in nature could be of the order of a kilometer or longer, even up to a few *km*. In place of applied potential on the plane electrode, it is the static potential on the clouds that develops rapidly under stormy conditions due to ionization or charge formation. The cloud potential increases rapidly by the movement of air and aerosols at high speed, explained in Section 4.1.3.3.

When the potential acquired by the clouds rises above the level required for the inception of partial breakdown (PB), first a wide spread streamer corona begins to take place in the air just below the cloud limits, the region of high field intensity. Occurrence of this wide spread streamer corona appears in the form of diffused glow of light as the streamer in atmospheric air is accompanied with low intensity of light. It is accompanied with a low intensity "flutter sound". However, the sound thus produced cannot be heard with naked ears because of its low intensity and the long distance.

The vigorous streamer activity is followed by the phenomenon of constriction resulting in the development of leader corona. The leader channels thus formed propagate downward towards the ground in pockets of air. Since the properties of air in these pockets are slightly different, the leader propagates towards the pockets having weaker insulating properties. Hence, the development of leader is in steps and accordingly it is known as "stepped leader". The potential gradient across the leader channels is known to be quite low around 1 kV/cm.

At the tip of each branch of stepped leader occurrence of streamer corona, a bunch of number of thin branches of local breakdown channels can be seen. In fact, it is the streamer corona at the tip that provides the required charged particles to the highly conductive leader channels directing the flow of charge towards the clouds. It can be understood by the principle of least resistance path followed by the charge to flow. This process causes further increase in potential of the clouds since it depends upon the magnitude of total charge acquired.

Towards the advance stage of the lightning strike, the field intensity at an appropriate object on the ground (the cathode) increases considerably. When it acquires the required magnitude, a similar process of partial breakdown begins to grow upwards in the air, giving rise to the development of anode directed streamer and even leader towards the last stage. However, the flow of charge/current in this part is downwards, as it provides the path of least resistance.

The lightning strike is accomplished only when both the cathode and the anode directed extended streamer/leader channels of partial breakdown meet together bridging the air gap between the clouds and the ground. The final jump of the lightning strike on the ground takes place with an unstable leader. The development of the anode directed breakdown channel from a suitable location on the ground is known as the "return stroke". The extent of the growth of return stroke in the air is much shorter as compared to the downward PB channels developed from the source, the clouds. The field intensity pattern in the long air gap is shown in the Fig. 4.9. At times it may happen that only streamer corona is able to develop at the grounded object, extending to a range of a few meters or as estimated by some, a few tens of meters. The whole process develops extremely rapidly, especially in the advanced stage of the lightning strike. The unstable leader towards final jump may propagate at a speed of 0.4 km/ms as explained in Section 3.3.5.1. The failure of the insulating properties of air gives rise to an impulse form of short circuit current to flow towards the ground. The total duration of this current could be in *ms*. In this way the clouds are able to evacuate their vast amount of acquired charge.

Different types of coronas or stable partial breakdown (PB) in air produce different types of audible noise (AN) due to sudden expansion of air in pockets under varying conditions of pressure, temperature, humidity, suspended particles and the ionized state. The stable streamer corona produces a low intensity flutter sound, whereas the stable leader corona produces a loud cracking sound, Section 3.3.2.3. An example is that this book's the author was engaged in a conversation on his copper wired telephone at a time when a severe storm, accompanied with thunder, was active in the area. Suddenly, a distinct flutter sound of increasing intensity followed by a single extremely loud cracking sound could be heard in the background during the conversation. The total duration of the two entirely different sounds heard consecutively would have been a few seconds. The author later learned that at that instant the lightning struck the ground only a few hundred meters away.

The flutter sound produced by streamer corona being of low intensity is difficult to hear in open air from a long distance. However, such a noise in ambience can be easily heard with the help of an amplifier. Thus the wide spread streamer corona noise was made audible on the telephone. The sound heard can also be described as the one produced when unloading a truck filled with crushed stone (gravel) by tilting it gradually. The increasing intensity of the sound of falling crushed stone is followed by a loud cracking sound when the gravel is emptied completely. This mechanism of sound is a symbolic representation of the sound produced by rapid development of wide spread streamer corona activity at the clouds producing flutter sound followed by the complete breakdown of air with unstable leader, causing the final jump of lightning strike with dart leader producing an extremely loud cracking sound.

4.2.3 Preference of Locations for the Lightning to Strike

Lightning seems unpredictable about where it is most likely to strike. However, it is most likely to strike sharp metallic objects.

At the tip of a metallic object, high field intensity is induced in the atmospheric air due to the potential at the source, the clouds in this case. The intensity and the gradient of this electric field is determined by the magnitude, type and polarity of the potential, the gap distance and, most importantly, by the shape of the electrodes (Chapter 2). Much research has been conducted to find the suitable shape of the conductor terminal, that could most attract the lightning to strike [4.10, 4.11].

The corona or the partial breakdown activity in air begins above its inception field intensity and grows with the increase in potential. The shape of the electrode facilitates different types of coronas: star, streamer or leader to be able to develop. The most suitable shape for the end terminal of the lightning conductor would be the one that could facilitate the growth of streamer corona in the air giving rise to a stronger return stroke with the development of leader. It is well known that a hemispherical rod shape of electrode is most suitable for streamer corona to grow. Although the pointed electrodes have a very small surface area, the intensity of field is extremely high in the air at their tips. This is the reason that traditionally pointed shape of electrodes are most popularly chosen for the end-terminals of the grounded lightning conductors. Sharp and pointed metallic objects such as antennas, umbrella rod stick tips, rucksack frames, poles, water pipe terminations, transmission line structures, wind turbines, and so on are known to be favorite objects for the lightning to strike upon. All tall buildings and structures are prone to lightning strikes. It has been observed that corners of buildings and structures are comparatively more prone to lightning strikes.

When no metallic object is present in the near vicinity of high field intensity under the clouds, lightning may strike tall trees or even sharp pine tree leaves. If nothing else, it may strike any living being or simply the ground.

A pre-ionized zone of the atmospheric air has weaker insulating properties. Ionized zones are quite often present over the electrical installations due to high intensity of electric field and the corona activity. This is the reason all-electrical installations are most prone to lightning strike. Laboratory investigations made by Daver [4.12] confirmed this phenomenon. Piles of biodegradable waste on the ground, such as cow dung, green agriculture waste, and so on, produce combustible gases having weaker dielectric properties. Such locations are also found to be prone to lightning strikes.

It has been recently observed that equipment producing high frequency communication channels in the atmosphere are more prone to lightning strike. One such example is that of people getting hit by lightning while using cell phone during thunder. On June 13, 1998, Lysa Selfon, sitting on the field level seats in a crowd of 66,000, was directly struck by lightning while attending a concert at RFK stadium in Washington DC. After a high intensity thunder activity started over the stadium, a lightning bolt hit her in the head and chest. It singed her hair, blew a hole in her right eardrum and left her with second degree burns over 20 percent of her body, mainly her face, neck and chest. She was clinically "dead" without a pulse for more than five minutes, but the first aid by CPR brought her back to life, Washington Post [4.13].

The worst burns on Lysa Selfon were on the chest, under the metallic wire of the right cup of her bra. Many questions were raised about how she alone could get hit by lightning in such a huge crowd? The only clue was that she had finished a conversation on a cell phone just seconds before getting hit by lightning.

Similar cases have also been reported from Mumbai, India. On June 15, 2002 Times of India, New Delhi reported the death of a person hit by lightning while he was talking on his cell phone in the presence of lightning in Mumbai. Consequently the "Consumer Guidance Society of India (CGSI)", Mumbai warned cell phone users against keeping their handsets on during lightning.

4.3 DELETERIOUS EFFECTS OF LIGHTNING

As discussed in the previous section, the phenomenon of lightning occurs due to the failure of insulating properties of atmospheric air. Breakdown of gaseous dielectrics is always accompanied with unstable leader followed by an arc. The characteristics of the free burning leader/arc column in air are determined by the situation depending upon the condition of the gas, potential (or the static charge) at the electrodes, the gap distance between them and the current/charge accompanied, Section 3.4.1. As the magnitude of charge or the accompanied current increases, it invokes a number of new features. The so called "plasma jet" along with self-generated gas blasts created by these jets are associated in these columns causing sudden expansion of the air. The temperature in the center of the leader/arc core may rise to the order of 30,000 K. This results in the sound of thunder when cloud-to-cloud lightning flashes take place.

Unlike thunder sound, a lightning strike on any object on the ground may give rise to a loud cracking sound. If the charge accompanied with lightning strike on the ground finds a path of least resistance to go inside the earth, no damage is done except the production of loud sound. However, if it does not find a path to go inside the earth, it may create havoc. The intensity of damage mainly depends upon the magnitude of the current accompanied with the stroke.

4.3.1 Loss of Life of the Living Beings

According to the Federal Centers for Disease Control and Prevention in the United States, nearly 100 people are killed by lightning each year in the United States. About 300 are struck but survive. In India, this figure is higher because the density of population is much higher. The chance of survival after getting struck is much lower in India due to the lack of awareness of how to treat lightning strike victims. A very large number of cattle also die due to direct lightning strike.

Neurological damage is the main cause of death. It could be caused when hit directly or indirectly due to the temporary interruption of the brain's vital oxygen supply during "cardiac standstill". A direct strike could also result in excessive burn injuries leading to death. An estimate reports that in the United States, only about one-quarter of people struck by lightning die. However, many survivors are left with a lasting injury, such as hearing loss, numbness or weakness of the arms and legs or damage to the automotive nervous system, which regulates the body temperature and digestion [4.13].

4.3.2 Fire Hazards Due to Lightning

Fire is the most common and dangerous hazard caused by lightning. Lightning often ignites fire in the jungles all over the world. These spread beyond control and cause heavy damage to life and property. Fire may ignite when the lightning strikes directly on trees, wooden structured buildings and residential houses. Gas and petroleum product storage tanks are highly vulnerable to catching fire when hit by a direct lightning bolt. All insulating materials are prone to catch fire or get damaged when hit by lightning.

A lightning strike injects anything between a few to hundreds of kA of impulse current (charge) in a few millisecond. Insulating materials, with high resistance on the order of hundreds and thousands of mega ohms, create intense temperature rise due to high I^2R losses getting converted into heat, before igniting fire. The materials that burn more easily catch fire first. Even on living creatures, which have fairly high resistance, the first item that gets charred is hair.

4.3.3 Blast Created by Lightning

Imagine the phenomenon of thunder taking place within a building! It turns out to be a bomb without any explosive material. The shock wave produced by the sudden expansion of air may damage the building structure, shatter the windowpanes, shift walls and dislocate items, in the process of equalizing the pressure towards the atmosphere. The more the structure is enclosed, the more severe is the intensity of the blast within the building.

Lightning is known to be attracted to striking upon sharp metallic objects terminating in the free air, for example, rods, tin shed-edges, pipes, and so on. Consider such a body projected from the terrace of a cement concrete building structure. If it terminates within the building, the charge injected by the lightning strike is easily conducted inside the building up to the level of its termination. Finding no conducting path beyond, it gets trapped within the structure. The surface resistivity of the concrete walls being much lower than the volume resistivity, the charge floats restlessly over the wall surfaces in search of a path of least resistance. It tries to attack all solid insulating materials with which it comes into contact. However, it is most attracted toward the path of least resistance, again the metals, which may be floating, for example, nails, screws, frames, and so on. Finding no other way of consuming electrical energy, the charge is injected into the air, an insulating material, through the floating metallic bodies to cause its sudden expansion, a "blast". Depending upon the situation, some charge injected through such metallic bodies may also be simultaneously consumed in the process of heating/burning those insulating materials, which get easily charred [4.14]. The shockwave produced due to the blast moves towards the atmosphere to equalize the pressure. On its way outside it shatters, lifts and throws items in its path.

The author had opportunity to examine several such incidents of lightning in the city of Kanpur, India. These incidents occurred in the early morning hours in the months of March, April and May, a season when thunder activity builds up locally in this region. Once, the author personally witnessed the phenomenon. In each episode, the floating metallic conductors that carried the lightning current inside the building could be located. An unfortunate discontinuation of the conductive path towards the ground, which gave rise to the blast, could be located in every such incident. Peculiar mechanical and thermal effects of lightning, such as explosion and charring of insulating materials on the floors and the walls could be identified as the "traces of lightning" left behind by the discharge current. Burning and charring of some small insulating materials did take place, but fortunately no major fire outbreak took place in these cases. No damage to any conducting metallic body could be observed in this process except some getting lifted and dislocated due to the shock wave. Conductors having much low resistance may get just slightly heated up when the full discharge current passes through them, since the duration of the whole phenomenon is in milliseconds.

4.3.4 Development of Transient Over-Voltage Due to Lightning Strike on the Electric Power System Network and Its Protection

As compared to other objects on the earth, the electric power system network is especially more prone to lightning strikes. This has two specific reasons:

1. **The pre-ionized zone:** Corona or Partial Breakdown (PB) in air is a continuous process on the whole power system network, that is, generation, transmission and distribution. Although attempts to prevent the corona activity have been made, it can be minimized but not eliminated. The corona process in

the air decomposes and ionizes the air. Thus, around the power system network an ionized atmosphere prevails known as the "pre-ionized zone". The insulating properties of this ionized zone are weaker as compared to the healthy air.

2. **Presence of metallic objects:** The transmission and distribution overhead lines of the power system stretch over long distances. These are interconnected with the help of large number of sub-stations, most of them in the open air. Depending upon the voltage level, the power grid sub-stations may occupy a huge area exposed to the atmosphere. All metallic bodies attract lightning as explained in Section 4.2.3.
3. **Presence of Streamer Corona:** The three types of stable PB in air are described as Star, Streamer and Leader coronas in Chapter 3. Out of these, the Star and the Streamer coronas are unavoidable on the power system network. In fact, their permissible limits are defined by various specifications. The Streamer corona extends itself in the air depending upon the shape of the electrode and its voltage level. Its presence on the live conductor makes it highly prone to lightning strikes.

Practically the whole power system network is protected from the direct lightning strikes. The transmission lines are provided with grounded wire on the top all along their lengths. The sub-stations are protected with the help of a ground conductor network. However, these schemes are unable to provide a fool proof lightning protection to the network. Lightning often strikes upon the live conductors at different voltage levels. It may damage the equipment directly or give rise to "transient over voltage" in the power system network. Although lightning surge diverters are installed all over the network, lightning strikes have often led to complete interruption of power supply for hours in vast areas all over the world. According to statistics, lightning in Japan causes more than 50% of failures of high voltage transmission lines.

The electrical breakdown of a dielectric equalizes the potential difference between the two electrodes through the highly conductive breakdown or the short circuit channel. The charge or the short circuit current is injected into the object at lower potential level from the electrode at higher potential. When the lightning bolt or the so-called, "dart leader" hits the live transmission line conductor, the clouds evacuate with their static charge through this highly conductive channel in a few milleseconds. An impulse form of current is injected into the object of the strike, which then distributes itself into different conductive paths available at the location in proportion to their impedance to ground.

Consider the lightning to strike upon a live conductor (phase) wire of a long distance high voltage transmission line. Let an impulse current I_{li} be injected at the location, as shown in the Figure 4.10.

Assuming equal impedance of the line on either side, the injected current distributes itself equally on the two sides. Considering the surge impedance Z_c distributed along the line, the injection of current gives rise to the development of a voltage equal to $\frac{1}{2} I_{li}Z_c$ superimposed upon the line voltage. This current accompanied with the voltage rush towards the loads through the sub-stations. The

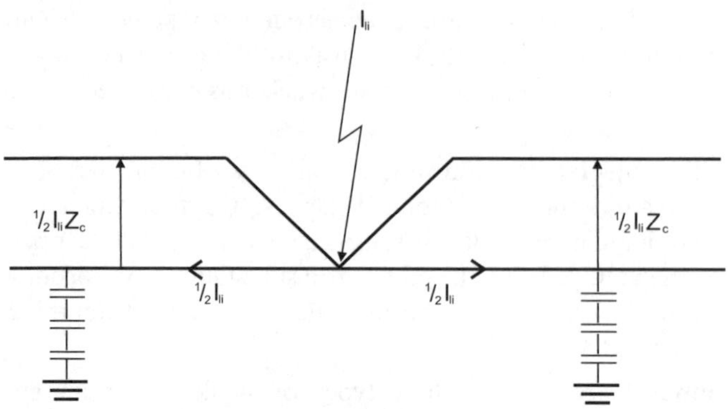

Figure 4.10 Lightning strike and the over-voltage on a transmission line.

overvoltage thus produced propagates in the form of a "traveling wave". As the whole phenomenon lasts for a very short duration of time, it is known as "transient over-voltage". The energy, injected by the lightning discharge current, is finally distributed and consumed at all the connected loads as the magnitude of power delivered increases momentarily.

The power system network is protected from such transient over-voltage traveling waves by installing "lightning arrestors", also known as "surge diverters" on all the overhead transmission lines terminating or beginning at the sub-stations or feeding a transformer. The transmission line insulator strings and bushings at the sub-stations are provided with horn-gaps to let the flash-over due to transient over voltage take place in the air and divert it directly to the ground. This way the solid dielectric insulators are saved from the high electric stress caused by transient over-voltages.

A new era in transient over-voltage protection was introduced in the 1970s with the advent of "Metal Oxide Varistors", (MOV). With time, a gapless surge diverter technique has developed. Metal oxide is a ceramic material having highly non-linear electrical characteristics similar to those of a back-to-back diode. It is developed by mixing zinc oxide (ZnO) with additive materials such as MnO_2, Sb_2O_3, Cr_2O_3. A granular layer of high resistivity oxide strongly bonded to each other surrounds the zinc grains having a low resistivity. The electrical properties of MOV are controlled by their grain boundaries. These lightning surge arrestors are characterized by fast transient voltage switching and letting it go to ground. With their high-energy absorption capability due to polycrystalline nature and inter-granular barriers, these devices have better long-term performance, [4.15 & 4.16].

4.4 PROTECTION FROM LIGHTNING

Loss of life is the most dreadful and irreparable damage caused by lightning. Economic loss due to damage and destruction of property in the United States alone is estimated to be billions of dollars (U.S.) every year. Loss of vegetation due to fire outbreaks is another cause for concern. Endeavoring to protect lives and property from the hazards caused by lightning has always been a motivation of scientific research.

Understanding the phenomena of lightning requires interdisciplinary knowledge of physics and high-voltage engineering. The subject of protection from lightning has become a highly specialized field of research involving many scientists, engineers and other professionals. Newer techniques for lightning protection are rapidly evolving with innovative ideas, experimental research, and development. Commercial exploitation of many such techniques is on going even without these having proved themselves. The phenomenon of lightning strike and its effects are so complicated that many of the results of research and development work in the laboratories fail in practice. Hence, only a brief introduction to well-established techniques, facts and figures are within the scope of this chapter.

4.4.1 Protection of Lives

It is rare to survive a lightning strike. However, there are incidents of people having survived even a direct hit or a side flash of lightning [4.13, 4.17]. The step and touch potentials of high discharge currents cause severe injuries.

Morris [4.17] has described a very interesting and revealing story of a young girl, Mary, hit by lightning. While trekking in Colorado's Roosevelt National Forest. Mary was struck with lightning just a few minutes after a hailstorm. A lightning bolt from the blue streaked through the clear sky and darted towards the metal frame of her rucksack. To her, it seemed the world exploded around her in bright, paralyzing light. The pulsing channel of lightning entered Mary's body with an ear-splitting crack, lifting her off the ground. Her muscles contracted violently, and her blood vessels went into spasms, cutting off the blood flow to her extremities. Her hair stood on end. As the charge surged down her legs, searing nerves and muscle tissue, the acrid smell of burning flesh filled Mary's nostrils. She fell to the ground and could feel nothing below her waist. The lightning strike left five raw coin-size burns in a neat semicircle on her back hips and foot. This must have been the spot where the rucksack metal frame would have rested on her body. Another similar story from [4.13] is described in Section 4.2.3.

High potential difference across the body of a living being causes injection of current. It can lead to ventricular fibrillation (unsynchronized muscle operation of the heart), causing respiratory arrest and neurological damage or even total burning of the body. In case the injury is not very severe, timely first-aid help may save the life even if the person is clinically dead for a few minutes [4.13]. Hence, it is strongly believed that only awareness and knowledge about lightning phenomena can protect and save the human lives.

Studies have revealed that often people are struck by lightning not during the peak intensity of a thunderstorm but before or after the storm peak, Gomes [4.18]. Therefore, one should not wait for the rain to start to seek shelter and should not leave the shelter immediately after the rain has stopped.

Personal safety measures under thunderstorm accompanied with lightning activity are recommended as follows:

- It is safest to take shelter indoors, i.e. inside a building provided with dependable lightning protection system. Often, however, the protection system is so old that it is no longer functional. Discontinuity of the lightning conductor and its ineffective earthing are the two most common causes for the malfunctioning of the protective system.
- When out in the open, one should never take shelter under a tree. Trees are highly vulnerable to lightning strike if metallic objects are not present in the vicinity. Lightning may split the tree and catch it on fire. Most of deaths of humans and cattle occur when lightning strikes the tree under which they take shelter to protect themselves from the rain. At the fag end of the strike, the dart leader is found to branch off. If a human directly meets one such branch, the survival is rare.
- Any metallic object carried by people out in the open attracts lightning. Objects like umbrellas, rucksack frames, ornaments and even cell phones may invite the lightning strike. Hence, it is advisable to avoid all metallic objects exposed from the body.
- If caught by thunder out in the open, it is best to sit on the ground with the feet close to each other, the head down, the ears covered with the hands. Covering the ears is advisable to protect from the very high intensity sound produced by lightning strikes. The close position of the feet minimizes the generation of step potential due to the flow of ground current. Shoes having a good insulating material sole also help.
- If caught in a vehicle, it is advisable to remain inside and not touch the metal body. The vehicle body works like a "Faraday Cage", which does not build potential inside. It is insulated from the ground by the tires. The charge injected by the lightning strike is most likely to discharge itself to the ground, finding the path of least resistance.
- When there is active thunder, it is advisable to avoid open water, for example, swimming pools, lakes, ponds, rivers or the sea. Because water conducts electrical currents.

4.4.2 Protection of Buildings and Structures

With the development of building and structure technology, lightning protection techniques have required newer approaches. For example, buildings and structures above 30 m in height may face lightning strikes on the sides even with protection system installed at the topmost location. Buildings with very large surface areas, that house valuable electronic gadgets or flammable materials, and so on, require special schemes for protection from lightning strikes.

Active or energized protection systems, for example, "Early Streamer Emission Lightning Conductor" with electrical triggering device, have been developed and even commercialized. Research is underway to develop techniques to divert lightning strikes to a desired location with the help of laser beams.

Large numbers of national and international standards have come up recommending the lightning protection schemes suitable for different situations. A list of

"Lightning Protection Codes and Standards" of different countries is provided by Richard Kithil of "National Lightning Safety Institute", (NLSI), USA in [4.19]. The code of practice followed in India is IS 2309 (2000), "Protection of buildings and allied structures against lightning". It is a very well written standard describing the effects of lightning and provides comprehensive lightning protection schemes for different situations.

However, the most effective, reliable, simple and widely-accepted method of lightning protection remains the lightning rod, invented by Benjamin the Franklin in 1752. It is also known as the franklin lightning rod. It comprises of a simple metal rod or conductor mounted on the top of a building and electrically connected to the ground in order to provide a low resistance path to the current injected by the lightning strike conducted safely to the earth.

This simple lightning protection system has three main components, that is, the air termination, the down conductor and the earth termination network described with their important features and development below.

4.4.2.1 Air Termination Network This may consist of vertical or horizontal conductors or a combination of both. The horizontal conductor layout specifications are well defined for various types of roofs and terraces. However, lightning tends to strike ridges and corners. Secondly, care must be taken that all metallic projections are bonded to form part of the air termination network.

The shape of the vertical conductor air termination, which would attract more lightning, has long been a field of research. There are differences of opinion about the effectiveness of the pointed versus the hemispherical shapes of the electrode. An interesting experimental work was performed by Mehndiratta [4.11] to investigate the suitable shape of the air terminal electrode. A flat inverted bowl shaped (Rogowski Profile) electrode of diameter 1.1 m was applied the high voltage *dc*, simulating it as the cloud. Various shapes of smaller electrodes were put below to measure the breakdown voltages, varying the gap distances between the electrodes. Breakdown voltages were measured with +ve as well as –ve polarity dc. As compared to the single needle and multiple needle shaped electrodes, lower breakdown voltages were measured with hemispherical rod shaped electrode of 2.5 cm diameter with both the polarities for longer gap distances. These results can be attributed to the development of stable streamer corona at the rod, unlike at the sharp and pointed electrodes as explained in Chapter 3. Just before the lightning strike, the development of streamer at the rod facilitates the return stroke to extend itself towards the propagating leader from the sky. Hence, a hemispherical rod shape of the air terminal is recommended by this work.

4.4.2.2 Down Conductor The down conductor is defined as the conductor connecting the air terminal to the earth electrode. The electrical continuity is an important criterion for any lightning protection system. A comprehensive list of recommended materials for all the component parts is provided in IS 2309. It suggests copper, aluminium and their alloys, galvanized steel, and so on. Because of the constant problem of corrosion, careful attention must be given to the metals and the joints and bonds involved. Any joint other than welded represents a discontinuity in the current conducting system and is susceptible to failure. Hence, the lightning protection system should have as few joints as possible. The joints should be mechanically and electrically effective. These should be clamped, screwed, bolted, crimped, riveted or welded, IS 2309.

4.4.2.3 Earth Termination System When the lightning current, having impulse form, flows through the conductor, a transient potential gradient is given rise around the electrode depending upon its resistance. Each down conductor is recommended to connect separately to the earth electrode. The combined resistance to earth of the whole lightning protective system should not exceed 10 Ω. However, a reduction in the value of resistance to earth is advantageous in the sense that the potential gradient build up around the earth electrode while discharging the lightning current will reduce proportionately, IS 2309.

A common earth termination network for the lightning protective system can be adopted with other services, for example, the electric power network. However, the resistance to earth for any of the individual services in this case should not exceed the highest value permissible. It must be mentioned here that the requirement for the power system network neutral grounding is not to exceed 1.0 Ohm.

4.4.3 The Protected Area

After the advent of the lightning rod by Benjamin Franklin in 1752, the extent of protection provided by this arrangement came into question first by a French scientist in 1779. There has been a wide discrepancy in this perspective. As Golde [4.20] reports, the suggestions regarding the "zone of protection" started in the year 1823. Mathematically, scientists have viewed the volume, rather than the area, under protection by lightning rod arrangement differently. Two widely accepted common practices, as recommended by many standards, are briefly described below.

4.4.3.1 Protected Volume Determined by a Cone In case of a single mast type vertical lightning conductor, the volume under protection is evolved by simple geometry. It is considered to have the shape of an inverted cone whose height is that of the mast, and the circle at the base having a radius equal to the height of the rod. Thus the angle of protection works out to be 45°, as shown in Figure 4.11(a). For stringent protective schemes the specifications recommend this protective angle should not exceed 30°. Such an arrangement may be required for the fire hazardous objects and so on.

If two or more vertical conductors are required to be installed to provide protection to objects spread in a large area, the protective angle within the space bounded by the conductors is permissible to be 45°, but towards the outside it must not exceed 30°, as shown in Figure 4.11 (b). Objects and structures, falling within the inverted cone/cones thus formed, are considered to be under protection from the direct lightning strikes.

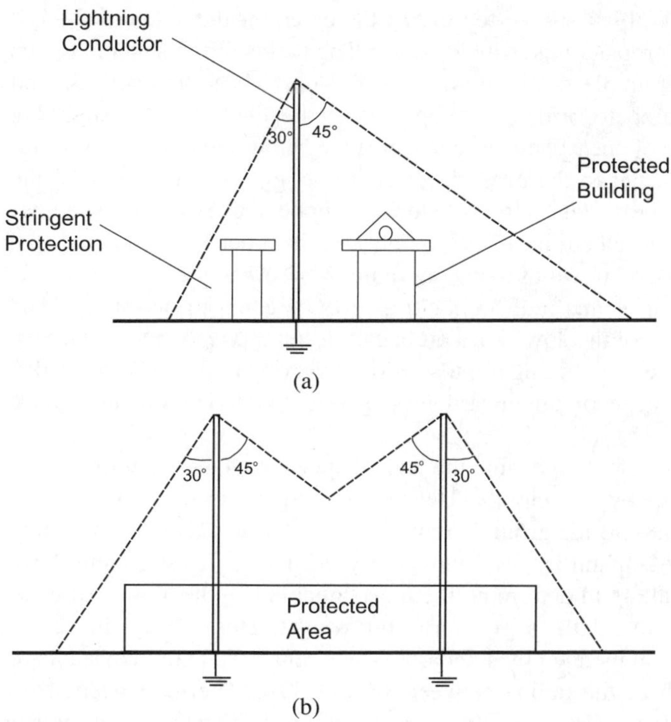

Figure 4.11 (a) Protected area by a single vertical lightning conductor, (b) Area protected by twin vertical conductors.

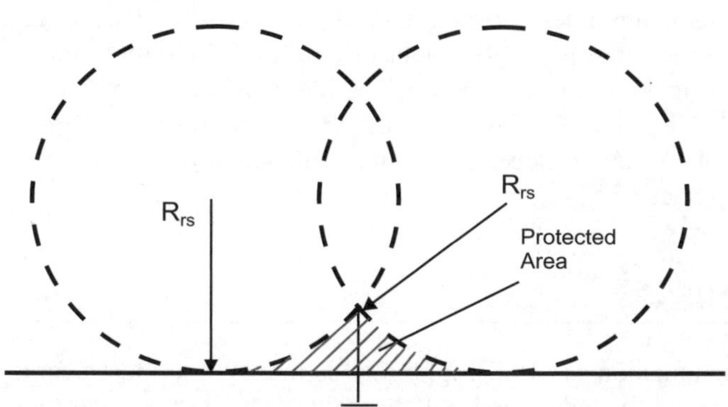

Figure 4.12 Protected volume derived by the Rolling Sphere.

4.4.3.2 Protected Volume Evolved by Rolling a Sphere In 1968 Armstrong and Whitehead [4.21] suggested a more scientific method to define the protective volume in simple three-dimensional sense, known as the "Rolling Sphere Method". Nauman [4.22] has described the procedure for the determination of protected volume derived by this method. An imaginary sphere of pre-determined radius R_{rs} is rolled over the protective arrangement. The volume derived underneath is considered to be under protection, as shown in Figure 4.12.

The center of the rolling sphere represents the extent or the reach of the last stage of the downward propagating stepped leader. The radius of the sphere "R_{rs}" is the nearest point, for example the lightning rod, from which a streamer initiated at the ground could jump to bridge the gap toward the final stage of lightning strike. Thus the radius of the sphere is determined by how close the downward propagating leader must get to the ground before attracting the streamer from the ground (the return stroke). Hence it also depends upon the extent of streamer required from the ground electrode to bridge the gap in order to accomplish the lightning strike. R_{rs} is also known as "striking radius" or "attractive distance". It is estimated that the length of this radius should always be equal to or smaller than the length of the last stage of the downward propagating stepped leader "l_{sl}", Nauman [4.22]. For the peak value of lightning impulse current varying between 3- and 15 kA the length of the last stage of downward propagating leader is estimated to be between 17 to 59 m.

The accumulation of charge and the gap distance together determine the potential effect built up by the clouds. Hence, the magnitude of charge in the clouds and their proximity to the ground are responsible for an accurate estimation of the above-mentioned quantities, R_{rs} and l_{sl}. Experimental investigations have been made by many authors to determine the most suitable length for the radius of the rolling sphere. However, it has given varying results. This is why the radius of the rolling sphere recommended by standards is in a range. The German standard recommends the length of the radius between 10 and 60 m. The longer length of the radius is suggested for protecting the low objects on the ground from higher magnitudes of lightning currents, that is, for lightning strikes of higher severity. The length of the radius reduces for taller objects as well as for the lightning strikes accompanying lower magnitudes of the current. More experimental investigations are required to determine the probable extent of streamer, the return stroke, just before the final jump in long air gaps. Since it is not possible to predict the magnitude of the current in a strike, for the design of the protective scheme length of the radius of rolling sphere should depend only upon the height of the object to be protected.

4.5 BALL LIGHTNING

Unlike lightning, the phenomenon of "Ball Lightning" is rare. While most people have not heard of, what's more understood ball lightning. The earliest literature reference of the "Red Whirling Balls" goes back to 1727 by James Thompson in his poem, "The Seasons", [4.23]. In some well known journals, for example, "La Nature", "La Meteorologie", in French and "Zeitschrift der Angewandte Meteorologie", in German, a vast number of articles on this subject have been published in the nineteenth and twentieth centuries. More than two thousand scientific papers have been written in various journals around the world on this subject as reported by Stenhoff [4.23]. Started in the year 1989, the "International Committee on Ball Lightning" organizes "International Symposium on Ball Lightning (ISBL)" every two years.

4.5.1 The Phenomenon of Ball Lightning

The natural phenomenon, as sighted by many individuals all over the world, is described as a cluster of luminous particles emitting orangish red color light, having a near spherical shape and size of a tennis ball to football or even greater, whirling and waving slowly and independently through air, often in horizontal direction above the ground, acquiring heights of up to a few meters and sometimes accompanied with low intensity hissing, heat, or a pungent odor, for example, that of ozone. It has been reported to last for a few seconds to a few minutes. Such sights have been found to pause for some time in their course of movement and at times rush off with great rapidity, [4.24]. The phenomenon is reported ending up in water with a hissing sound or simply weakening off in the air. Its explosive decay has also been reported when penetrated by a conductor, [4.25].

The phenomenon of ball lightning, also known as "Fire Ball", has been reported to take place in open air. It is seen entering houses through windowpanes or chimneys. It has even been sighted inside a flying aircraft and sailing ship. Almost 96% reports of the sightings of ball lightning state that the events occurred during or after heavy thunderstorms, often accompanied with rainfall, [4.26]. Such publications also report that ball lightning was seen following a lightning strike on the ground. However, there are also reports of occurrence of this phenomenon in the absence of lightning discharge [4.24].

4.5.2 Injurious Effects of Ball Lightning

Ball lightning is reported to damage insulating bodies/materials by overheating or even burning, for example, dress fabric, plastics, and so on. Large numbers of deaths of living beings, including humans, have been reported due to electrocution and burning caused by ball lightning. Surprisingly, the majority of them relate to eighteenth and nineteenth centuries. Comparatively fewer reports have been published in the twentieth century, [4.23]. When encountered directly, injuries to attributed to ball lightning include electric shock; paralysis; loss of consciousness; burns; singeing of hair; loss of hearing and even vision; shattered limbs, face, fingers and thumbs. The last type of injury could have been caused due to people running in panic after having seen such a peculiar object. Animals that have allegedly been killed by ball lightning include farmyard animals, such as horses, cows, sheep and pigs and also domestic pets such as cats and so on. A premature demise of ball lightning is reported with explosion, particularly when encountered with destructive interruption.

James [4.27, 4.23] reported in 1992 an interesting incident that happened in Angust 1978. This occurred at an altitude of some 3900 m in the Caucasus Mountains. He reported the ball lightning or the fire-ball as a "blob". As he describes, an amorphous yellow "blob" made its way into a tent occupied by climbers. Floating one meter above the floor, it suddenly dived into a sleeping bag whose occupant screamed in pain. The ball jumped out and went on to circle over the other bags, entering one after the other, accompanied with anguished howls from the victims. When the climbers were flown to hospital, their injuries were described as "worse than burns", with pieces of muscle ripped off. One of the climbers was instantaneously killed. It was discovered later that he was sleeping in the bag lying on a rubber mattress, an extra insulation from the ground. Such incidents have been characterized as "aggressive" or "attacking" ball lightning in the literature.

4.5.3 Models and Physics of Ball Lightning

When lightning strikes on the open field, the discharge is likely to have longer retention on the ground surface in the absence of a conducting path to penetrate into the ground. The ground has a vast surface area unlike a particular object of strike. Every ground may have its own soil composition too. It may be rich in silicon, ferrous or other compounds. Depending upon the area, terrain and weather conditions, the soil offers different volume and surface resistivities. Water retention in the soil affects its resistivity to a great extent, which in turn depends upon its composition and weather conditions.

Under some conditions suitable for the development of lightning, the charge may spread over the surface, acquiring a low superficial charge density on the ground. This facilitates relatively slow interaction between the soil compounds and electric charge. It gives rise to a favorable condition for the development of metastable state of the soil molecules described as the prime source for the formation of ball lightning by many authors, [4.28].

There have been continuous attempts by the researchers over the last 150 years to provide an acceptable physical model for ball lightning. Models based on ionized gas and plasma are divided into two groups. One involves a selfs contained energy source (plasmoids) and the other involves external energy (discharge), [4.23].

In 1960, Hill opined that this phenomenon is not a plasma state of gas; rather it is a concentration of strong inhomogeneous distribution of space charge (like polarity charge) in the form of a highly ionized state of gas in molecular form, [4.29].

Continuing with the same line of thoughts, the Finkelstein-Rubenstein model was proposed in 1969–70 along with their experimental laboratory evidence [4.30]. They based their model on the Townsend theory of ionization. Their explanation for the ball lightning structure comprises the central region where multiplication of charge particles due to impact ionization by accelerated electrons in electric field occurs. In the interactive region, a radial increase in temperature takes place and the outermost region is surrounded by air.

The most recent model has been described by Abrahamson and Dinniss in "Nature", 2000 [4.31]. It suggests that the extreme heat generated at the location of lightning strike on the open ground can sometimes turn the soil into silicon and silicon compounds with oxygen and carbon. The process is similar to the technique used in industry to extract pure silicon from sand. The tiny "nanoparticles" of silicon link together into chains, which are lifted above the ground by surface air currents. These particles then burn slowly, giving off heat and light.

The models describing the ionized state of particles derived from external sources of energy appear to be more convincing. Naugebauer in 1937 [4.28] described the phenomenon based on quantum mechanics theory. Through quantitative analysis of a macroscopically neutral sphere composed of free electrons and positive ions, he showed that such a mass of ionized gas could exist in metastable equilibrium. The quantum mechanical exchange energy of the electrons, a weak attractive force between electrons of opposite spin, hold it together. This could be made responsible for the formation of a cluster of such particles.

In physics, the metastable is described to be a particular excited state of an atom, the nucleus of another system that has a longer lifetime than the ordinary excited state. For certain gas molecules the lifetime in the excited state, known as "Metastable State", may extend to a few tens of milliseconds. A metastable state of a molecule may thus be considered to be a kind of temporary energy trap, a somewhat intermediate state of energy. Metastables have a relatively higher potential energy and therefore ionize neutral particles on collision. These excited molecules may emit photons of electromagnetic radiation on returning to the ground, the lowest energy level. This type of incandescence excitation is by thermal energy. If the cause of excitation is a photon, the process is called photo-luminescence and if it is an electron, it is called electro-luminescence. In the absence of very large amounts of power input availability for the reported visual brightness of 40–100 W bulb, Powell & Finkelstein (1970) suggested that the ball lightning glow is because of electro-luminescence, an unusually efficient conversion process of electrical energy to light. Partially ionized metastable state or gaseous plasma state of a matter is known to be very rich in collecting phenomenon, for example, linear waves, and the solitons or the solitary waves that retain their shape and speed after colliding with each other. The occurrence of ball lightning may be attributed to one of these nonlinear phenomena.

As of today, no satisfactory reproduction of the phenomenon has been made in any laboratory in spite of several attempts.

4.5.4 Ball Lightning without Lightning Strike

Like every individual case of lightning strike, every occurrence of ball lightning is also peculiar in its own form. It is reported in general to take place after the injection of charge on the ground through the natural lightning strike. However, ball lightning can take place due to other sources of charge under suitable conditions of its occurrence, Arora [4.32], also reported by Talbot [4.24].

Wide scale recurrence of this otherwise rare phenomenon took place during unusually warm and prolonged dry spell of summer of the year 2002 over the districts of Mirzapur, Lucknow and Kanpur in the state of Uttar Pradesh in northern India during June, July and August. The affected area spread to about 200 by 300 km. Millions of people had the rare opportunity to see ball lightning due to its repeated occurrence.

In the absence of knowledge and scientific explanation, residents mass as well as the media remained perplexed for weeks. Fear prevailed during this period in the region. Confused descriptions like "Muhnochwa" (one who scratches the face), extra terrestrial invasion by UFOs, attack by insect etc. were reported by the media before the author had an opportunity to identify it as the phenomenon of ball lightning, Arora [4.32]. The actual description reports and a video film, shot coincidently by a video-photographer, helped to reveal the occurrence.

4.5.4.1 The Weather and Climatic Conditions It was a prolonged and dry summer. Temperatures during the day soared to above 42°C and around 28°C in the night. There were no rains, no thunder. Fast and hot winds blew all throughout the day but a lull prevailed during the evening and the night hours when the phenomenon was repeatedly observed. Dust storms occasionally occurred.

The district of Mirzapur falls in the Vindhyachal mountain range and has a rocky terrain with sand and gravel. Even under normal weather conditions, the soil resistivity in this region is known to be very high. Under the adverse conditions the soil resistivity could increase many fold its normal value of 1000 Ω-m in such a region. The electrical earthing/grounding systems are rendered to be near ineffective due to lack of moisture in the soil. Such conditions make it difficult for the current/charge to find passage inside the ground. Instead, it is forced to remain on the surface or just underneath for a considerable time since the surface resistivity, depending upon the particular soil conditions, becomes lower than the ground resistivity.

4.5.4.2 The Man Made Sources of Charge/Current In the absence of injection of charge from the natural source of lightning, occurrence of the phenomenon of ball lightning was unconceivable at the first instance. A deeper visualization of the situation led to the presence of multiple possibilities of man-made sources of electric charge injected on the surface of the ground in this area from the electric power distribution network. These are described in the following paragraphs.

The single-phase power supply to electric traction at 25 kV by two-phase 132 kV transformers introduces a highly unbalanced load system. The return path for the current from the moving locomotives completes the circuit through the rails to the substation located at a distance of every 40 km along the track. Thus the rails are responsible for providing the return path for the huge amount of load current, on the order of a few hundred amperes, continuously drawn from each substation. Indian Railways has a very wide network of electrified track. Measurement of current supplied to the load (locomotives on the track) at 25 kV and the return current at the ground bus on such a transformer revealed that only about 55 to 60% current returned. In the total absence of grounding or earth plates in this system, 40 to 45% of the load current simply spreads over the surface of the ground all along the electric traction track length over the vast area. These measurements also revealed that the farther the locomotive was from the transformer (substation), the less percentage of the load current returned to the source. Thus the electric traction is a major source of injection of charge on the ground.

Power theft is a common feature in this part of India. Estimation revealed that at certain locations even 40% of the power supply is stolen and accounted for as losses. All single-phase power thefts are accomplished by providing the return path for completing the circuit by simple unscientific ground connection instead through the neutral wire. Consequently, it contributes large amounts of charge/current to the surface of the ground.

The current due to unbalanced load carried by neutral grounding in three-phase power distribution system is another source that could flow on the surface of the ground under adverse conditions of the soil. The last source of electric charge on the ground from the power distribution network is by the ground wire provided for safety in the single-phase power supply. It continuously carries the charging/leakage current of the insulation system to the ground. The contribution by both these sources is, however, not significant.

REFERENCES

[4.1] "Extra-terrestrial invades UP", Sunday Times of India, Aug. 18, 2002, Lucknow, India.

[4.2] Arora, Ravindra and Gomes, Chandima, "The Lightning Phenomenon", Published by NAM S & T Center, New Delhi and National Science and Technology Commission, Colombo, Sri Lanka (2009), Daya Publishing House, Delhi, India.

[4.3] Sharma, Asit, "Read the clouds", Young World, The Hindu (daily), July 29, 2005 p. 2.

[4.4] Cloud Forms, "Webster's Ninth New Collegiate Dictionary" (1991), Merriam–Webster Inc., Publisher, Springfield, MA, USA.

[4.5] Hekkila, W.J., "Outline of a magnetospheric theory", Journal of Geophysics Res., Vol. 80, p. 2496 (IEEE Spectrum, June 1995).

[4.6] Dooling, Dave, "Stormy Weather in Space", IEEE Spectrum, June 1995.

[4.7] "Ultraviolet Radiation", Integrated Environmental Management, Inc., USA (1999–2008).

[4.8] Jayaratne, Chandana, "Cloud Charging Mechanisms", UNESCO Regional Training Workshop on Lightning Physics, April–May, 2003, Colombo, SriLanka, the author is with Department of Physics, University of Colombo.

[4.9] Arora, Ravindra, "A new approach to the mechanism of lightning strike", International Roundtable on Lightning Protection, Colombo, Srilanka, May 2007.

[4.10] Verma, N.K., "Investigation of Lightning Protection", MTech thesis, Dept. of Elect. Eng., IIT Kanpur (March 1996).

[4.11] Mehndiratta, Vikram, "Experimental investigations of breakdown voltage of atmospheric air for different electrode configurations under weakly and extremely nonuniform electric field conditions with high voltage dc", MTech thesis, Dept. of Elect. Eng., IIT Kanpur (May 2008).

[4.12] Davar, Vivek, "An investigation on effect of pre-ionization on the Lightning Air Terminal", MTech thesis, Dept. of Elect. Eng., IIT Kanpur (April 1994).

[4.13] "Lightning, Cover Story", Washington Post Health, Washington DC, USA, June 23, (1998).

[4.14] Arora, Ravindra, "Traces of Lightning Incidences on Structures", Proceedings, 24th International Conference on Lightning Protection (ICLP), Sep. 1998, paper 95, Staffordshire University, UK.

[4.15] Schei, A. et al., "Metal Oxide Surge Arrestors in ac system", Part 1, Electra, Vol. 128, (Jan 1990).

[4.16] Eda, K., "Zinc Oxide Varistors", IEEE Electrical Insulation Magazine, Vol. 5, No. 6, pp. 28–41, (1989).

[4.17] Morris, Deborah, "Hit by Lightning", Drama in Real Life, Reader's Digest, Vol. 139 No. 832, Aug. 1991, pp. 72–77.

[4.18] Gomes, Chandima, "Human Safety against Lightning", article in the book "The Lightning Phenomenon", Daya Publishing House, Delhi (2009), pp. 132–138.

[4.19] Kithil, Richard, "A review of dynamic changes in USA lightning codes and standards", International Lightning Detection conference, Tuscon, Arizona, (Oct. 2002).

[4.20] Golde, R.H., "Lightning Protection", Edward Arnold Ltd. Pub., London (1973).

[4.21] Armstrong, H.R., et al., "Field and Analytical Studies of Transmission Line Shielding", IEEE Trans. Vol. PAS-87, No. 1 (1968), pp. 270–281.

[4.22] Naumann, W., "Die Konstruktion der Schutzraeume im Blitzschutz", Elektropraktiker, Berlin, 40 (1986) 8, pp. 258–263, (German).

[4.23] Stenhoff, Mark, "Ball Lightning", An unsolved Problem in Atmospheric Physics, Kluwer Academic/Plenum Publishers (1999).

[4.24] Talbot, Michel T., "Ball Lightning" new information, site: www-bprc.ohio-state.edu/bdaye/balligh.html.

[4.25] Tessan, M.de, and Hebd, C.R., "Sur la foudre en boule", Seances Acad. Sci., 49 (1859).

[4.26] Rayle, W.D., "Ball Lightning Characteristics", NASA Tech. Note D-3188, Washington, D.C. (1966).

[4.27] James, A., "Fatalities attributed to ball lightning", Proceedings of fourth TORRO (Tornado and Storm Research Organization) Conference, edited by M. Stenhoff, Oxford Brookes University, Oxford, (1992) U.K.

[4.28] Naugenbauer, T., "Zu den Problem des Kugelblitzes", Zeitschrift d. Physik, 106 (7, 8) 474 (1937).

[4.29] Hill, E.L., "Ball Lightning as Physical Phenomenon", Journal of Geophysics Research, 65, 1947 (1960).

[4.30] Finkelstein, D. and Rubenstein, J., "Ball Lightning", Physical Review, 135, A390 (1964).

[4.31] Abrahamson and Dinniss, "Phenomenon of Ball Lightning", Nature (2000).

[4.32] Arora, Ravindra, "Ball Lightning without Lightning Strike", Proceedings, 8th International Symposium on Ball Lightning (ISBL04), National Central University (NCU), Chung-li, Taiwan, (Aug 2004).

CHAPTER 5

ELECTRICAL PROPERTIES OF VACUUM AS HIGH VOLTAGE INSULATION

The idea of vacuum as insulation is quite old, going back to 1897 when R.W. Wood first described discharges in vacuum while investigating the production of X-ray tubes [5.1]. The desire to produce X-ray tubes operating at high voltages impelled the investigations to focus on the dielectric properties of vacuum. Ever since, the vacuum as insulation has gradually found its application in electronic valves, microwave tubes, electron microscopes, Klystrons, photocells, particle accelerators and separators, controlled nuclear fusion devices, and so on. On the other hand, vacuum insulation is applied in high voltage apparatus such as electrostatic generators, low-loss or loss free capacitors, power circuit breakers as well as for outer space applications.

Compared to dielectric gases, an ideal vacuum has the highest possible electrical breakdown strength because there are no charge carriers to conduct the current. However in practical conditions, the electric field intensity in vacuum generates charge carriers by literally pulling them out from the electrodes and dielectric materials, Bugaev et al. [5.2]

At extremely low gas pressures, the impact ionization process by electrons becomes inadequate to cause a breakdown. It is because the "mean free path" of an electron, defined as the distance an electron can travel without colliding with another particle, is very long. In vacuum better than 10^{-4} Torr (1.333×10^{-2} Pa), less than $3 \cdot 10^{12}$ molecules per cm^3 are estimated to be present. The length of the mean free path of an electron in this condition is in the order of meters. In such a vacuum, an electron may cross a gap of a few *cm* between two electrodes without making any collision. Therefore, unlike in gases, in vacuum the initial process of breakdown cannot be due to the formation of electron avalanche. The process of multiplication of charged particles by collision in the space between the electrodes is far too insufficient to create avalanches. However, if a gas cloud is able to form in the vacuum, the usual kind of breakdown process takes place. Thus investigations of the breakdown in vacuum have oriented toward the mechanisms of the formation of gas clouds in vacuum.

High Voltage and Electrical Insulation Engineering, First Edition. Ravindra Arora, Wolfgang Mosch.
© 2011 Institute of Electrical and Electronics Engineers. Published 2011 by John Wiley & Sons, Inc.

An excellent review of the historical development of the work on breakdown phenomena in vacuum is made by Farrall [5.1]. A vast bibliography of references on electrical discharges in vacuum from 1877 through 1979 is given by Miller [5.3]. From decades of experimental work, it has been established that the "field electron emission process" occurs at isolated metallic microprotrusions, also known as "whisters". It was understood that the field in the gap having such whisters could be geometrically enhanced to about 3×10^4 kV/cm, which is responsible for the appearance of pre-breakdown currents as well as for maintaining the development of the succeeding discharge processes [5.4]. However, the work by Cox in the mid-1970s provided the first direct evidence of the existence of an alternative emission mechanism, as reported by Latham [5.5]. This alternative mechanism suggests that the pre-breakdown currents that flow between vacuum insulated high voltage electrodes frequently originate from a nonmetallic emission mechanism. These are associated with some kind of insulating/semiconducting surface oxides or impurity concentrations. From the technological point of view, the microscopic conditions of electrode surfaces continue to play an important role in providing the required charged particles.

Breakdown in vacuum is a rather complicated phenomenon involving a large number of processes at high electric field intensities. The "field emission" process, that is, the electron emission from metallic surfaces in the presence of strong electric fields, established itself through a considerable amount of work performed in the twentieth century. After this a new process, a form of complex nonmetallic emission mechanism has made its breakthrough to explain the pre-breakdown conduction. Both of these mechanisms are described below.

5.1 PRE-BREAKDOWN ELECTRON EMISSION IN VACUUM

When the voltage across a very small gap (a few *mm*) in vacuum is increased sufficiently, a relatively steady conduction current begins to flow. A general observation made for small gaps is that the pre-breakdown current flow has been found to consist mainly of electrons [5.6]. For longer gap spacings (>1 cm), small pulse currents of millisecond durations and charges of the order of micro-Coulombs (micro-discharges) are measured. Such micro-discharges may exist without a steady current or superimpose themselves on the steady current. On raising the voltage further, the micro-discharges eventually give rise to a steady current [5.7]. The phenomenon of electron emission is categorized in two main mechanisms, the metallic and nonmetallic, described in the following sections.

5.1.1 Mechanism of Electron Emission from Metallic Surfaces

Large number of mechanisms are described in the literature that can produce electron emission from metallic surfaces under various conditions. These are, for example: thermionic emission; field assisted thermionic, also known as Schottky or T-F emission; field or cold emission; photoelectric emission; secondary emission caused by electron bombardment of an electrode and emissions caused by positive ion bombardment and metastable atoms [5.7]. Out of these, the most important mechanisms of electron emission in vacuum are due to field, field assisted thermionic, and thermionic emission. As reported by Meek and Craggs [5.6], the general interrelation of these three phenomena has been discussed by many authors described below.

The current density "J", from an electrode as a function of electric field "E" at the tip, is given analytically by the Richardson-Schottky equation for field assisted thermionic emission, and by the Fowler-Nordheim equation for field emission. These are suitable for consideration for pre-breakdown currents as follows:

(a) field assisted thermionic emission:

$$J = \frac{4\pi me(kT)^2}{h^3}\exp\left(\frac{-(\Phi - e^{3/2}E^{1/2})}{kT}\right) \tag{5.1}$$

where	m	-	the electron mass	$(9.106 \times 10^{-31}$ kg)
	e	-	the electron charge	$(1.602 \times 10^{-19}$ C)
	k	-	Boltzmann's constant	$(1.380 \times 10^{-23}$ J/K)
	T	-	the absolute temperature	(K)
	h	-	Planck's constant	$(6.630 \times 10^{-34}$ J.s)
and	Φ	-	work function	(eV)

Equation 5.1 represents the case when the electrons are emitted over the lower potential barrier.

At higher field intensity, an extended form of Schottky emission law prevails giving the current density as:

$$J = \frac{\pi q}{\sin \pi q}\frac{4\pi me(kT)^2}{h^3}\exp\left(\frac{-(\Phi - e^{3/2}E^{1/2})}{kT}\right) \tag{5.2}$$

where $q = \dfrac{C}{kT}$ and $C = \dfrac{he^{1/4}E^{3/4}}{2\pi m^{1/2}}$

(b) field emission with temperature correction:

The emitted current density for electron tunneling through the barrier is,

$$J = \frac{\pi l}{\sin \pi l}\frac{e^2E^2}{8\pi^2 h\Phi t^2(Y)}\exp\left\{\frac{-8\pi(2m\Phi^2)^{1/2}}{3heE}v(y)\right\} \tag{5.3}$$

where $l = \dfrac{kT}{d'}$ with $d' = \dfrac{eEh}{4\pi(2m\Phi)^{1/2}t(y)}$ and $y = \dfrac{(e^3E)^{1/2}}{\Phi}$

The functions $t(y)$ and $v(y)$ are slow varying and were tabulated by Miller in 1966 [5.8]. These are frequently regarded as constants. When T tends to zero, the term $(\pi l/\sin \pi l)$tends to be equal to one. Equation 5.3 then reduces to the well known Fowler-Nordheim form,

$$J = AE^2\exp\left\{-\frac{B\Phi^{3/2}v(y)}{E}\right\} \quad A/cm^3 \tag{5.4}$$

where $A = \dfrac{1.541 \times 10^{-6}}{\Phi t^2(y)}$, $B = 6.831 \times 10^9$

E is in V/cm and Φ in eV

and $y = 3.795 \times 10^{-4} \dfrac{\sqrt{E}}{\Phi}$

Equation 5.4 can be rewritten as,

$$\log\left[\frac{J}{E^2}\right] = -\log\left[\frac{1}{A}\right] - \frac{B\Phi^{1.5} v(y)(1/E)}{2.3026} \tag{5.5}$$

Since Φ is a constant and the terms $t^2(y)$ and $v(y)$ are also essentially constants, a plot of log (J/E^2) against $(1/E)$ is a straight line having a negative slope. This straight line test is often performed to assess the applicability of the Fowler-Nordheim relation to experimental data. Such typical plot of field emission currents for different gap spacings between clean parallel tungsten electrodes given by Alpert et al. in 1964 [5.9] are illustrated in Figure 5.1. The Fowler-Nordheim theory provides the means for calculating the total energy distribution of emitted electrons and is widely accepted. Both of these equations (5.3 and 5.4), have been verified by measurements of the total current emitted from a small hemispherical end of a field-emitting microscopic tip.

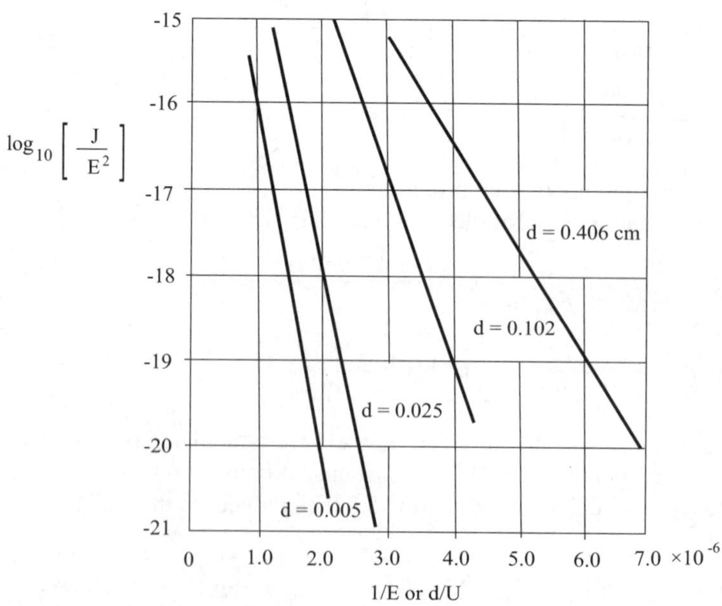

Figure 5.1 Typical Fowler-Nordheim plots of the field emission current between clean parallel tungsten electrodes for various gap spacing (d in cm), Alpert et al. [5.9].

From the form of equations (5.3) and (5.4), it is clear that small changes in local work function or electric field due to surface irregularity or any other cause would result in large changes in the emitted current. Small changes in the local fields can be caused at high current densities by the formation of a negative space charge outside the emitter surface. The problem of calculating field emission current densities in the presence of space charge is very difficult as it involves complex geometries and irregular boundary conditions.

For gap spacing of less than 5 mm and voltages up to 100 kV, the majority of investigators have reported the presence of strong field dependent currents, mostly identified as electron currents. Initially these electron currents were thought to be due to field emission from micro-protrusions on the surface having enhanced local fields, but subsequent experimentation revealed many unusual phenomena. For instance, the electron currents were observed at low cathode field intensities of the order of 10^2 kV/cm, but according to equation (5.4), representing the standard field emission law, cathode fields of the order of 2 to 4×10^4 kV/cm were required for appreciable emitted current densities. Theoretical estimation of the requirement of such high impractical field intensities for field emission are possibly derived when micrometer, μm, or smaller dimensions of irregularities on the electrode surface in uniform fields are directly inverted (μ *to* M). Such an estimation of the enhanced field intensity in the dielectric appears to have been concluded simply mathematically without taking into consideration the field equalization effect by the comparatively very large electrode surface.

Several theories are put forth to explain the phenomenon of enhanced cathode emission that occurs for various reasons. The reasons given are, for example, oxide films and associated trapped positive surface charges, absorbed particles on cathode surface, local hot spots, field-assisted thermionic emission from low work function regions and cathode microprojections (protrusions).

Many physicists have investigated in-depth the problem of surface protrusions on metals. Three basic types of protrusions are characterized. These are micropoints caused by surface diffusion effects, gross surface deformations due to arc damage, local welding, cracking or microparticle bombardment and impurities in the materials. These protrusions may get blunted by heating or by ion bombardment, Chatterton in [5.6].

5.1.2 Non-Metallic Electron Emission Mechanisms

Pre-breakdown conduction currents between vacuum-insulated high voltage electrodes frequently originate from a nonmetallic mechanism. These are associated with some form of insulating/semiconducting oxide layer on the surfaces or impurity concentrations. The microinclusions present on electrode surfaces can stimulate strong electron emission and significantly reduce the breakdown strength of the gap.

The phenomenon of "microdischarge", identified as one of the sources of pre-breakdown charge transfer, was first studied by Clifford at the University of London in 1950, Hawley in [5.7]. In vacuum, the microdischarges follow as self-extinguishing pulses of charge and their initiation is voltage/field dependent. A number of analytical and simulation works have been described by various authors in order to predict the conduction ion currents, Chatterton in [5.6]. In these theories the source of ions has also been attributed to the electrode surface impurities or contaminations. This was confirmed during the experiments on the measurement of threshold voltage for microdischarges.

As described by Halbritter [5.10] and Hawley in [5.7], large surface area electrodes, especially in high voltage vacuum equipment, are contaminated with oxides, adsorbates and dust. Most electrodes, even those of noble metals, are prone to form a thin layer of oxides or ionic complexes on their surfaces. Oxidation in vacuum takes place much more rapidly than under normal pressure conditions. Growth of a molecular layer of oxide on copper at a pressure of about 0.01 Pa of oxygen has been shown to be almost instantaneous. At a pressure of 1 mPa of oxygen, the estimated time for a layer of oxygen molecules to cover a surface is 0.06 s. Typical thickness of oxide layer coatings in such a case is of the order of 5 nm and that of adsorbates 10 nm.

If in the oil diffusion pump rubber O-rings and vacuum grease are used, the organic vapours are bound to be present. This gives rise to organic contamination that is deposited on the electrode surfaces. Porous oxide layer on the large surface areas results in drastic enhancement of H_2O and hydrocarbon adsorbates.

Even in completely sealed-off systems, the electrode surfaces may still carry some contamination. It has been discovered that when glass is heated to its working temperature, such as when sealing the electrodes into a closed cell, fluxes are vaporized from the glass. These are redeposited in the cool inner surfaces in the form of spherical particles up to a μm diameter. Any electrode surface in a sealed system may therefore carry such contamination resulting from particles that contain sodium, potassium and boron as well as traces of aluminium and silicon, Hawley in [5.7].

The oxides, adsorbates and dust contaminants on the electrode surfaces undergo changes when electric field is applied. For example, oxides and adsorbates undergo chemical reactions initiated by the impact of electrons, photons and ions. The "electron stimulated adsorption" (ESA), is accompanied with the desorption of H_2, CH_4 etc., Halbritter [5.10].

These contaminants on the electrode surfaces limit the maximum field intensity withstood by the system in three ways: (1) adsorbates and dust enhance the field emission of electrons; (2) oxides, adsorbates and dust enhance the secondary electron emission; and (3) oxides, adsorbates and dust show stimulated desorption of molecules and ions under the impact of electrons, photons or ions. Because of enhancement in electron emission process through these mechanisms, more electric field energy is converted into kinetic energy of electrons and ions leading to an increase in surface energy of the metal. Thus the electric field required for electron emission in the gap may reduce to the level of 10 kV/cm in contrast to the field intensity requirement theoretically estimated to be of the order of 10^4 kV/cm for the field-assisted electron emission process, as reported in the literature.

According to Latham [5.5], it was work by Cox (1975–79) that provided the first direct evidence of the existence of an alternative emission mechanism to the field emission. Cox used an anode probe-hole technique mounted within a scanning electron microscope. This way, a 10 μm diameter microregion of the electrode surface was able to be resolved that contained an emitting site. His observations consistently failed to reveal the presence of any metallic protrusions within the isolated microregion. Instead, the topography was typically characterized by the presence of insulating nodular features, possibly oxide structures, which were often in the vicinity of surface cracks or grain boundaries. Contour maps of current density distribution in the plane of the anode probe hole indicated emission sites to consist of several subcenters distributed over a distance of several μm on the plane of the cathode. The findings of Cox were later confirmed by Athwal and Latham in 1981. They also made topographical investigations of emission sites in scanning electron microscope using an anode probe technique to obtain still better spatial resolution of 2 to 3 μm. With the help of an ancillary X-ray spectrometer facility, it was further demonstrated that insulating inclusions, identified as emitting sites on copper electrodes, frequently contained a "foreign" element (Al, Ag).

Several theoretical interpretation models have been developed to provide the basis for the nonmetallic electron emission mechanism. But this phenomenon is so diverse that a range of physical mechanisms could occur with the complex surface condition existing on typical high voltage electrodes. The "Hot Electron Model", proposed by Latham and his co-workers in 1983 [5.5] is supposed to have provided the first physical interpretation of nonmetallic emission process. In 1986, they proposed to call this process "field induced hot electron emission" (FIHEE) mechanism. It involves a composite metal-insulation emission regime to explain the physical origin of pre-breakdown currents, Latham [5.11].

Based on experimental observations with a scanning electron microscope, it is assumed that the emitting microregime is an insulating, rather than semiconducting microinclusion, as shown in Figure 5.2 (a). The microinclusion is in intimate contact with the metal substrate of the cathode. On applying an external field, the electric stress penetrates the inclusion through the metal substrate. First it initiates a dielectric switching mechanism and subsequently accelerates or heats the electrons, which manage to cross the metal insulation interface. Only a very few of these electrons would be required in order to rapidly establish an electron avalanche through regenerative ionization process, also known as scattering process.

According to this model, the electrons initially tunnel from the metal through the interface potential barrier of region I, as shown by the band structure representation in Figure 5.2 (b). The electrons are then rapidly heated to cross the major portion of the insulating layer at the bottom of the conduction band in region II. Finally, the electrons are accelerated or heated through several eV in the high field surface region

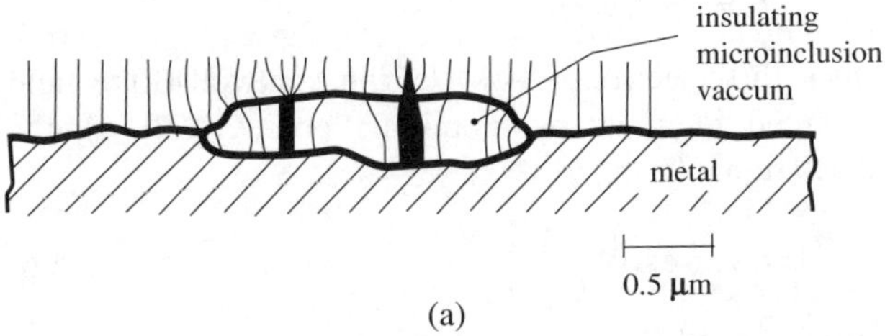

(a)

Figure 5.2 (a) A schematic representation of insulating microinclusion emission regime with conducting channels and the associated microscopic field enhancement, Latham, [5.5, 5.11].

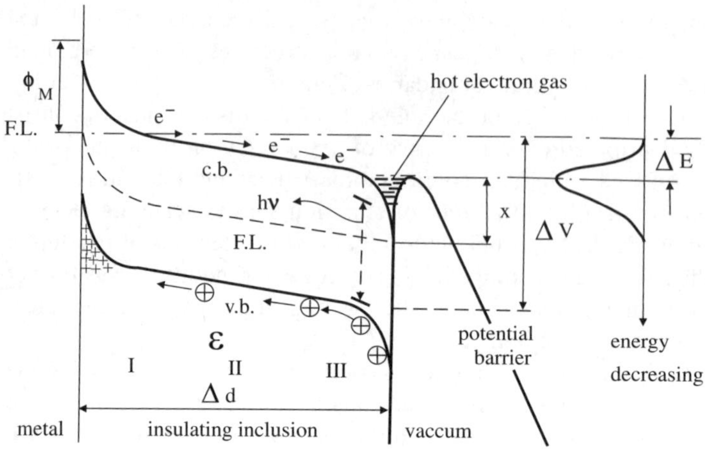

(b)

Figure 5.2 (b) A band structure representation of an insulating microparticle in "switched on state" of a conducting channel yielding electron emission. F.L.-Fermi Level, c.b. -conduction band, v.b.-valance band, Latham [5.5].

III to overcome the surface potential barrier at some energy ΔE below the Fermi level of the cathode substrate. In the high field region III, where the population of hot electrons exists, the electron-photon scattering processes occur, giving rise to electroluminescence phenomenon. Electrons and holes recombine via interband trapped states to release an optical photon "$h\nu$", as illustrated in Figure 5.2 (b). The holes created in this region migrate towards the metal-insulator interface and create a positive space charge domain that assists electrons to tunnel from the metal. It is precisely this physical presence of a tiny fraction of the metal insulator interface, found on the electrode surfaces, that allows the initial tunneling of electrons from the metal.

One of the most attractive features of this hot electron model described by Latham is to provide an alternative physical explanation to the well-known linearity of the Fowler-Nordheim plots of the pre-breakdown current-voltage characteristics. Assuming that the population of hot electrons in region III of Figure 5.2 (b), is similar to those existing on a metal surface at an enhanced temperature T_e, given by the relation,

$$T_e = 2e\,\Delta U/3k$$

Assuming further that these hot electrons also obey the Maxwell-Boltzmann statistics, so that they are emitted by a quasi thermionic process following a Richardson-Dushman type of relation,

$$J = KT_e^2 \exp\left[\frac{-e\chi}{kT_e}\right] \tag{5.6}$$

where K is a constant, appropriate to the present emitting regime and χ is the height of the surface potential barrier, that is, the electron affinity of the insulator.

From the dielectric considerations, ΔU is approximately given by $(\Delta d/\varepsilon d)U$ which leads to,

$$T_e = (2e\ \Delta d/3k\varepsilon d)U \tag{5.7}$$

and from equations (5.6) and (5.7), we obtain,

$$J = K\left[\frac{2e \cdot \Delta d}{3k\varepsilon d}\right]^2 U^2 \exp\left[\frac{-\chi}{(2\Delta d/3\varepsilon d)U}\right] \tag{5.8}$$

Equation (5.8) is similar to equation (5.4) having the well known Fowler-Nordheim form. The plot of log(J/U^2) versus $1/U$ yields a straight line with a slope,

$$m = (3\varepsilon d\chi/2\Delta d) \tag{5.9}$$

and an intercept,

$$C = \ln\left[K\left(\frac{2e \cdot \Delta d}{3k\varepsilon d}\right)^2\right] \tag{5.10}$$

Considering $d = 0.5\ mm$, $\varepsilon_r = 3$, $\chi = 4\ eV$ and $\Delta d = 0.1\ \mu m$ as being realistic practical values for these parameters, equation (5.9) predicts m to be approximately equal to 7×10^4. This is in good agreement with the value of $m \approx 10^5$, obtained from the typical single-site experimental plot shown in Figure 5.3.

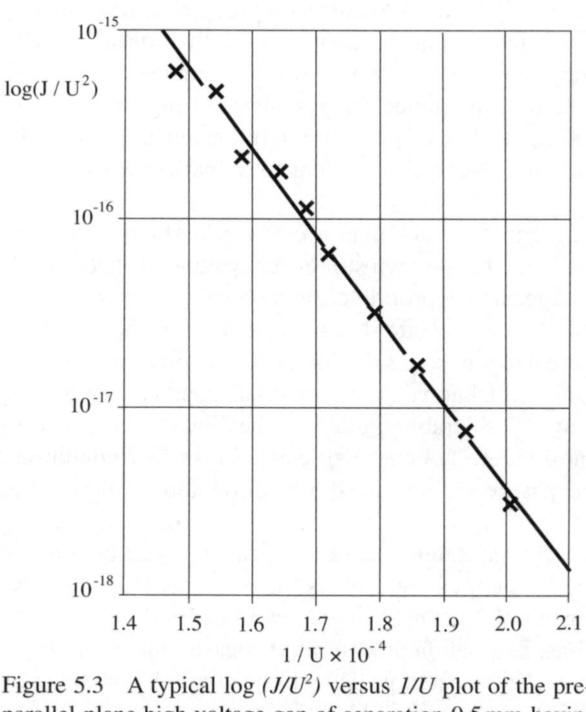

Figure 5.3 A typical log (J/U^2) versus $1/U$ plot of the pre-breakdown characteristics of a parallel-plane high voltage gap of separation 0.5 mm having only one electron emission site, Latham [5.5].

Further, K is assumed to have the value $1.2 \times 10^6 \text{A/m}^2\text{K}^2$, normally used in the Richrdson-Dushman equation. The emitting area of this site, responsible for the plot of Figure 5.3, is calculated from equation (5.10). It works out to be approximately equal to $10^{-4} \mu\text{m}^2$, a value physically realistic according to Latham.

This relatively new theoretical model, involving highly precision values, provides an insight into the origin of pre-breakdown emission. However, the exact physical nature of the microstructures, responsible for the nonmetallic emission process, remains doubtful. Improved higher resolution techniques are urgently needed to be developed for identifying and analyzing the microstructure of the sites. By now, it is well recognized that the pre-breakdown electron emission originates from microsized particulate structures, randomly located on electrode surfaces. The breakdown events generally result from such emission centers or sites, when they become unstable.

5.2 PRE-BREAKDOWN CONDUCTION AND SPARK BREAKDOWN IN VACUUM

The previous section established that the field emission from metallic surface as well as the electron emission from nonmetallic microinclusions on the metal surfaces are responsible for pre-breakdown conduction in vacuum. As the voltage is raised beyond the onset of the pre-breakdown phenomena, various conduction processes intensify. For example, in very small gaps (mm or lower range) with extremely clean electrode surfaces, where the field emission process is predominant, the conduction current increases and bright blue spots of light appear on the anode. This light emission is described to be due to "transition radiation", caused by the impinging electron beams into the dielectric (vacuum). At higher voltages, the local hot regions or the anode itself can get extremely heated up and begin to radiate due to thermionic activity, eventually leading to a spark breakdown of the gap.

On the other hand, in long gaps (greater than a *cm*) the field emission pre-breakdown currents are low. While the breakdown still occurs with a localized spark channel, field emission theories appear inappropriate for such gaps. Therefore, it is in this gap region that the micro-sized particulate structures, located on the electrode surfaces, could dominate the breakdown process. However, the precise limits of the field emission theories are not yet known. Even at small gaps (1–2 mm), the microinclusions, if present, may considerably influence the breakdown process. Experiments performed with most carefully controlled electrode surface conditions and for very small as well as very large gap spacings, may throw more light on this phenomenon.

It was proposed through early measurements on uniform and weakly nonuniform field conditions performed by various researchers that there existed a critical cathode field, "E_c", for each material, at which breakdown occurred due to field resistive heating at the protrusions. Estimation of E_c through measurements has been made for different electrode geometries, materials and gap spacings. A partial list of E_c, prepared by Chatterton in [5.6] is given in Table 5.1.

TABLE 5.1 Estimated critical breakdown field intensity, "E_c", measured experimentally for different materials, Chatterton in [5.6]

Material	Gap spacing (cm)	E_c (kV/cm)	Authors
W	0.0051–0.4	$6.5 \pm 1.0 \times 10^4$	Alpert et al. (1964)
W	0.035–0.125	$6.9 \pm 1.0 \times 10^4$	Kranjec and Ruby (1967)
Ni	0.01–0.1	$9.7 \pm 1.0 \times 10^4$	Kranjec and Ruby (1967)
Cu	0.003–0.1	$6.9 \pm 1.0 \times 10^4$	Bloomer and Cox (1968)
Cu	0.1	$6.3 \pm 0.8 \times 10^4$	Davies and Biondi (1968)
Mo	0.003–0.1	$5.4 \pm 1.0 \times 10^4$	Bloomer and Cox (1968)
Stainless Steel	0.003–0.1	$5.9 \pm 1.0 \times 10^4$	Bloomer and Cox (1968)
Stainless Steel	0.038–0.228	$7.8 \pm 1.3 \times 10^4$	Hackam and Salman (1974)

As seen from this table, there is a strong effect of electrode material on the critical breakdown field intensity in vacuum. As mentioned, such high impractical values of "E_c", could be the result of direct inversion of micro-dimensions into mega units of voltage.

Breakdown and pre-breakdown conduction current measurements have been widely performed but on very small gap lengths. However, in the high voltage power apparatus, for example, circuit breakers using vacuum as insulation involve relatively large and broad area electrode surfaces as well as longer gap spacings. These also involve other peculiar conditions of arc, electrode phenomena, and particle effects. In this respect, a large number of experimental works have been performed and published after 1990 [5.2, 5.12–5.15].

Pre-breakdown conduction currents and breakdown voltages with *ac* on needle and hemispherical rod-plane gaps up to 20 mm in vacuum were measured by Schmidt [5.16]. Pre-breakdown peak conduction currents as a function of the applied alternating 50 Hz voltage, measured on a 10 mm gap with needles/rod of different tip curvatures are shown in Figure 5.4. These experiments were performed in a stainless steel vacuum chamber with electrolyte copper electrodes. A vacuum of the order of 10^{-5} mbar (1 mPa) was maintained with the help of a turbomolecular pumping system. Satisfactory measurements of conduction currents could be performed only at a spacing of 10 mm and above because of the strong effect of radius of the electrode on pre-breakdown currents at smaller gaps. The electrode surfaces were polished with 400 grade emery paper and then washed with methanol.

As seen from the figure, the conduction current also increases exponentially in vacuum before the breakdown. The pre-breakdown currents measured were relatively steady but of different wave forms, depending upon the electrode radius and gap lengths. It can be seen that the magnitude of the conduction current depends upon the electric field intensity at the tip of the electrode. Since the electric field intensity at the sharp needle tip is highest, it has led to the same patterns and magnitudes of conduction current at lower voltages.

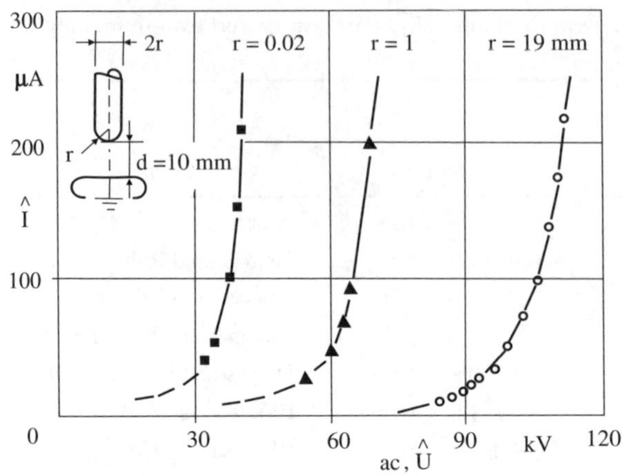

Figure 5.4 Pre-breakdown conduction current $\hat{I}$ as a function of the applied ac voltage $\hat{U}$ for different dimensions of needles and rod at constant gap distance, Schmidt [5.16].

On the same set of experimental arrangement described above, the peak breakdown voltage as a function of increasing gap distance up to 20 mm were also measured, as shown in Figure 5.5 (a). The electrodes were conditioned by a few hundred flashovers before measuring the breakdown voltage. These curves show that the breakdown voltages with sharp needle-like electrodes having a diameter of 2 mm do not appreciably change beyond a gap length of about 3 mm. This is a case of extremely nonuniform field configuration. On the other hand, the hemispherical rod with 38 mm diameter represents a weakly nonuniform field for the given gap spacings. The breakdown voltages in this case continuously increase with increasing gap distance.

From the measured breakdown voltage curves and the electrode geometry, Schmidt [5.16] also calculated the maximum/minimum field intensities at breakdown, $\hat{E}_{b\max}$ and $\hat{E}_{b\min}$ in the electrode system i.e. at the tip of the rod/needle as well as on the plane electrode, Figure 5.5(b). The Schwaiger factor, η, being very small for sharper tip electrode, results in much higher maximum field intensity at the tip of the needle electrode. These curves show that for small gap spacings ($d \leq 5mm$), the high maximum field intensity required for breakdown in both cases decreases rapidly with increasing gap distance, while the breakdown field intensity remains nearly constant for gap lengths above 5 mm. These calculated values also indicate that the required critical field intensity for field emission is achieved at the cathode (the plane) at very small gap lengths of a few *mm*.

Breakdown strengths of vacuum at different pressures in weakly nonuniform field under practical conditions were measured with *ac* power frequency and lightning impulse voltages by Khan [5.17]. These measurements involved stainless steel spheres of diameters 50 and 25 mm, hemispherical brass rod of diameter 18.5 mm and rounded edge stainless steel plane of 85 mm diameter. For each measurement, the glass vacuum chamber as well as the electrodes and the connecting shanks were first thoroughly cleaned with liquid soap, distilled water and then with acetone without bare hands touching any part. Figure 5.6 (a) shows the *ac* 50 Hz peak

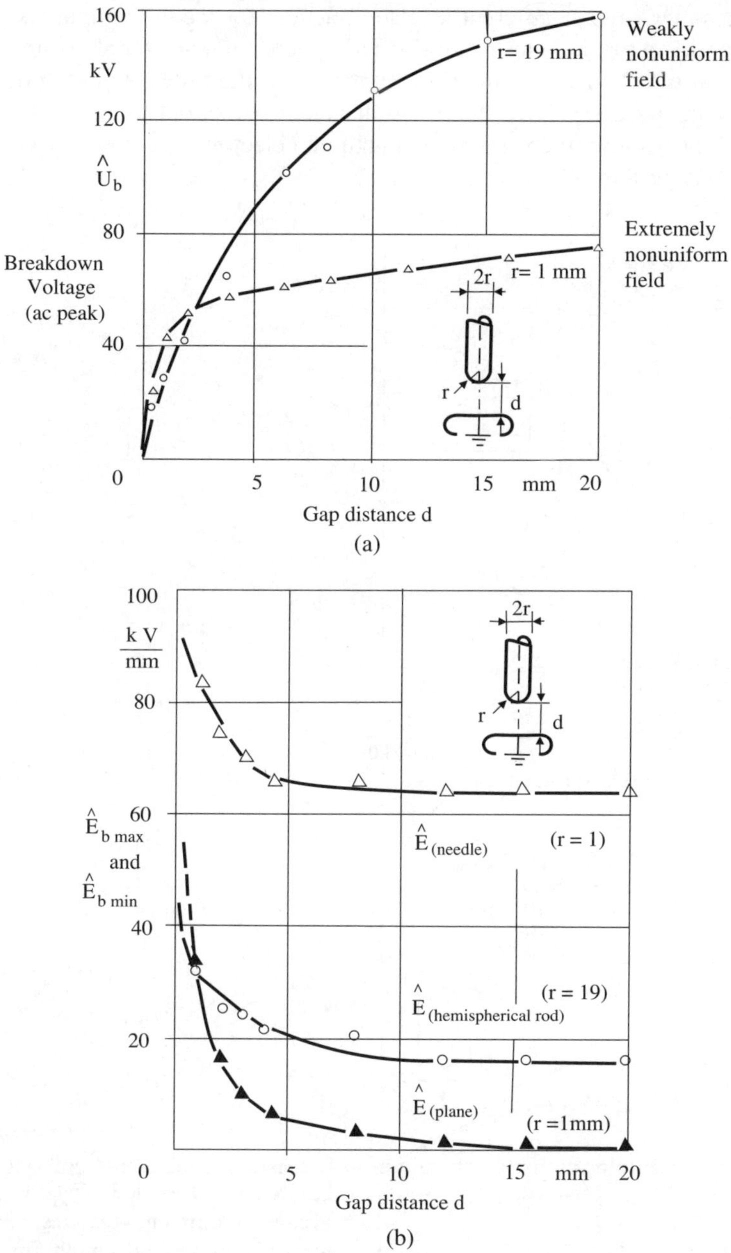

Figure 5.5 (a) ac Peak breakdown voltage for copper hemispherical rod/needle-plane electrodes at 1.0 mPa (7.5×10^{-6} Torr) with increasing gap distance. (b) Calculated field intensities at breakdown at the hemispherical rod/needle tip, $\hat{E}_{b\max}$, and at the plane, $\hat{E}_{b\min}$, for increasing gap distance, Schmidt [5.16].

breakdown voltages measured at different levels of vacuum for increasing gap distance up to 3 mm between two identical stainless steel spheres of diameter of 50 mm. The level of vacuum maintained in power interrupters is of the order of 10^{-6} Torr. The value of Schwaiger factor, η, for these gap distances works out to be greater than 0.95. Figure 5.6 (b) shows the variation in maximum electric strength at breakdown in the electrode system.

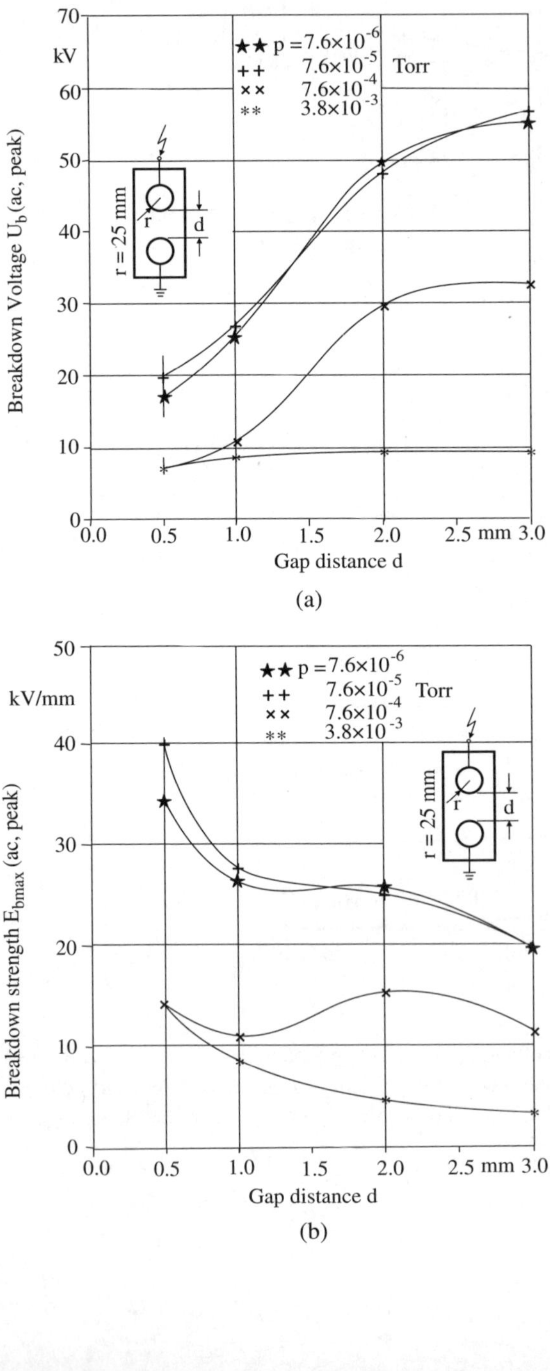

Figure 5.6 Breakdown in vacuum at different pressures measured in weakly nonuniform field between identical spheres of 50 mm diameter for increasing gap distance, (a) ac peak breakdown voltage, (b) maximum ac peak electric strength in the electrode system at breakdown, Khan [5.17].

Figure 5.7 shows the results of standard lightning, *li*, impulse breakdown voltage, U_{b-50}, measured at 7.6×10^{-6} Torr for both polarities on the same identical sphere electrode system as above. Unlike with *ac*, these measurements could be performed up to increasing gap distance of 30 mm. The estimated value of η at this gap distance is ~0.5. This confirms that the field between the electrodes at these gap distances was weakly nonuniform.

The measured results by Khan [5.17] are consistent with the measured results of Schmidt [5.16] at the pressure of ~7.5×10^{-6} Torr between hemispherical rod and plane also representing weakly nonuniform field. Both these measurements reveal that the breakdown strength of vacuum is very high, more than 30 $k\hat{V}$/mm for very short gap distances of a few *mm* between the electrodes. It reduces rapidly for longer gap distances, and acquires almost a flat characteristic for gap distances of 1 cm and above.

Breakdown characteristics with standard lightning and switching impulse voltages, *li* and *si*, are of interest not only for studying the physical process of discharge mechanisms in longer gap distances between broad area electrodes but also for the technical application of vacuum insulation in power systems. The electric strength of a gap in vacuum is mainly determined by the weakest point, which is difficult to establish on broad electrode surfaces.

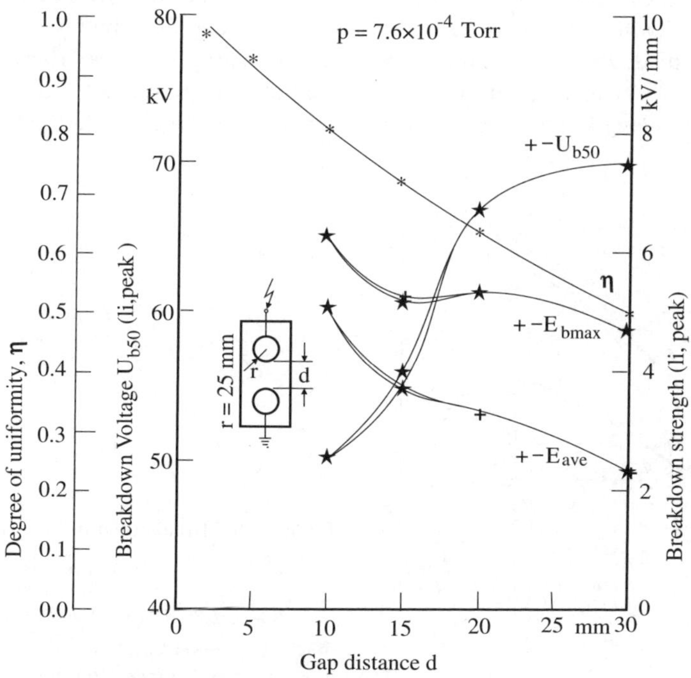

Figure 5.7 Characteristics of standard *li* impulse breakdown voltage, U_{b-50}, for positive and negative polarity, maximum field intensity at breakdown, $E_{b\max}$, and Schwaiger factor with increasing gap distances, between identical spheres at contant pressure of 7.6×10^{-6} Torr, Khan [5.17].

Breakdown in long vacuum gaps (cm range) with standard lightning impulse voltage 1.2/50 μs, were measured by Bender and Kärner [5.18]. A time lag in the range of 10 to 60 μs was discovered while measuring the pre-breakdown conduction currents as a function of the gap spacing for different electrode materials. The time lag measured had a tendency to increase with the gap spacings. It was also observed that the range of pre-breakdown conduction current peak values significantly depend on electrode materials. These values increase in the order of aluminium, stainless steel and copper.

Investigations mentioned above were performed in a vacuum of less than 0.1 mPa. The residual gases analyzed were mainly water vapor with small amounts of nitrogen and oxygen. The Borda profile shaped uniform field electrodes of stainless steel, copper and aluminium were chemically polished and finally cleaned with distilled water and methanol to obtain the desired surface conditions.

The 50% li breakdown voltage, U_{b-50}, is difficult to determine for an electrode arrangement in vacuum. This is because of the changing surface conditions of the electrodes due to spark after each impulse. In the experiments by Bender et al., the 50% breakdown voltages were estimated by the standard "up and down" method after thorough conditioning of the electrodes by sparks/flashovers.

Figure 5.8 shows the U_{b-50} li breakdown voltages measured with increasing gap distance in uniform field for identical Borda profile shaped electrodes of different materials. These electrodes were conditioned by *ac* sparks up to 130 kV. The extremely high breakdown voltages measured for Al electrodes are contrary to earlier investigations made with high direct as well as alternating voltages. These results contradict the effect of material as such on the breakdown strength. Al is normally considered to have a more rough surface than stainless steel and copper.

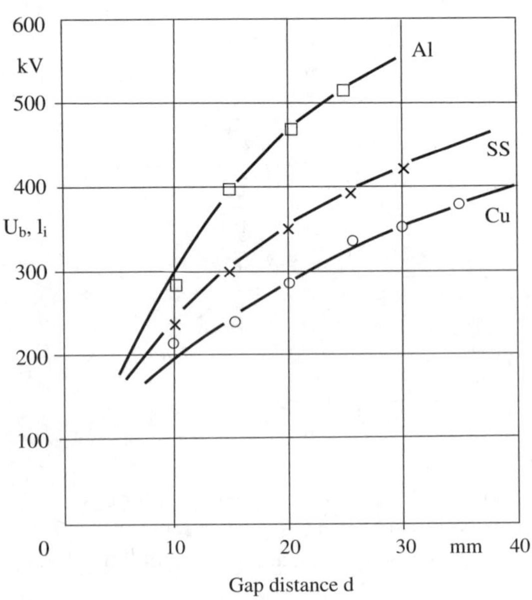

Figure 5.8 50% li breakdown voltage, U_{b-50} as a function of increasing gap distance after conditioning (with ac 130 kV) for Borda profile uniform field electrodes, Bender and Kärner [5.18].

In this case, better results could have been obtained because of excellent surface finish of the Al electrodes for investigations.

The measurements by Schümann et al. [5.19] also confirm a strong influence of the surface roughness on the breakdown strength in vacuum. Smoother surface conditions lead to higher electric breakdown strengths. Surface roughness of both, the cathode as well as anode electrodes affect the breakdown voltage.

Breakdown strength of gaps between different electrode arrangements in vacuum is a highly complex phenomenon. It is influenced by a wide range of parameters such as electrode materials, shape, size, surface area, surface finish, conditioning state of the surface and the vacuum level and its arrangement. A comparison of experimental results of different work is therefore not advised, Schümann et al. [5.19].

5.2.1 Electrical Breakdown in Vacuum Interrupters

For voltage levels up to 66 kV of power distribution system, there has been a rapid increase in the application of vacuum circuit breakers since 1980s. Compact size, reliability, long service life, low maintenance requirement and complete environmental compatibility are the advantages over oil and SF_6 gas circuit breakers. The high breakdown strength of short gap distances together with the high arc current breaking capability have enabled the design of compact arc quenching chambers using vacuum technology. Investigations have revealed that the electrical breakdown in vacuum interrupters is strongly influenced by the stability of the arc, the contact design, their material and the effect of particles in the gap, Farrall [5.13].

5.2.1.1 High Current Arc Quenching in Vacuum The short circuit current ratings of 36 kV vacuum circuit breakers have increased to 50 kA and above. Copper-chromium is the commonly-used contact material in these breakers. The arc initiated at contact separation could have two different states of existence, the diffused and the constricted modes. At low currents, the arc in vacuum is diffused. Charge carriers produced by cathode spots distribute randomly over the contact surface of the breaker. Under such conditions, the dielectric recovery in the gap that could withstand the transient switching over voltage is accomplished within ~10 μs, Dullni and Schade [5.20].

At higher currents, the arc is constricted and locally heats a thin layer of the electrode surface well above its melting temperature due to excessive heating leading to the vaporization of the contact material. The density of metal vapor produced by the arc determines the dielectric strength recovery and hence the breakdown process of the gap after the current zero. The liquid droplets and residual plasma play a minor role in this process. By melting the contact material and due to the effect of the plasma pressure, craters of μm dimensions are formed on the contacts in the vicinity of the anode spots. The molten metal is pushed aside and metal droplets are ejected into the vacuum.

Dullni and Schade [5.20] concluded through their investigations that the high current interruption process in vacuum is mainly determined by the molecular metal vapor formation by the arc itself. The molar mass of a gas is defined as the ratio of the mass of one molecule of the gas to that of an hydrogen atom under similar conditions. The vapor density in turn is expressed as the ratio of the mass of n molecules of the gas to the n molecules of hydrogen. Hence the vapor density is equal to half of the molar mass. If the vapor density after current zero exceeds $10^{22}\,m^{-3}$, the probability of breakdown due to transient voltage restrike is 100%. If the current is reduced so that the mean vapor density after current zero is $\approx 3\times10^{21}\ m^{-3}$, there exists a finite probability of a successful interruption. Thus the main parameter determining the recovery of arc is the presence of metal vapor density. Therefore, it is essential to keep the heat load of the contacts as low and uniform as possible. One way to achieve this is by superposition of the magnetic field produced by high current with strong axial component. It keeps the arc in a diffused state even at high current values. Another way is by construction of such contact geometries that produce radial magnetic fields. These drive the arc away towards the periphery of the contact electrodes by magnetic forces distributing the heat of the arc all over the contact surface.

5.2.1.2 Delayed Re-Ignition of Arcs High voltage vacuum circuit breakers occasionally go through the phenomenon of delayed re-ignition of arcs after successful interruption of the short circuit currents. Such re-ignition of arc could occur due to the loss of dielectric strength with a delay after extinction of the arc at current zero.

Investigations by Gebel and Hartmann [5.21] revealed that this phenomenon is caused by particles that adhere to the contact electrodes and vapor shield surfaces. These particles are released due to the strong mechanical shocks the vacuum bottle goes through during the breaker operation.

Measurements of the delay in re-ignition after current interruption were carried out as a function of the mechanical and electrical stresses exerted on the vacuum circuit breaker bottles. A correlation between the occurrence of strong, shock-like mechanical acceleration peaks of $>10^4\,m/s^2$ and the increase in the probability of delayed re-ignition could be established. By simple mechanical damping of the shock wave amplitudes on the breaker structure, the probability of the delayed re-ignition of the arc reduced by as much as 50%. Because of the standard careful baking and conditioning processes in modern production lines of industrial manufacture, such a phenomenon is rare at nominal voltage, Gebel and Hartmann [5.21]. It could occur due to high restrike voltages across the electrodes after successfully quenching arcs.

5.2.1.3 Effect of Insulator Surface Phenomena The dielectric properties of vacuum not only depend upon the electrode surface but are strongly influenced by the insulator material surfaces used in the apparatus, Gollor and Rogalla [5.22].

The triple junction between the electrode, vacuum and the solid dielectric formed at the cathode and the anode have inevitable microscopic irregularities responsible for primary electron emission. The secondary electron emission, triggered by high energy particle impact, may also contribute to charge carrier multiplication leading to enhancement of surface charge on the insulator. The accumulation of charge results in increased local electric field intensity and the occurrence of surface discharge or tracking phenomenon. It may lead to a flash over along the surface of the insulator at much lower voltage compared to the breakdown strength of the gap in vacuum.

Tracking resistant solid dielectric materials are desirable for spacers. Such materials have low gas desorption or the outgassing rates from their surface. The suitable materials are, for example; alumina-filled epoxy, polymethylmethacrylate (PMMA), polytetrafluroethylene (PTFE). Ceramic materials such as alumina is used very widely in vacuum interrupters. It has low outgassing rate, good thermal and mechanical properties. Since it has excellent radiation resistivity, it has found its application for space borne electrical systems. The other materials used are polymers, silicones polyurethanes, polycarbonates and polyamides for special purpose application.

5.2.2 Effect of Conditioning of Electrodes on Breakdown Voltage

Millikan and Sawyer in 1918 first observed that electric breakdowns (spark breakdowns) of frequent recurrence on an electrode gap in vacuum increase the breakdown voltage gradually until it reaches a plateau. The electrode gap is then said to be "conditioned". This phenomenon has since been further investigated and confirmed by other researchers.

Since the breakdown strength of vacuum is relatively very high, smaller gap distances have been investigated in scientific work. For smaller gap distances, the condition of the electrode surface and its roughness play a more important role. As accepted in a general sense, the conditioning is a process through which the micro-emission sites are supposed to be destroyed or blunted. For improving the capacity of the gaps in vacuum to withstand the voltage, the process of conditioning of electrodes is of immense practical importance.

Various scientists have proposed and studied a number of methods for conditioning electrodes. According to Hawley in [5.7], the method of conditioning that has been found to give the most consistent results is to treat the electrodes by means of hydrogen glow discharge, the Faraday glow described in Section 3.2.5. Other methods of conditioning consist of allowing the pre-breakdown currents in the gap to flow for some time or to heat the electrodes in vacuum to a high temperature besides the slow method of treating the electrodes with repeated spark breakdowns.

For hydrogen-conditioned electrodes, the number of spark breakdowns required to reach a plateau are about 10, whereas for untreated electrodes it can be as many as 10,000. As Hawley describes further, in some cases the conditioning is not permanent. If the voltage is removed for a considerable time or the electrodes are exposed to the atmosphere, then the next breakdown value measured is again much lower. This phenomenon is understandable, considering the fast growth of molecular layers of oxide on some electrode materials in vacuum as well as in atmospheric conditions. Hence it is important for equipment that has a vacuum chamber that after the electrodes are conditioned, they are not exposed to the atmosphere.

Bender and Kärner [5.18] also investigated the effect of different conditioning procedures on the breakdown voltage of copper electrodes as shown in Figure 5.9.

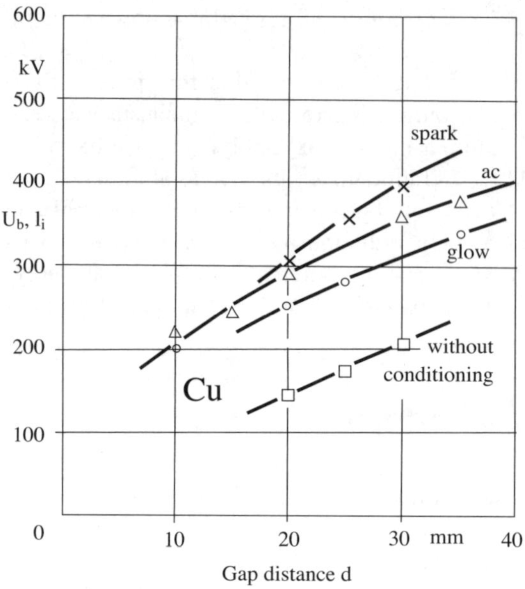

Figure 5.9 Effect of different conditioning procedures on the electric strength of copper electrodes, Bender and Kärner [5.18].

Argon glow discharge conditioning was performed with argon in the Paschen's minimum at alternating conduction current of 200 mA for 20 minutes. Under the high conductivity of the medium, the electrode surfaces are bombarded with heavy positive ions that remove the emission sites and adsorb gas layers in the process. This improves the breakdown voltage. The second method of conditioning was by subjecting the gap to a number of breakdown sparks by *ac* up to 80 kV. The third method chosen was by spark breakdown with breakdown currents up to 2 kA initiated by high impulse voltage. Out of these, the third method is time consuming, especially for broad electrodes. This method is also presumed to give rise to new emission sites on the surface due to heavy erosion by the arcs.

As shown in Figure 5.9, Bender and Kärner [5.18] believed that more intensive *ac* and Faraday glow discharge conditioning may further improve the impulse breakdown voltage characteristic.

5.2.3 Effect of Area of Electrodes on Breakdown in Vacuum

As in other high voltage insulation systems, in vacuums the effect of area of the electrodes on the breakdown voltages has been observed. For example, lower breakdown voltages are measured if the area of the electrodes is increased for the same gap distance in uniform field.

The effect of area of electrodes, using a single ceramic spacer in the gap in vacuum, was experimentally investigated by Juchniewicz et al. [5.23]. Two identical stainless steel plane electrodes were given Rogowski profile. The cylindrical ceramic spacer (insulator) had a diameter of 8 mm and a length of 5.7 mm. Thus, for a gap spacing of 5.7 mm, the maximum diameter of electrodes with which the experiments

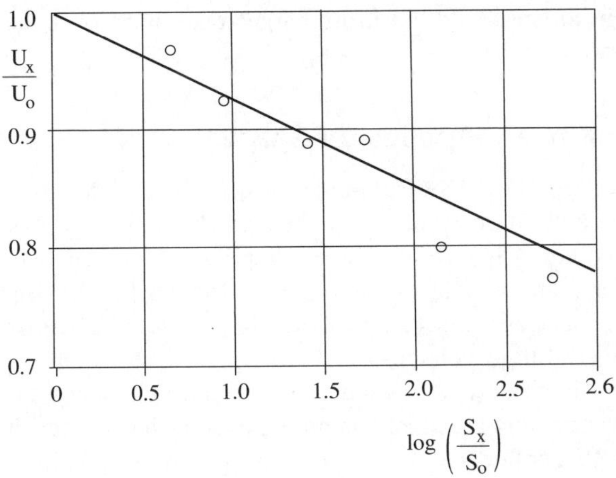

Figure 5.10 Effect of electrode area on breakdown voltage for Rogowski profile stainless steel electrodes, Juchniewicz et al. [5.23].

could be performed was 190 mm. A pressure lower than 8 mPa was maintained and the experiments were performed with ac 50 Hz. Special care was taken to achieve the conditioning of the electrodes. The measured results are shown in Figure 5.10. The related breakdown voltage is shown as a function of related electrode area. The voltage U_o represents the breakdown voltage with the electrode area having the same diameter as that of the spacer S_o (8 mm).

The straight line in Figure 5.10 can be related as follows:

$$\frac{U_x}{U_o} = \left[\frac{S_x}{S_o}\right]^{-a} \tag{5.11}$$

where the value of a in this case is equal to 0.036, representing a good correlation with the measured values under given conditions.

The experimental investigations made by Schümann et al. [5.19] in 2003 also confirm that the effect of area on breakdown voltage is valid for electrode arrangements in vacuum. The breakdown strength decreases with increasing electrode diameter or the area. The effect is attributed to the enhanced probability of the presence of micro-irregularities and particles on larger area of the electrode surface that could initiate a breakdown.

5.3 VACUUM AS INSULATION IN SPACE APPLICATIONS

There are two separate areas in which the performance of vacuum has been investigated for its application in outer space. Vacuum as insulation has found increasing application in spacecrafts and satellites for supplying power at high voltages since 1980s. Additionally, the outer surface of the spacecraft bodies coated with dielectric materials for their good thermal insulation properties give rise to peculiar problems in outer space. In low earth orbits at pressures of the order of 25 Torr and below where the ambient plasma environment could exist, discharges take place over the insulated surface of the spacecraft.

5.3.1 Vacuum-Insulated Power Supplies for Space

Increasing requirement of higher levels of electrical power in space-borne electrical systems—satellites, space ships and space shuttles—has led to the application of vacuum as insulation at high voltage. Spacecraft power supplies are designed to supply power at voltages higher than even 15 kV, Gollor and Rogalla [5.22]. The gadgets, for example, analyzers, spectroscopes, cameras, traveling wave tubes, radar systems, and so on, require power at high voltages.

Unlike on earth, the dielectric properties of vacuum as insulation in space are determined by the space environment and the local contaminations of the spacecraft structures. The specific aspects that influence are temperature, pressure, radiation, charge particle intensity, outgassing products and effluents. Gas desorption and material outgassing processes from surfaces are strongly temperature dependent. Due to the drastic thermal cycles in the space environment such phenomenon becomes important. The internal pressure increases when the satellite moves from the night to the daylight zone. The equilibrium state between material outgassing, desorption and gas diffusion into the space as a rule is responsible for the fact that in practical satellite operation the internal gas pressure often varies in the range of 10^{-4} to 10^{-6} Pa although the external pressure in the space is orders of magnitude lower, Gollor and Rogalla [5.22].

As the power and voltage demand levels in spacecrafts increase, a vacuum insulated system is advantageous due to its light weight. For studying the earth's atmosphere from the space a carbon dioxide gas laser may require to operate at voltages higher than 100 kV. There are two variations for the high voltage power supplies in this field. These are an open construction in which the space vacuum becomes part of the total insulation system, or application of a complete encapsulated high voltage system. The latter is considered to be safer as the first one may have the risk of uncontrollable discharges and flashover behavior in the space environment. In space, there is always a presence of radiation and charged particles. There may also be impact of meteoroids, dust particles and debris that could trigger surface discharge. However, the open construction of 15 kV TWT amplifier used on the European Radar Satellite, ERS-1, has shown a successful operation. It has demonstrated that the space vacuum insulation is an interesting alternative to achieve reliable, flexible and lightweight high voltage power supplies for space applications, Gollor and Rogalla [5.22].

5.3.2 Vacuum Related Problems in Low Earth Orbit Plasma Environment

The outer surface of the spacecraft body is coated with various thermal insulating materials to maintain the desired temperature inside. For such coatings, dielectric materials are often used because they also possess good thermal insulation properties. The coating layer over the body forms a capacitance in between the outer space

Figure 5.11 An astronaut busy in working on the electrical gadgets during space walk, NASA.

and the spacecraft body. The spacecraft body acts as a ground for its electric power supply equipment inside. The negative end or neutral of the power supply is usually connected to the body. Therefore, the spacecraft body behaves like a cathode and the plasma environment an anode. In case of solar arrays, the negative end of the array itself has negative potential, a cathode, Mengu Cho et al. [5.24]

As described in Section 3.2.5, Faraday glow discharge takes place in the ambient pressure range of 22 to 1×10^{-4} Torr when cathode rays gain energy from the applied electric field. In the low earth orbits of satellites and spacecrafts, they must be encountering the ambient pressure in this range and hence the problems of high conductivity plasma due to glow discharge. Such problems are also faced by spacecrafts returning back to the earth after their journey in space, while crossing the ambient pressure regions mentioned above.

5.4 CONCLUSION

A fairly vast amount of experimental and analytical work has been carried out to investigate the phenomena of electron emission in vacuum. The emphasis on old and classical mechanisms of electron emission from metallic surfaces has gradually shifted towards nonmetallic electron emission mechanisms. Nonmetallic contamination on electrode surfaces such as oxide layers, adsorbates and dust not only enhance the electron emission but also affect the breakdown strength. Modern experiments conducted under high resolution scanning electron microscopes have revealed significant knowledge in this respect.

The pre-breakdown conduction as well as breakdown mechanisms get influenced to a great extent by physical and experimental conditions. The most significant of these are the gap distance, electrode material—their size and geometry—treatments of their surfaces, presence of microinclusions and adsorbates; conditioning of the electrodes and above all, the residual pressure in the gap. However, breakdown in vacuum is a highly complex phenomenon. Comparison of experimental results of different research work is therefore not advisable.

Initial research in the past was more concentrated on very small, *mm* range, gap distances. Investigations with broad area electrodes and longer gap spacings, applicable to high voltage power systems picked up in the late 1970s. Fairly vast research work has been performed since the 1990s. Application of vacuum in high voltage interrupters and power supplies for spacecrafts has given rise to tremendous research work in these areas.

REFERENCES

[5.1] Farrall, G.A., "Electrical Breakdown in Vacuum", IEEE Trans. EI-20 (1985) 5, pp. 815–841.

[5.2] Bugaev, Sergei P. and Proskurovsky, Dmitry I., "Electrical Breakdown and Insulation in Vacuum", IEEE Trans. on Dielectrics and Electrical Insulation, Vol. 2 No. 2, April 1995.

[5.3] Miller, H. Craig, "Electrical Discharges in Vacuum 1877–1979", IEEE Trans. on Electrical Insulation, Vol. 25 No. 5, October, 1990.

[5.4] Fursey, G.N., "Field emission and vacuum breakdown", IEEE Trans. EI-20 (1985) 4, pp. 659–670.

[5.5] Latham, R.V., "Pre-breakdown electron emission", IEEE Trans. EI-18 (1983) 3, pp. 194–203.

[5.6] Meek, J.M., Craggs, J.D., "Electrical Breakdown of Gases", John Wiley and Sons (1978), Chapter 2, Vacuum Breakdown by Chatterton, P.A.

[5.7] Alston, L.L., "High Voltage Technology", Oxford University Press (1968).

[5.8] Miller, H.C., "Values of Fowler-Nordheim Field Emission Functions: v(y), t(y) and s(y)", Journal of Franklin Institute, Vol. 252, No. 6, December 1966, pp. 382–388.

[5.9] Alpert, D., Lee, D., Lyman, E.M. and Tomaschke, H.E., Journal of Vacuum Science and Technology, (1964) 1, 35.

[5.10] Halbritter, J., "On Contamination of Electrode Surfaces and Electric Field Limitations", IEEE Trans. EI-20 (1985) 4, pp. 671–681.

[5.11] Latham, R.V., "A New Perspective of the Origin of Prebreakdown Electron Emission Process", IEEE Trans. EI-23 (Feb. 1988)1, pp. 9–16.

[5.12] Fröhlich, K., Kärner, H.C., König, D., Lindmayer, M., Möller. K. and Rieder, W., "Fundamental Research on Vacuum Interrupters at Technical Universities in Germany and Austria", IEEE Trans. on Electrical Insulation, Vol. 28 No. 4, August, 1993, pp. 592–606.

[5.13] Farrall, George A., "Electrical Breakdown in Vacuum Interrupters", IEEE Trans. on Electrical Insulation, Vol. 28 No.4, August, 1993, pp. 580–591.

[5.14] Wetzer, J.M., "Vacuum Insulation: Fundamental and Applications", IEEE Trans. on Dielectrics and Electrical Insulation, Vol. 6 No.4, August 1999, p. 393.

[5.15] Jiyan, Z., König, D., Miller, H.C., "Vacuum Insulation, Fundamentals and Applications", IEEE Trans. on Dielectrics and Electrical Insulation, Vol. 9 No. 2, April, 2002, p. 163.

[5.16] Schmidt, Bernd-D., "AC Breakdown of vacuum gaps at high voltages", 3rd ISH, Milan (1979), paper 31.09.

[5.17] Khan, Md. Abdul Goffar, "Investigation of Insulating Properties of Vacuum under High voltage", Ph.D. Thesis, Department of Electrical Engineering, Indian Institute of Technology, Kanpur, India, December 1995.

[5.18] Bender, H.G. and Kärner, H.C., "Breakdown of large vacuum gaps under lightning impulse stress", IEEE Trans. on Electrical Insulation, Vol. 23 No. 1, February 1988, pp. 37–42.

[5.19] Schümann, Ulf, Giere, Stefan and Kurrat, Michael, "Breakdown Voltage of Electrode Arrangements in Vacuum Circuit Breakers", IEEE Trans. on Dielectric and Electrical Insulation, Vol. 10 No. 4, August 2003, pp. 557–562.

[5.20] Dullni, E. and Schade, E., "Investigation of High Current Interruption of Vacuum Circuit Breakers", IEEE Trans. on Electrical Insulation, Vol. 28 No. 4, August 1993, pp. 607–620.

[5.21] Gebel, R. and Hartmann, W., "Mechanical Shocks as Cause of Late Discharges in Vacuum Circuit Breakers", IEEE Trans. on Electrical Insulation, Vol. 28 No. 4, August 1993, pp. 468–472.

[5.22] Gollor, M. and Rogalla, K., "High Voltage Design of Vacuum-insulated Power Supplies for Sapce Applications", IEEE Trans. on Electrical Insulation, Vol. 28 No. 4, August 1993, pp. 667–680.

[5.23] Juchniewicz, J., Mazurek, B., Tymann, A., "On some size effects of high voltage vacuum insulation", 3rd ISH, Milan (1979), paper 31.02.

[5.24] Mengu Cho, Naoki Miyate, Masayuki Hikita and Susumu Sasaki, "Discharge over Spacecraft Insulator Surface in Low Earth Orbit Plasma Environment", IEEE Trans. on Dielectric and Electrical Insulation, Vol. 6 No. 4, August 1999, pp. 501–506.

CHAPTER 6

LIQUID DIELECTRICS, THEIR CLASSIFICATION, PROPERTIES, AND BREAKDOWN STRENGTH

From the point of view of molecular arrangement, a liquid can be described as "highly compressed gas" in which the molecules are very closely arranged. This is known as kinetic model of the liquid structure. A liquid is characterized by free movement of the constituent molecules among themselves but without the tendency to separate. However, the movement of charged particles, their microscopic streams, and interface conditions with other materials cause a distortion in the otherwise undisturbed molecular structure of the liquids. The well known terminology describing the breakdown mechanisms in gaseous dielectrics—such as impact ionization, mean free path, electron drift, and so on—are, therefore, also applicable for liquid dielectrics.

Liquid dielectrics are accordingly classified in between the two states of matter; that is, gaseous and solid insulating materials. Wide range of application of liquid dielectircs in power apparatus also characterizes this intermediate position of liquid dielectrics. Insulating oils are used in power and instrument transformers, power cables, circuit breakers, power capacitors, and so on. Liquid dielectrics perform a number of functions simultaneously, namely:

- insulation between the parts carrying voltage and the grounded container, as in transformers
- impregnation of insulation provided in thin layers of paper or other materials, as in transformers, cables and capacitors, where oils or impregnating compounds are used
- cooling action by convection in transformers and oil filled cables through circulation
- filling up of the voids to form an electrically stronger integral part of a composite dielectric
- arc extinction in circuit breakers

High Voltage and Electrical Insulation Engineering, First Edition. Ravindra Arora, Wolfgang Mosch.
© 2011 Institute of Electrical and Electronics Engineers. Published 2011 by John Wiley & Sons, Inc.

- achieve higher capacitance with liquid dielectrics having high permittivity in power capacitors

A large number of natural and synthetic liquids are available that can be used as dielectrics. These possess a very high electric strength and their viscosity and permittivity vary in a wide range. The appropriate application of a liquid dielectric in an apparatus is determined by its physical, chemical and electrical properties on the one hand, and on the other, it depends upon the requirements for the specific functions to be performed.

The application of liquid dielectrics in power apparatus has been gradually declining in the recent past by the countries adopting more advance technologies. The developments in SF_6 gas and vacuum technology since the 1970s are responsible for this trend.

6.1 CLASSIFICATION OF LIQUID DIELECTRICS

Dielectric materials can be divided into two broad classifications: organic and inorganic. Organic dielectrics are basically chemical compounds containing carbon. Earlier, under organic chemistry only those compounds, which were derived from either plant or animal organism, were considered. This concept underwent changes with concurrent developments in chemical technology. Carbon compounds, in general, are now called organic. Among the main natural insulating materials of this type are petroleum products and mineral oils. The most important and widely used organic liquid dielectrics for electrical power equipment are mineral oils. The other natural organic insulating materials are asphalt, vegetable oils, wax, natural resins, wood, and fiber plants (fibrins).

A large number of synthetic organic insulating materials are also produced. These are nothing but substitutes of hydrocarbons in gaseous or liquid forms. In gaseous forms are fluorinated and chlorinated carbon compounds. Their liquid forms are chlorinated diphenyles, besides some nonchlorinated synthetic hydrocarbons. The chlorodiphenyles, although possessing some special properties, are not widely used because they are unsafe for humans and very costly.

Among halogenfree synthetic oils are polymerization products, the polyisobutylenes and the siliconpolycondensates. Polyisobutylene offers better dielectric and thermal properties than mineral oils for its application in power cables and capacitors, but it is many times more expensive. Silicon oils are top grade, halogenfree synthetic insulating liquids. They have excellent stable properties, but because of being costly, have so far found limited application for special purposes in power apparatus.

Among inorganic liquid insulating materials, highly purified water, liquid nitrogen, oxygen, argon, sulphurhexafluoride, helium etc. have been investigated for possible use as dielectrics. Liquefied gases, having high electric strength, are more frequently used in cryogenic applications. Water and water mixtures are being actively investigated for use as dielectrics in pulse power capacitors and pulsed power modulators, and son on, because of their high relative permittivity, low cost, easy handling and disposal.

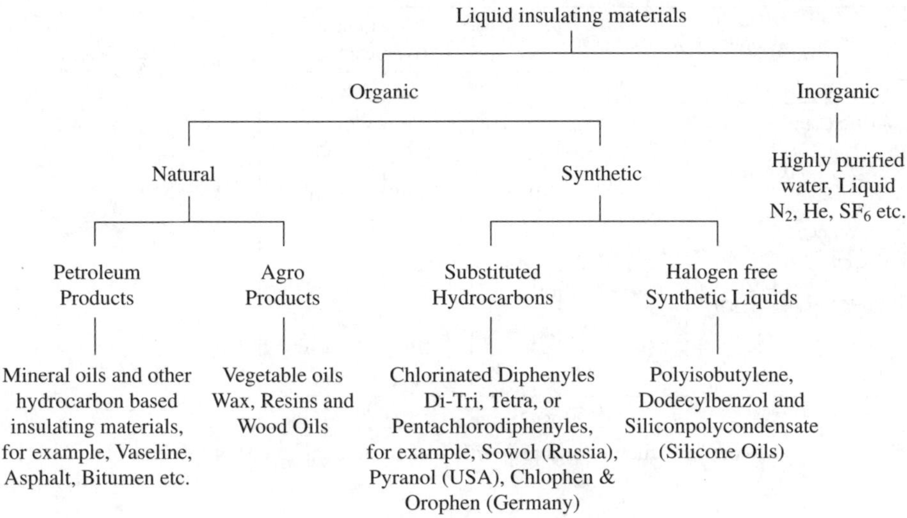

Figure 6.1 Classification of liquid insulating materials.

To summarize the classification of liquid dielectrics explained above, a schematic is drawn in Figure 6.1. The insulating liquids, commonly applied in high voltage apparatus, are classified as illustrated in this figure.

6.1.1 Mineral Insulating Oils

Amongst the liquid dielectrics used in power equipment, mineral oils are most important. Mineral oils are the suitably refined mixtures of different hydrocarbons obtained by fractional distillation of natural petroleum. The properties of individual oils strongly depend upon their chemical compositions. These oils mainly consist of saturated hydrocarbons of paraffinic and naphthenic (C_nH_{2n}) structures, besides having unsaturated aromatic hydrocarbons (C_nH_{2n-6}) in different proportions. The molecular structure and the range of percentage proportions of hydrocarbon bases are illustrated in Figure 6.2. An oil base is known as naphthenic or paraffinic when the content of either of them exceeds the other. If their contents are of equal amounts, the oil is known as mixed oil. Further, an oil is known as a weak aromatic when the aromatic content is less than 5%, and highly aromatic if its proportion exceeds 10%. Naphthenic mineral oils are more in use as dielectrics because they do not have a wax type consistency, which reduces their fluidity at low temperatures.

Electrical and other important properties of different insulating oils are brought together in Table 6.1.

Mineral oils are used in transformers, CTs, PTs, oil filled cables, condensers and circuit breakers. In the case of cables and condensers, high quality and low viscosity oils are preferred to achieve smoother circulation, hence a better convection of heat.

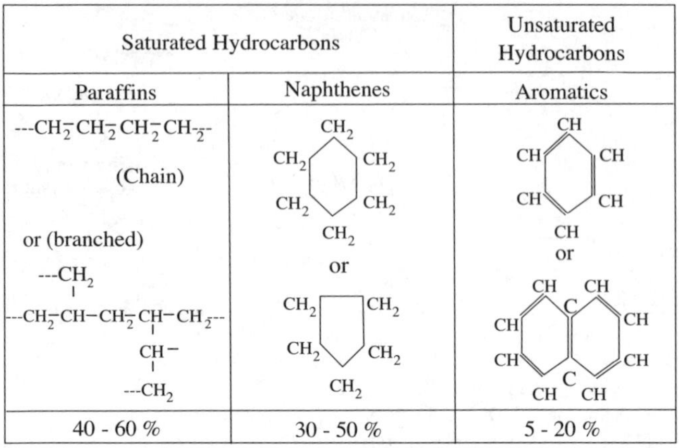

Figure 6.2 Molecular structure and proportion of basic constituent of mineral oils.

6.1.1.1 Mineral Insulating Oil in Transformers "Transformer oil", the petroleum-based mineral insulating oil is refined specifically to meet the requirements of electrical insulation and heat transfer for its application in transformers. The transformers in electric power transmission and distribution systems are expected to provide reliable and efficient service for their life expectancy of at least forty years. The quality of the oil in a transformer plays a very important role in satisfactorily performing its functions.

Petroleum-based transformer oils are the most widely used electrical insulating liquids in the world for the past century. The amount of oil contained in transformers varies significantly. It mainly depends upon the power rating of the transformer; for example, a 25 kVA transformer may contain about 20 gallons whereas a 400 MVA may contain more than 10,000 gallons (U.S.) of oil. According to an estimate made by Rouse [6.1], there were more than 30 million transformers in service just in the United States in 1998. The total volume of oil contained in these was estimated to be of the order of a billion gallons (U.S.). The different quality insulating oil sales in U.S. were 74 million gallons in 1996 as reported by "National Petroleum Refiners Association" in September 1997.

The crude oils are complex mixtures of many individual hydrocarbon compounds. The relative proportions of these hydrocarbons could be quite variable in crude oils from different sources. Like the crude oils, the compositions of acceptable transformer oils also vary substantially. The limits for concentration of the various hydrocarbons in transformer oils are not clearly specified by the standards. Some specified important physical and electrical properties are listed in Table 6.1. For transformer oils, viscosity and breakdown strength are the principal parameters for heat transfer and electrical insulation respectively required for design calculations.

6.1.2 Vegetable Oils

Besides mineral oils, some vegetable oils have also found their application in electrical equipment for their suitable properties. A large number of vegetable oils are

TABLE 6.1 Nominal/standard values of some properties of pure insulating liquids in high voltage applications [6.2, 6.3 and 6.5]

Insulating liquids	Relative permittivity ε_r	Loss tangent $\tan\delta$	Dynamic viscosity* Pa.s	Density at 20°C g/(cm^3) (kg/m^3 × 10^{-3})	Electric strength E_b kV/cm	Application
Mineral insulating oils (Transformer oil)	at 20°C 50 Hz ≈ 2.0	at 50 Hz 20°C ≤ 10^{-3} 90°C ≤ 4 × 10^{-3}	at 40°C 0.0067–0.0143 Tr, oil at 27°C 0.0243 (IS–335)	<0.895	350 … 500 Minimum value for transformers ≥300 For Circuit Breakers ≥175	Power Transformers CTs, PTs Circuit Breakers Bushings Cables and Condensers
Linseed oil	at 20°C 50 Hz ≈ 3.2	at 50 Hz 20°C > 10^{-3}	at 40°C ≈ 0.0260 at 100°C ≈ 0.0067	0.930	—	—
Castor oil	at 20°C 50 Hz between 4.2 and 4.5	at 50 Hz 20°C, <10^{-2}	at 40°C ≈ 0.2684 at 100°C ≈ 0.0192	0.96–0.97	175–250	Condensers
Chlorinated Diphenyles	at 20°C 50 Hz between 4–6	at 20°C, 50 Hz between 10^{-4}–10^{-3}	at 20°C ≈ 0.0600 at 90°C ≈ 0.0040	1.400–1.550	250–500	Transformers Condensers (prohibited in some countries)
Silicone oils	at 20°C 50 Hz ≈ 2.6	at 20°C 50 Hz < 10^{-4}	at 20°C 0.0096–0.9700	0.960–0.970	300–400	Cables Condensers Bushings

* Dynamic Viscosity (cP) = Kinematic Viscosity (cSt) × Density (g/cm^3), cP-centipoise, cSt-centistokes, Pa.s Pascal second, Pa.s = (N/m^2).s = 10 P.

available, for example castor, linseed, rapeseed, soya, groundnut, corn, olive, sunflower, mustard, clove almond, mangoseed, cottonseed oils, and so on. They are basically fatty acids accumulated in vegetable seeds. Chemically, these are ester compounds produced from sebacic acids and glycerine. On the other hand, the volatile vegetable oils, for example turpentine oil, have a strong odor and are produced from leaves, wood and roots of special plants. The higher the molecular weight of these oils, the more is the specific resistance and lower the dielectric loss tangent, ($\tan\delta$) [6.2].

Linseed, soya and castor oils are important components for the production of "oil modified alkaline resins". Such resins incorporate the advantages of oils to improve their elasticity as against the hard dried resins. Soya oil with epoxyresin is known as "softener" for some synthetic materials. Castor oil, which has hydroxide contents of about 5%, is an important polyisocyanide reagent. The chemical composition of this unsaturated oil is given as $C_{18}H_{32}OHCOOH$, [6.3], which has a high relative permittivity between 4.2 and 4.5. Castor oil has therefore found wide application as an impregnating agent in power capacitors. Using castor oil considerably reduces the size of the capacitors for a desired value of capacitance. Some of these oils are often mixed with suitable solvents in order to improve the soaking or impregnating properties. Turpentine oil can dissolve lac, and thus serves as a solvent. It also affects the drying and used for lac polishing because of high percentage of peroxide contents.

6.1.3 Synthetic Liquid Dielectrics, the Chlorinated Diphenyles

Among synthetic liquid dielectrics, chlorinated diphenyles have very good electrical properties for their suitable usage in power capacitors. Replacing two to five hydrogen atoms of each molecule by chlorine atoms produces these. Accordingly di, tri, tetra or penta-chlorodiphenyles are produced; for example, trichlorodiphenyle ($C_{12}H_7Cl_3$) and pentachlorodiphenyle ($C_{12}H_5Cl_5$). These dielectrics are also known by different commercial names in different countries; for example, Sowol (Russia), Pyranol (the United States), Chlophen, Orophen (Germany), and so on. Because of the electronegative effect of chlorine atoms, the chlorodiphenyles have comparatively higher relative permittivity number of the order of 5.6 at 20°C. The relative permittivity of paper is of the order of 6.8, therefore the impregnation of paper with these oils gives rise to a more uniform distribution of field compared to natural mineral oils that have relatively lower permittivity of the order of 2 to 2.3. For a desired capacitance the total size of the high voltage condenser is reduced considerably because of smaller volume of insulation thus required.

The chlorodiphenyles are highly stable and do not decompose easily under normal conditions. They are oxidation, flame/fire resistant and do not show other forms of aging. They do not endanger any explosion and cause no damage to metals. By mixing trichlorobenzol ($C_6H_3Cl_3$) their viscosity can be brought down to a level of 0.004 Pa.s at 90°C, which is desirable for the use in transformers as a cooling agent.

However, the chlorodiphenyles cause serious problems as they do have a number of undesirable properties. They appear to dissolve organic matter, but to a limited extent, and cause damage to such materials. They also get decomposed by electric arcs and above all, are injurious to health. By inhalation or by direct contact, these can cause serious health hazards. Compared to mineral oils, synthetic liquids are about ten times costlier. Because of the above mentioned behavior, the use of chlorodiphenyle insulating hydrocarbons is prohibited in some countries. In other countries, where these liquids are used for transformers and power capacitors, their use is restricted to completely sealed units in order to prevent any spillage. Moreover, a very strict control of their disposal is observed.

6.1.3.1 Halogen Free Synthetic Oils Among pure hydrocarbon synthetic liquid dielectrics, polyisobutane and dodecylbenzole have found their application in power cables and condensers. Polyisobutane, a chemically stable compound, is a polymerization product of butane of the series polyolefin. It is used as a high viscosity impregnating agent for mass impregnated power cables and condensers.

Dodecylbenzole is a low viscosity insulating material from the series Alkalinebenzole. It is used for low-pressure oil filled cables and also for condensers.

The most important halogen free synthetic insulating liquids are pure polydimethylsiloxane, as shown below,

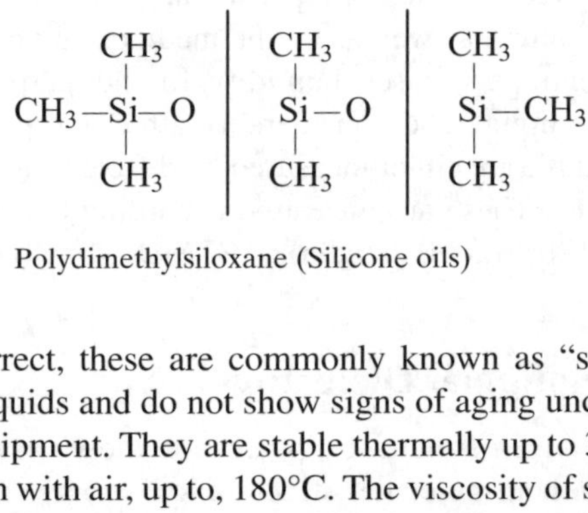

Polydimethylsiloxane (Silicone oils)

Although incorrect, these are commonly known as "silicone oils". Silicone oils are very stable liquids and do not show signs of aging under the working conditions of electrical equipment. They are stable thermally up to 300°C and chemically, that is under oxidation with air, up to, 180°C. The viscosity of silicone liquids depend upon their chain structure/length, "*n*", which is only marginally affected by the temperature of the liquid. Hence silicone oils with their viscosities in a very wide range are available, Table 6.1. Their relative permittivity is about 2.6 at 20°C but reduces to 2.3 at temperatures around 200°C. The dielectric loss tangent ($\tan\delta$) varies only between 10^{-4} and 2×10^{-4} in this wide temperature range of 20 to 200°C. Since silicone oils have a very low surface tension, these are water resistive, that is, water collects as separate drops instead of getting mixed with the oil. One can say in other words that silicone oils have a property to reject water. Therefore, even a thin layer of silicone oil on insulators protects them from water conductive contaminants. Silicone oils are used in power cables and condensers where they are required to be in service over a wide range of temperatures. An extensive use of silicone oils is restricted only because of their high cost.

6.1.4 Inorganic Liquids as Insulation

Extremely purified water has a relative permittivity of the order of 80 and comparatively possesses a high conductivity. Application of water as an insulation is therefore suitable only for high frequency pulse lines. Due to electrical discharge in water, high energy impulse waves and divergent currents are produced. By providing proper shape to the metal plates, these high energy impulse waves can be successfully used in communication for pulse-power capacitors, pulse-forming transmission lines, and switches, Zahn et al. [6.4]. Ethylene glycol is added to water as it maintains the high permittivity with a large resistivity and also allows low temperature operation without freezing.

There is a significant space charge injection into highly purified water when high voltage is applied. It causes anomalous voltage-current characteristics and distortions in the electric field. The nature (+ve or −ve) and the magnitude of space charge strongly depend upon the electrode materials. Thus, by appropriate choice of electrode material combinations and voltage polarity, it is possible to have uncharged, unipolar charged (negative or positive), or bipolar charged water. The recent charge injection analysis, terminal voltage-current measurements, and Kerr electro-optic field mapping measurements by Zahn et al. [6.4] have shown significant space charge effects on the electric field distribution in water.

In Table 6.1, nominal/standard values of some important properties of insulating liquids, used in high voltage equipment, are brought together as given in [6.2, 6.3 and 6.5]. Single values of these properties are not possible as they strongly depend upon the temperature, as well as on the methods and conditions of measurements. Moreover, the entrapped gases, humidity, foreign particles and above all the aging conditions of the liquid also considerably affect the properties. The electric strengths given in this table are either measured by different investigators or required by various specifications. These are measured according to VDE-0370 (Germany), BIS-335 (2005) and 1866 (2000) (India), ASTMD-3487 (USA) or IEC-296, as described in the text.

6.1.5 Polar and Nonpolar Dielectrics

The dielectrics, which have permanent dipoles, are known as polar. In such dielectrics, the asymmetry in their molecular structure leads to a permanent displacement of the positive and negative charge centers even without an external electric field. However, the field vectors in these isotropic materials are distributed in a way such that from outside the material behaves electrically neutral.

The polar dielectrics have a high relative permittivity ε_r. The relative permittivity number determines the extent of their being polar. Typical examples of polar dielectrics are, among liquefied gases; ammonia (NH_3) and hydrochloric gas (HCl), and among liquids; water (H_2O), propylene carbonate and $C_2H_5NO_2$. PVC and epoxy resins are the solid dielectrics belonging to the polar group. Highly purified water ($\varepsilon_r = 80$) and propylenecarbonate ($\varepsilon_r = 69$) are known as strongly polar liquids. In these liquids the enhancement of dissociation process by the applied electric field is very limited. Hence the dissociation rate in these can be considered to be constant. The polar liquids are often good solvents and are commonly used in electrochemistry.

The dielectrics, which do not build dipoles appreciably without an external electric field, are known as nonpolar. The dipoles formed in such dielectrics on applying an electric field are not permanent. In other words, in nonpolar dielectrics the atoms return back to their original state on removing the electric field. The liquids included in Table 6.1 may be regarded as nonpolar liquids.

6.2 DIELECTRIC PROPERTIES OF INSULATING MATERIALS

The electrical properties of liquid and solid dielectrics are described in the following sections.

6.2.1 Insulation Resistance Offered by Dielectrics

The *dc* resistance offered by an insulating material represents the concept of insulation resistance of a dielectric. It is generally described as specific insulation resistance "ρ_{ins}", which is reciprocal of the dc conductivity κ_{dc},

$$\rho_{ins} = \frac{1}{\kappa_{dc}} \quad \Omega \cdot m \tag{6.1}$$

Consider a direct voltage U_{dc} applied across two uniform field electrodes separated by a block of insulating material having an area A and length d as shown in Figure 6.3. From the equivalent circuit diagram constituting a capacitance C and a *dc* resistance R_{dc} in parallel, the following relation can be derived,

$$R_{dc} = \rho_{ins} \frac{d}{A} \tag{6.2}$$

$$i_{dc} = \frac{U}{R_{dc}} = \frac{U \cdot A}{\rho_{ins} \cdot d} \tag{6.3}$$

for a uniform field where $E = \frac{U}{d}$,

$$\begin{aligned} i_{dc} &= \frac{E \cdot A \cdot d}{\rho_{ins} \cdot d} = \frac{E \cdot A}{\rho_{ins}} \\ &= \kappa_{dc} \cdot A \cdot E \end{aligned} \tag{6.4}$$

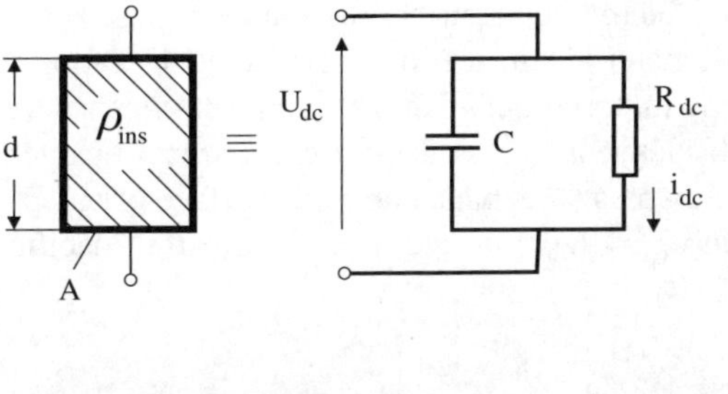

Figure 6.3 An insulating material and its equivalent circuit diagram.

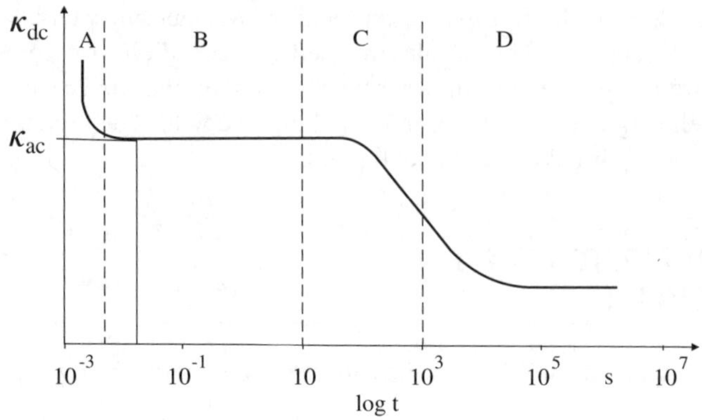

Figure 6.4 A schematic of dc conductivity of an insulating oil with respect to the time of applied voltage.

Like κ_{dc}, the specific insulation resistance also strongly depends upon the temperature.

The specific insulation resistance, or the *dc* conductivity is a function of time for which the voltage is applied. Schematic illustration of the variation of κ_{dc} with respect to the time of application of the direct voltage for an insulating oil is shown in Figure 6.4. The *dc* conductivity, which is very high initially (region A), is determined by the orientation of dipoles in the dielectric present even before the application of voltage.

In region B, the conductivity is determined by the movement of free charge carriers under the influence of applied electric field. The magnitude of conductivity in this region also represents the *ac* power frequency conductivity "κ_{ac}". The region C in the figure represents the development of space charge in front of the electrodes. The steady ion current due to dissociation is depicted by region D, which is achieved after considerable amount of time of application of the voltage.

In every dielectric matter, including those which have a low concentration of free charge carriers, a conductive current through the dielectric is always present on applying a voltage across the volume of the dielectric. Small currents may also flow along the surfaces of the dielectric. Accordingly, two different conductivities, the volume and the surface conductivities of a dielectric, are distinguished. Reciprocal of these are the specific volume and surface insulation resistances, respectively. The specific insulation resistance ρ_{ins} described above represents the specific volume resistance, "ρ_v", of the dielectric and it has a unit of Ω·m. The specific surface resistance, "ρ_s", has accordingly a unit of Ω only. These are properties of the material also known as volume and surface resistivities, respectively.

Surface Resistance "R_s" is defined as the dc electrical resistance between two electrodes in contact with an insulating material surface. It is derived as the ratio of the voltage applied to that portion of the material between the two electrodes to the current flowing between the two electrodes. Since it is the current that creeps over the surface of the dielectric, it is also known as "creepage current". Surface or creepage currents of solid dielectrics depend upon their surface resistivity or the specific surface resistance, "ρ_s", which is determined by the ratio of *dc* voltage drop per unit length to the surface current per unit width of the material.

The surface resistivity of solid dielectrics may have strong variation in its value influenced by the condition of the surface of the material. Temperature, moisture, pollution, composition of the pollution layer, hydrophobicity of the material, and so on, affect the performance of the surface considerably. The surface or the creepage currents of solid dielectrics depend upon their effective surface resistivity.

6.2.2 Permittivity of Insulating Materials

An important electrical parameter of insulating materials is the capacitance offered by them between given electrode systems. The capacitance thus formed is always accompanied with some losses determined by the permittivity "ε" and the specific insulation resistance ρ_{ins} of the material.

The permittivity of insulating materials ε is defined as the product of absolute permittivity of free space (vacuum) "ε_o" and the relative permittivity "ε_r" of the material or the medium,

$$\varepsilon = \varepsilon_o \varepsilon_r$$

The absolute permittivity of free space, ε_o, is constant and has a value,

$$\varepsilon_o = 8.854 \cdot 10^{-12} \quad F/m$$

On the contrary, the relative permittivity of a material "ε_r" is not a constant. It depends upon the thermal conditions of the material as well as the magnitude and frequency of the applied voltage. "ε_r" is often mentioned as dielectric number or permittivity number in the literature. "ε_r" of a dielectric is defined as the quotient of the capacitances C_x to C_o. Where C_x, the capacitance of a condenser, when the given material constitutes as the dielectric and C_o, the capacitance of the same condenser when vacuum constitutes as the dielectric,

$$\varepsilon_r = \frac{C_x}{C_o} \tag{6.5}$$

The phenomenon of polarization in a dielectric material under the influence of an external electric field can be explained on the basis of the displacement of charges making up electric dipoles. These charged particles have mass and inertia. Considering the effect of frequency of the applied field, it can be predicted that the polarization will decrease with increasing frequency. In other words, as the frequency of the applied field increases, the inertia of the particles will tend to prevent the particle displacement from following the field (in time). Hence, it can be concluded that the displacement will have in phase as well as out of phase components. Accordingly, for the mathematical analysis taking into account the losses in the dielectric due to polarization process at high frequencies prevalent in the communication network, it was necessary to introduce the complex relative permittivity "$\overline{\varepsilon}_r$" described as,

$$\bar{\varepsilon}_r = \varepsilon_r' - j\varepsilon_r'' \tag{6.6}$$

where

ε_r' is the real part of the complex relative permittivity, which is related to the stored energy within the medium. Hence it is also called the *ac* capacitivity, and,

ε_r'' is the imaginary part of the complex relative permittivity, which is related to the dissipation (or loss) of energy within the medium. At a given frequency, the imaginary part leads to absorption loss if it is positive and gain if it is negative. ε_r'' is also known as the dielectric loss factor or the "loss index" of *ac* signals in a medium, the so called "Lossy Transmission Line". It can be shown that ε_r'' is a positive quantity.

6.2.3 Polarization in Insulating Materials

Polarization in insulating materials is basically a phenomenon of interaction between the applied external electric field and the charge carriers, the atoms, ions or molecules present in the dielectriç. Not only does the applied electric field give rise to the polarization, but in turn, polarization modifies the microscopic field within the dielectric. Thus, it is a process of reversible displacement between the positions of positive and negative charges in the molecular structure caused by the applied electric field. The interaction takes place by a force, exercised on the basic structural elements (charge carriers) of the dielectric.

The effect of polarization is analytically described by the relative permittivity ε_r. Consider a uniform field electrode system in vacuum applied an electric field E. The electric flux density $\vec{D}_o$ is given by,

$$\vec{D}_o = \varepsilon_o \cdot \vec{E} \tag{6.7}$$

For the same electrode and the magnitude of the applied electric field on replacing the vacuum by an insulating material (solid, liquid or gas) the electric flux density in the dielectric $\vec{D}_{ins}$ is increased as given by,

$$\vec{D}_{ins} = \varepsilon_o \cdot \varepsilon_r \cdot \vec{E} \tag{6.8}$$

The increase in electric flux density "$\vec{D}_p$" is caused by polarization in the dielectric. "$\vec{D}_p$", therefore, represents the property of the material and is given by,

$$\vec{D}_p = \vec{D}_{ins} - \vec{D}_o = \varepsilon_o(\varepsilon_r - 1)\vec{E}$$

or

$$\frac{\vec{D}_p}{\vec{D}_o} = (\varepsilon_r - 1) = \chi \tag{6.9}$$

The quotient D_p/D_o is known as "dielectric susceptibility" or "polarization capacity" of a dielectric. It is denoted by χ.

Most of the gaseous dielectrics have their ε_r nearly equal to one. Hence polarization is not an important phenomenon in gases, but in liquid and solid dielectrics it plays a significant role determining the conductivity and the losses. Different types of polarization mechanisms are described under the following three main categories.

Displacement Polarization

Within a molecular bond, the positive and negative charges of individual molecule or atom are rendered to oscillate in synchronism under the influence of an applied electric field. Similar oscillations also take place between the nucleus and the electron shell of an atom, building dipoles. The displacement caused by the oscillations is proportional to the applied electric field and on removing the electric field the atoms return back to their original state. Where only this type of polarization mechanism is present, the dielectric materials are described to be "nonpolar" and these have no dipole moment. Classic examples of such dielectrics are the gases, which have ε_r nearly equal to one.

Space Charge or Boundary Surface Polarization

The heterogeneous dielectrics, for example partially crystalline insulating materials, have the positive and negative ions as free charge carriers. Without an external field, the dielectric is in neutral condition and the positive and negative charges neutralize each other. However, under the effect of an external electric field, the charge carriers in the dielectric move towards opposite polarity electrode surfaces, giving rise to a macroscopic dipole. Like the mechanism described in the previous case, in this case too on removing the applied electric field, the dielectric returns to its original neutral state. Hence it is known as a type of displacement polarization.

Dielectrics having this type of polarization mechanism are also described as nonpolar, that is, these do not build a dipole without an external field. The boundary surface polarization is commonly present in heterogeneous insulating materials, such as impregnated paper insulations, hard pressed boards, taped insulations used in electrical machines and even at voids in solid homogeneous dielectrics. Such materials normally have a low value of ε_r.

Orientation Polarization

Orientation polarization is the main polarization mechanism in liquid dielectrics. Usually, the polarization depends upon the applied electric field intensity. However, in some materials a permanent polarization is "frozen" due to the permanent dipoles present in the dielectrics even without an external field. Such dielectrics have an asymmetry in their molecular structure and are described as polar.

On applying an external electric field, the dipoles arrange themselves according to the field lines. The effect of applied external field on the dipoles is determined by their dipole moments. Depending upon the bond between the positive and negative charges in the molecules of the material, the orientation of dipoles takes place. The effective field intensity established in the material is the quantity induced by the external source and the sources within the material itself. The orientation of dipoles may be in the form of an arrangement in an element or they may simply straighten out like a chain, which usually depends upon the local field in which the molecules are situated. The local microscopic field may not be necessarily equal to the macroscopic applied electric field.

6.2.3.1 *Effect of Time on Polarization* The displacement of charge or orientation of dipoles is accompanied with a time lag. Depending upon the frequency of the voltage applied, a time constant determines the extent of polarization. Not only the frequency but also the temperature affects the polarization. The effects of variation of voltage with time on polarization are described in the following.

6.2.3.1.1 Polarization under Direct Voltage Consider a uniform field parallel plate condenser of area A and gap distance d, having vacuum in one half of its volume ($\varepsilon_{r1} = 1, \kappa_1 = 0$) and let the other half filled with a solid dielectric ($\varepsilon_{r2} = 2, \kappa_2$) as shown in Figure 6.5 (a). The equivalent circuit diagram of this condenser is shown in Figure 6.5 (b).

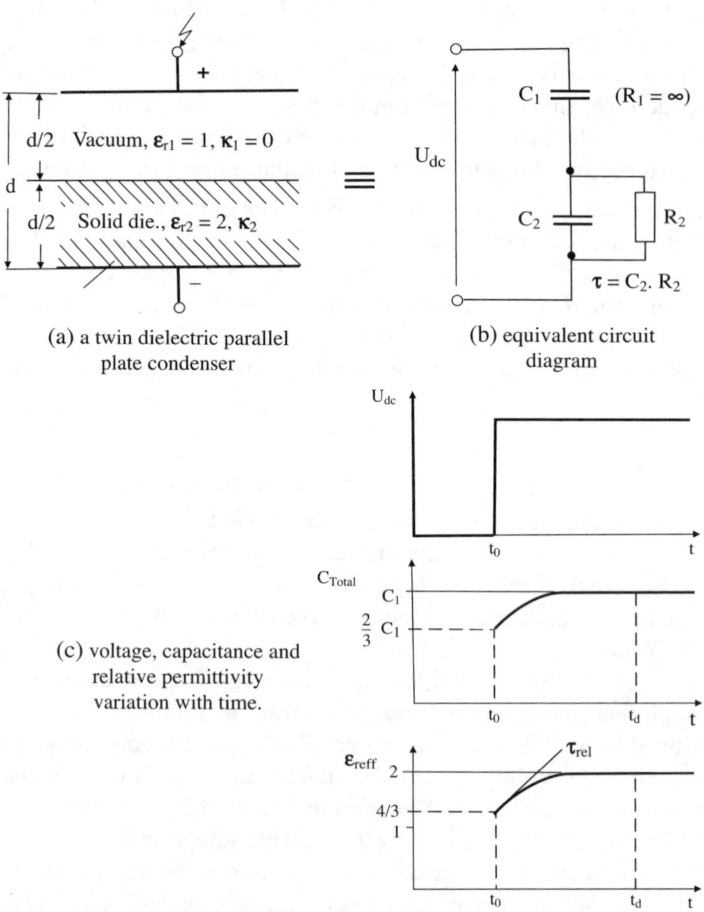

(a) a twin dielectric parallel plate condenser

(b) equivalent circuit diagram

(c) voltage, capacitance and relative permittivity variation with time.

Figure 6.5 Polarization under direct voltage.

On applying a direct voltage across this condenser at time $t = t_o$, at first no current flows through the condenser because of the time lag of ions present in the solid dielectric. In other words, there is no space charge polarization present immediately after the application of voltage. The total capacitance of the condenser at the instant t_o is given by,

$$\frac{1}{C_{Total}} = \frac{1}{C_1} + \frac{1}{C_2}$$

or

$$C_{Total} = \frac{C_1 \cdot C_2}{C_1 + C_2} \tag{6.10}$$

Since ε_{r2} is equal to $2\varepsilon_{r1}, C_2$ is then equal to $2C_1$. Putting this value in equation (6.10).

$$C_{Total} = \frac{2}{3} C_1 \tag{6.11}$$

as shown in Figure 6.5 (c). The capacitance of the parallel plate condenser shown in Figure 6.5 (a) is given by,

$$C = \frac{\varepsilon \cdot A}{d} \tag{6.12}$$

Using this relation for the equivalent capacitance, the combined or the effective relative permittivity at $t = t_o$ is determined as follows,

$$\frac{1}{C_{Total}} = \frac{1}{C_1} + \frac{1}{C_2}$$

or

$$\frac{d}{\varepsilon_{r(eff)} A} = \frac{d/2}{\varepsilon_{r1} \cdot A} + \frac{d/2}{\varepsilon_{r2} \cdot A}$$

or

$$\varepsilon_{r(eff)} = \frac{2\varepsilon_{r1} \cdot \varepsilon_{r2}}{\varepsilon_{r1} + \varepsilon_{r2}}.$$

or

$$\varepsilon_{r(eff)} = \frac{4}{3} \tag{6.13}$$

The conductivity of the solid dielectric, represented by the resistance R_2 in the equivalent circuit diagram, is shown in Figure 6.5 (b). On applying the voltage, a displacement of charge within the solid dielectric takes place. The capacitance C_2 is discharged over R_2 with a time constant τ equal to $C_2 R_2$. Let C_2 be completely discharged by the time t_d, where t_d is greater than $(t_o + 3\tau)$. Under this condition, the total applied voltage stands across C_1. The effective capacitance of the given condenser, after a time lag of t_d, is therefore C_1, Figure 6.5 (c). For a time $t \geq t_d$, the following relationship is derived,

$$C_{Total} = C_1$$

or

$$\frac{\varepsilon_{r(eff)} A}{d} = \frac{\varepsilon_{r1} \cdot A}{d/2}$$

or

$$\varepsilon_{r(eff)} = 2\varepsilon_{r1} = 2 \tag{6.14}$$

This process can be interpreted by the phenomena of displacement of space charge in the solid dielectrics. The space charge polarization leads to an increase in effective relative permittivity, and hence also the capacitance. However, the polarization is effective only when the time required for the formation of dipoles as well as their displacement is provided. The time required for this process is known as "Relaxation time". In Figure 6.5 (c), the tangent to the curve showing increase in $\varepsilon_{r(eff)}$ with time, represented by τ_{rel} is a measure of the time required for polarization process, hence it can also be denoted by τ_p.

The relaxation times required by different types of polarizations vary considerably from each other. For example, the relaxation time is of the order of 10^{-15} s for electron polarization, whereas for ion polarization it is 10^{-12} s. For orientation polarization it is 10^{-8} s, and for space charge or boundary surface polarization, the longest relaxation time of the order of 10^{-3} s is required.

6.2.3.1.2 Polarization under Alternating Voltage On applying alternating field to a dielectric, the dipoles or the charges must change their direction every half cycle. If the frequency of alternating field is very low, in other words, if the duration of the half cycle is very long compared to the required "relaxation time" of the dipoles, then the polarization caused by orientation of dipoles would have its maximum effect. Let this be represented by ε_{r1} in Figure 6.6. In the equivalent circuit diagram shown in this figure for the region of the frequency ($\omega << \omega_e$), C_p represents the increase in capacitance caused by polarization, whereas R_p represents the pure resistance required to interpret the time dependence of polarization. The time constant is given by,

$$\tau_p = C_p \cdot R_p$$

At very high frequencies, where the duration of a half cycle is very short compared to the relaxation time, the dipoles are not in a position to follow the change in the field intensity. This reduces the polarization caused by orientation of dipoles that ultimately diminishes to zero at very high frequencies as illustrated by the right hand side of the curve in Figure 6.6. It is represented by a lower constant value of relative permittivity, ε_{r2}. The total effective capacitance in this region of very high frequency is C, as shown in the equivalent circuit diagram.

The transition domain of frequency is also known as the "dispersion domain". The relative permittivity ε_r depends strongly upon the frequency. Corresponding to the relaxation time τ_p, the eigen frequency "ω_e" of the dipoles is defined as:

$$\omega_e = \frac{1}{\tau_p} \tag{6.15}$$

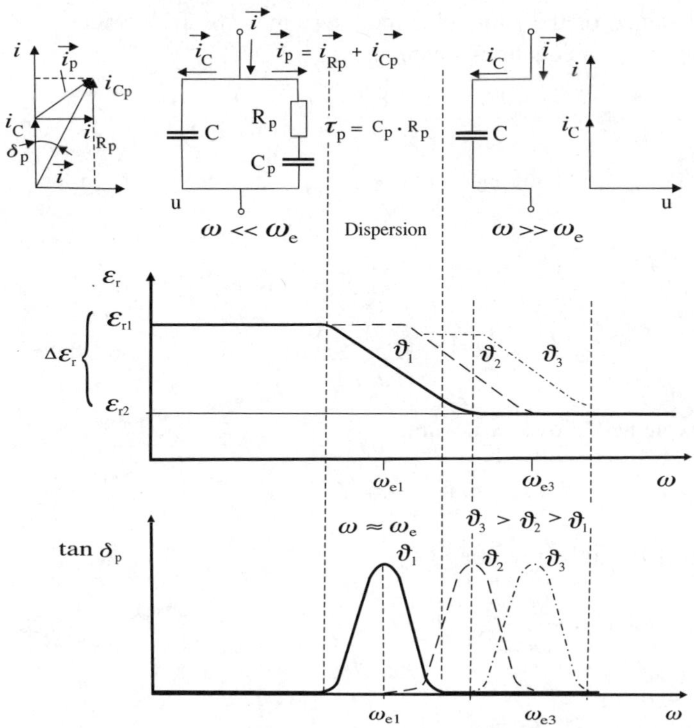

Figure 6.6 Polarization under alternating voltage and the effect of frequency and temperature on ε_r and polarization losses.

The eigen frequency ω_e of a dielectric is its characteristic property and may vary considerably for different dielectrics. For example, it is 10 kHz for hard pressed boards, whereas for highly purified water, it is of the order of 100 MHz. Eigen frequency of dielectrics is also a function of temperature. It is influenced by the binding energy of the material, which in turn is affected by temperature. Consequently, at higher temperatures the eigen frequency of a dielectric increases, as shown in Figure 6.6. From this figure the following can be concluded,

for

$$\omega << \omega_e, \quad C_{Total} = C + C_p$$

and for

$$\omega >> \omega_e, \quad C_{Total} = C$$

Taking into account the polarization losses, the total admittance offered can be calculated as,

$$Y_{Total} = j\omega C + \frac{j\omega C_p}{1 + j\omega R_p C_p} \tag{6.16}$$

The total admittance offered across the terminals can also be given by,

$$Y_{Total} = j\omega\varepsilon_r \cdot C_o \tag{6.17}$$

where C_o is the capacitance of the same electrode system with air or vacuum as dielectric. By equating the two equations given above, we get,

$$\varepsilon_r = \frac{C}{C_o} + \frac{C_p}{C_o} \cdot \frac{1}{1 + j\omega R_p \cdot C_p} \tag{6.18}$$

Putting $R_p \cdot C_p = \tau_p = 1/\omega_e$ in the above equation,

$$\varepsilon_r = \frac{C}{C_o} + \frac{C_p}{C_o} \cdot \frac{1}{[1 + j(\omega/\omega_e)]} \tag{6.19}$$

or

$$\varepsilon_r = \frac{C}{C_o} + \frac{C_p}{C_o} \cdot \frac{[1 - j(\omega/\omega_e)]}{1 + (\omega/\omega_e)^2}$$

From Figure 6.6 as explained above, we obtain,

$$\frac{C}{C_o} = \varepsilon_{r2} \quad \text{and} \quad \frac{C_p}{C_o} = \Delta\varepsilon_r$$

Therefore, equation (6.19) can be rewritten as,

$$\varepsilon_r = \varepsilon_{r2} + \frac{\Delta\varepsilon_r}{1 + (\omega/\omega_e)^2} - j(\omega/\omega_e) \cdot \frac{\Delta\varepsilon_r}{1 + (\omega/\omega_e)^2} \tag{6.20}$$

or

$$\bar{\varepsilon}_r = (\varepsilon_r' - j\varepsilon_r'')$$

Equation (6.20) is similar to equation (6.6), proving that $\bar{\varepsilon}_r$ is a complex quantity having real and imaginary parts. The complex quantity $\bar{\varepsilon}_r$ given by equation (6.20) for the case explained in Figure 6.6 is plotted in Figure 6.7. Like ω_e, the relative permittivity ε_r is also a function of temperature.

From the equivalent circuit diagram given in Figure 6.6 for the region $\omega << \omega_e$ (for example power frequency region), the total vector current $\vec{i}$ is given by,

$$\vec{i} = \vec{i}_c + \vec{i}_p \tag{6.21}$$

where the polarization current $\vec{i}_p$ constitutes of real $\vec{i}_{Rp}$ and reactive $\vec{i}_{cp}$ parts. The total reactive current is then given by,

$$\vec{i}_{Total(reactive)} = \vec{i}_c + \vec{i}_{cp}$$

and total active current is $\vec{i}_{Rp.}$

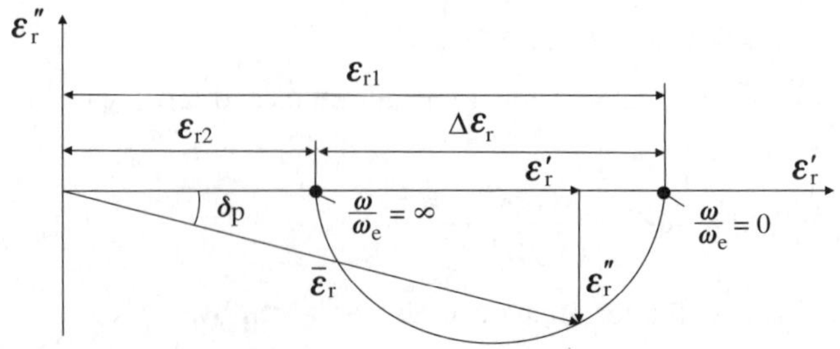

Figure 6.7 Plot of complex relative permittivity variation with frequency of applied voltage.

The angle between total vector current and the total reactive current in the dielectric is known as angle δ_p, which is a measure of the active power loss in the dielectric due to polarization. The tangent of angle δ_p is equal to the ratio of total active to reactive currents, that is, the ratio between the real and imaginary parts of the current in the dielcctric due to polarization, as shown by the vector diagram for the currents in Figure 6.6,

$$\tan\delta_p = \frac{i_{Rp}}{i_c + i_{cp}} \tag{6.22}$$

From Figure 6.7 and equation (6.20)

$$\tan\delta_p = \frac{\varepsilon_r''}{\varepsilon_r'}$$

or

$$\tan\delta_p = \frac{\Delta\varepsilon_r(\omega/\omega_e)}{\varepsilon_{r2}[1+(\omega/\omega_e)^2]+\Delta\varepsilon_r} \tag{6.23}$$

From Figure 6.7, it is clear that $\delta_{P,}$ hence $\tan\delta_p$ has maximum value in the dispersion domain of the frequency. For $(\omega/\omega_e) << 1$ and $(\omega/\omega_e) >> 1$, $\tan\delta_p$ tends to be equal to zero. Accordingly a schematic plot of $\tan\delta_p$ with respect to the frequency is shown in Figure 6.6. However, since ε_r and ω_e, both are functions of temperature, with increasing temperature $(\theta_{2,}\ \theta_3)$ the loss tangent maximum due to polarization in a dielectric also shifts towards higher frequencies with the increase in eigen frequency.

It can be concluded with this analysis that the peak active power loss due to polarization in a dielectric takes place at its eigen frequency. This is important knowledge, which must be taken into consideration while exercising choice of the dielectric for use in high frequency high voltage equipments. The eigen frequency of the dielectric in such an apparatus should in no way be close to its working frequency.

6.2.4 Dielectric Power Losses in Insulating Materials

Besides the capacitive charging currents, real or active currents are also present in the dielectrics. These currents are caused due to different types of conductivities and polarizations present in the materials. As shown in the previous sections, these currents not only depend upon the frequency and magnitude of the applied voltage but also upon the thermal conditions of the dielectric. The conductive currents present in an insulating material determine its dielectric power loss property. Each conductive current mechanism causes currents of different characteristics. These currents contribute to the total dielectric current, "i_{ins}", as depicted schematically for alternating voltage in Figure 6.8.

The dielectric loss tangent, "$\tan\delta$" is defined as the quotient of active to reactive power loss in a capacitor or in a volume of dielectric. It is derived from Figure 6.8 as follows,

$$\tan\delta = \frac{\textit{Active Power}}{\textit{Reactive Power}} = \frac{u\cdot i_{ins}\cos\phi}{u\cdot i_{ins}\sin\phi} = \frac{i_{R(Total)}}{i_{C(Total)}} \tag{6.24}$$

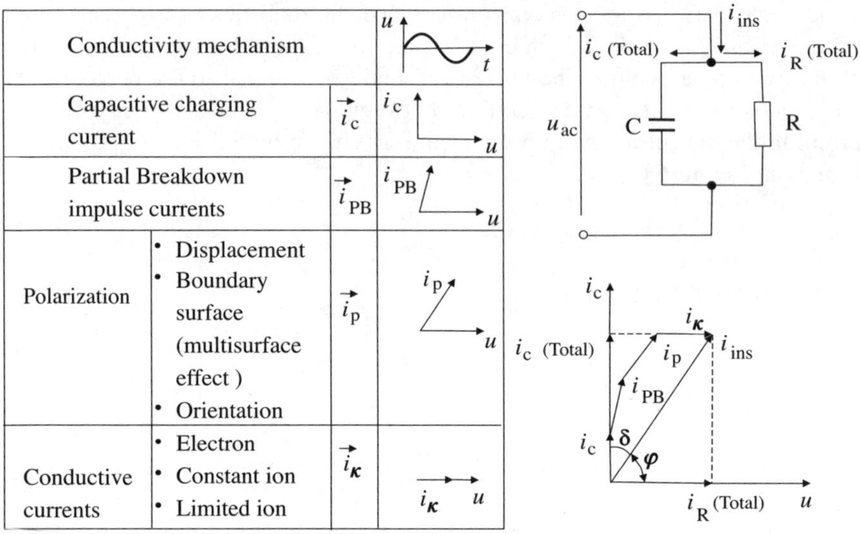

Figure 6.8 Conductive mechanisms in insulating materials for alternating voltage with equivalent circuit and vector diagrams.

The dielectric loss tangent ($\tan\delta$) represents the complete power loss in a dielectric, hence it is a parameter with which the power losses in a capacitor can be estimated. However, $\tan\delta$ is a function of frequency and magnitude of the applied voltage as well as the temperature of the dielectric, because these affect the conductivity κ and the polarization processes in the dielectrics.

When alternating voltage of rms magnitude U and frequency ω is applied to a condenser having total effective capacitance C, the total capacitive conductive current $I_{C(Total)}$ is given by,

$$I_{C(Total)} = \frac{U}{1/\omega C} = \omega \cdot C \cdot U \tag{6.25}$$

The active part of the total insulation current is $(\omega CU \cdot \tan\delta)$, hence the active power loss "P_{ac}" is given by,

$$P_{ac} = \omega \cdot C \cdot U^2 \cdot \tan\delta \tag{6.26}$$

where $\omega = 2\pi f$ and if f is in Hz, C in F and U in volts, P is given in Watts. In practice, the capacitance C of the given test object is generally measured.

Considering a parallel plate condenser of area A and gap distance d having dielectric with relative permittivity ε_r forming a uniform field E, equation (6.26) can be rewritten as,

$$P_{ac} = \omega \cdot \varepsilon_o \varepsilon_r \frac{A}{d} \cdot (E \cdot d)^2 \tan\delta$$

or

$$P_{ac} = \varepsilon_o \cdot \omega \cdot E^2 \cdot V \cdot \varepsilon_r \tan\delta \tag{6.27}$$

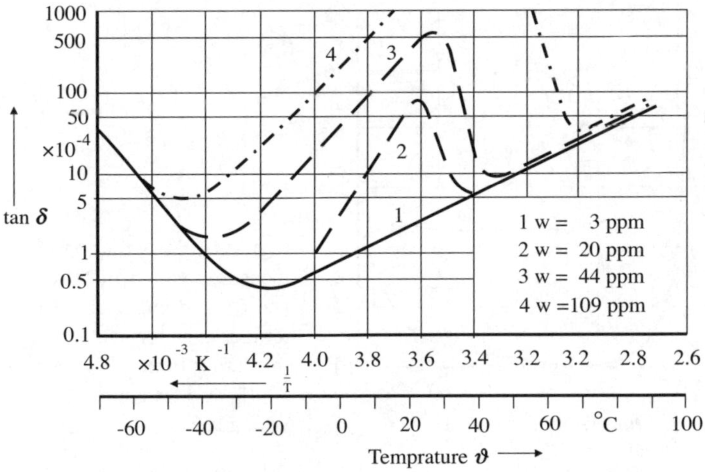

Figure 6.9 Loss tangent "tan δ" of transformer oil having different ppm moisture contents with varying temperature at 50 Hz constant voltage, Holle [6.6].

where *V*, the volume of the dielectric, is given by (*Ad*) in this case. Equation 6.27 shows that the dielectric losses depend upon the applied field intensity and its frequency, volume of the dielectric, its relative permittivity, and the loss tangent.

On applying a direct voltage to the dielectric, the losses depend only upon the magnitude of the applied voltage U_{dc} and the *dc* resistance, "R_{dc}" offered by the dielectric, determined by equation (6.3) and (6.4) as follows,

$$P_{dc} = \frac{U_{dc}^2}{R_{dc}}$$

or

$$P_{dc} = E^2 \cdot V \cdot \kappa_{dc} \tag{6.28}$$

Variation in loss tangent of a transformer oil having different moisture contents, measured for a wide range of temperatures by Holle [6.6] is shown in Figure 6.9. These measurements were conducted on a standard test cell (guard ring capacitor) applying a constant 50 Hz voltage.

Figure 6.10 shows the variation of $\tan\delta$ and ε_r of synthetic insulating liquid "Clophen A 50", with respect to temperature measured at 50 Hz constant voltage.

As described earlier, the loss tangent is a function of the applied voltage. Increase in $\tan\delta$ of transformer oil samples, measured at constant temperatures for increasing 50 Hz voltage/field intensity magnitudes is shown in Figure 6.11, for two different moisture contents, Bayer et al. [6.5]. These measurements were made at Schering Institute of HV Engineering, University of Hannover. Extra care was taken to design the measuring arrangement, which was developed to be PB free for upto 200 kV. As seen from these curves, the increase in temperature as well as moisture content in oil, both result in a premature increase in $\tan\delta$ values with increasing voltage.

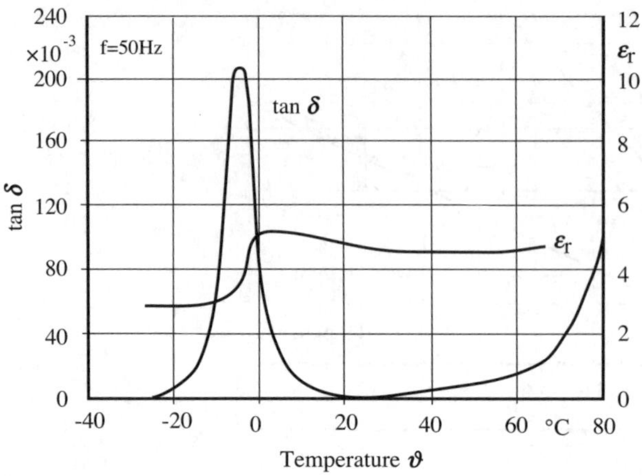

Figure 6.10 Variation in measured values of $\tan\delta$ and ε_r with increasing temperature for synthetic insulating liquid, "Clophen A 50" at constant 50 Hz ac voltage.

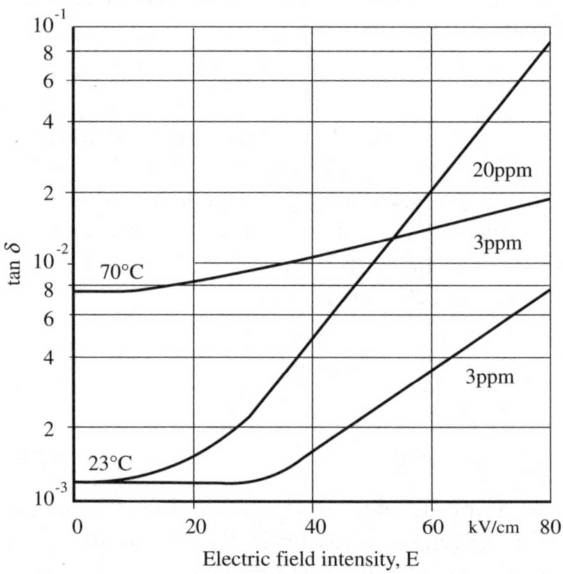

Figure 6.11 Loss tangent ($\tan\delta$) of transformer oil measured with increasing 50 Hz voltage/field intensity at different constant temperatures and moisture contents in oil, Bayer et al. [6.5].

6.3 BREAKDOWN IN LIQUID DIELECTRICS

A very large number of external factors affect the breakdown strength of liquid dielectrics. For example, electrode configuration, their material, size and surface finish; the type of voltage, its period of application and magnitude besides the temperature, pressure, purification of the liquid and its aging condition. Dissolved water, gas or the presence of any other form of contamination considerably affect the breakdown strength. It is, therefore, not possible to describe the breakdown mechanism by a single theoretical analysis, which could take into account all known observed factors affecting the breakdown. Hence, there is no unified physical theory of breakdown in liquids. However, the measured breakdown strength of any dielectric is distinguished into two broad categories known as the "intrinsic" and the "practical" breakdown strengths.

Before describing these, it would be interesting to investigate the mechanisms of pre-breakdown conduction in insulating liquids. The electric conduction primarily determines the dielectric properties. The trend of research in this area has been more practical, involving the commercial liquids used in the industry, Umemura et al. [6.7]. The electric conduction and dielectric properties of insulating liquids used in practice are generally affected by the amount of ionic impurities that they contain.

6.3.1 Electric Conduction in Insulating Liquids

The conduction mechanism in a dielectric liquid is strongly affected by the degree of its purification. A schematic of typical characteristic of conduction current density J versus the applied *dc* voltage U in a pure liquid is shown in Figure 6.12. At low voltages, the current is directly proportional to the voltage, region 1, representing the ohmic behavior. On raising the applied voltage, and thus the field intensity, the ionic current gets saturated, region 2. At still higher applied voltage, the current density increases rapidly until the breakdown occurs, region 3.

Similar behavior is observed generally in both polar and nonpolar liquids. In some cases the saturation region may be completely missing. This characteristic has an analogy to what has also been commonly observed in gases. In liquids, which have not been highly purified, when subjected to field up to about a few kV/cm the conduction is primarily ionic. The dissociation processes of impurities as well as the injection of charge by the electrodes through electrochemical reactions affect ionic conduction. The electrons do not take part in the passage of current through such liquids. In nonpolar liquids, electrons exhibit a transitory existence. They recombine with positive charges, transfer to surfaces of the vessel or become attached to impurities and form negative ions, Schmidt [6.8].

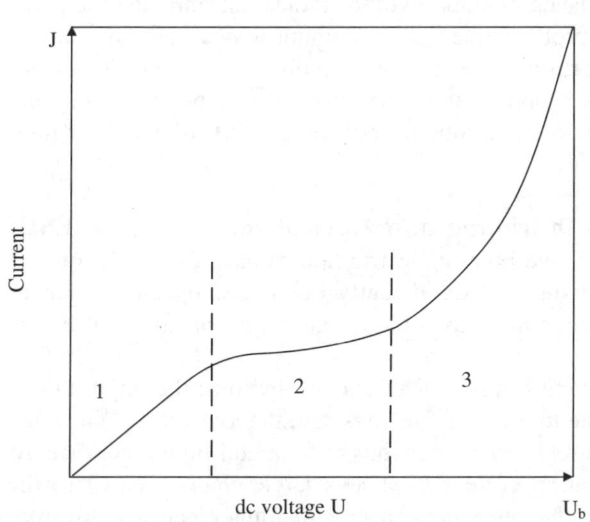

Figure 6.12 Typical voltage—current ($U - J$) characteristic in dielectric liquids.

The concentration of charge carriers and the viscosity of insulating liquids are both strongly affected by the temperature, which in turn influence the conductivity of the liquid. According to Van't Hoffsch law, the conductivity, "κ" within a certain range of temperature follows the relation,

$$\kappa = \kappa_o \exp(-F/kT) \tag{6.29}$$

where k is Boltzmann constant, T absolute temperature, κ_o and F are the material constants. F is known as "activation energy" and is expressed as kcal/mole.

However, Van't Hoffsch law is valid only in the region where the conduction current follows the ohmic behavior, that is, region 1 in Figure 6.12. It thus depends upon the particular liquid and its impurity contents (for example, moisture content etc). For some aromatic insulating liquids, Umemura et al. [6.7] have discussed the applicability of the above equation at different temperatures and have also estimated the values of the constants. According to them, the conduction phenomenon of the liquids can be divided into two temperature regions. In the low temperature regions, that is, room temperature and lower, the conductivity is governed mainly by ionic carrier mobility. However, in the high temperature region, it is characterized by thermal dissociation of impurities as well as carrier mobility whereas the viscosity of the liquid governs the carrier mobility.

Variation in transformer oil conductivity for a wide range of temperature and different moisture contents measured by Holle [6.6] in near uniform field are illustrated in Figure 6.13. Equation (6.29) is proved to be valid only in the temperature region above the room temperature. The conductivity curve in this region represents a straight line on logarithmic scale (curve 1). It is evident from this figure that the conductivity of oil increases as the water ppm contents rise.

In Figure 6.14 the conduction current characteristics in extremely nonuniform field are shown for negative direct voltage applied to a needle-plane electrode configuration in transformer oil and liquid nitrogen, measured for different radii of curvature, "r_t" of the needle tip by Takashima et al. [6.9]. For these experiments, the needle constituted of a polished platinum wire of radius 0.25 mm and the plane of 50 mm diameter Rogowsky profile curved plate of stainless steel. The current was measured through the plane electrode for a constant gap distance of 4 mm. The sharp bend measured in the curve is supposed to represent the development of a strong space charge around the needle possibly due to inception of PB in extremely nonuniform field.

6.3.1.1 Liquid Dielectrics in Motion and Electrohydrodynamics (EHD)

The behavior of transformer oil and other dielectric liquids used for insulation and cooling in power system equipment is significantly influenced by the motion or circulation of the fluid. The motion in fluids could be natural or forced by the action of circulating pumps.

The total conduction current has two distinct contributions: the impurities in the liquid, which may dissociate to produce free ions, and the ions (charge) injected from the electrodes due to electrochemical reactions at the metal-liquid interface. At low electric field, the dissociation current is always larger than or equal to the injected current. However, the charges generated by streaming electrification from

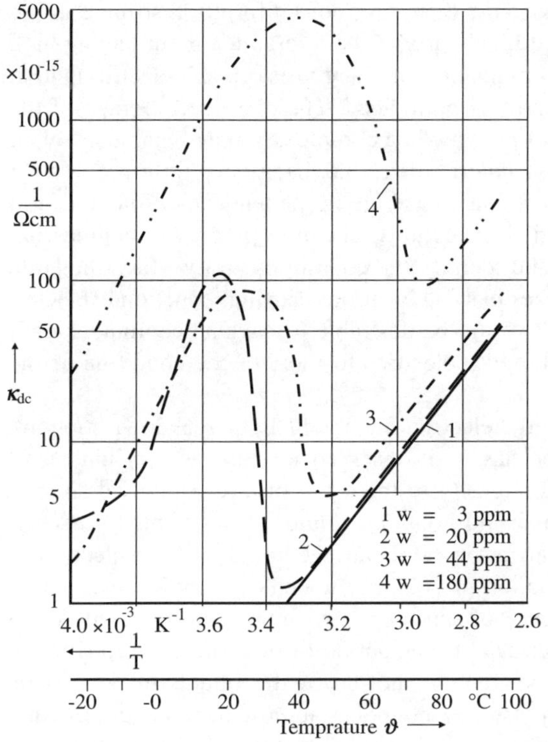

Figure 6.13 Direct current conductivity of transformer oil for different water contents (ppm) with respect to the reciprocal of absolute temperature, Holle [6.6].

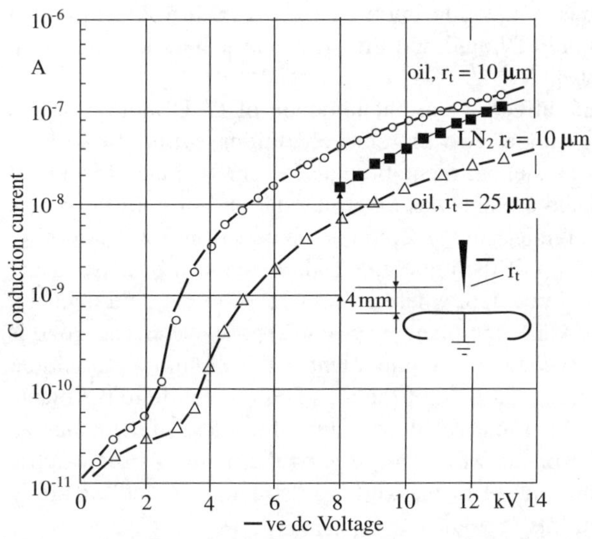

Figure 6.14 Conduction current in oil and liquid nitrogen for negative needle-plane electrode gap of 4 mm with increasing voltage. Oil at 20°C and LN_2 at 77.3 K, Takashima et al. [6.9].

the electrodes are much higher and may, therefore, not uniformly distribute in the liquid. The charges accumulate at critical parts of the hydraulic circuit having high velocity or turbulence. Such an accumulation of charge distorts the electric field at locations where the dielectric integrity is prejudiced. The dielectric strength of the liquid is also altered by the action of the flow in a complex, but predictable, manner, Nelson [6.10]. However, as Nelson reports, there has been very little progress in identifying the trace of contaminants that contribute to the tendency of nonuniform distribution of charge in the liquid dielectric fluids. The most productive engineering approach would be to investigate and identify the conditions under which the large amounts of charge accumulation takes place at locations causing reduction in electric strength of the dielectric. Hence, it would be desirable to have a "continuous contaminant sensing mechanism" in liquid dielectrics to monitor the condition of the equipment.

Since the highly purified liquid dielectrics have very high breakdown strength, a peculiar conduction phenomenon has been observed in these at very high field intensities. On increasing the field intensity to the order of a few hundred kV/cm, the conduction current in both polar and nonpolar liquids is augmented predominantly by large amount of charge carriers injected into the liquid from the electrodes. Beside this, the field aided dissociation process of molecules is intensified. Under certain given conditions, the presence of space charge, say of density q, gives rise to a Coulomb's force of density qE. Due to the action of this force, hydrodynamic instability is caused, developing convection motion in the liquid. Investigations made by many authors, on both polar and nonpolar liquids, have concluded that whenever conduction in an insulating fluid is accompanied with a significant magnitude of charge injection, the convection motions occur. These are known as "electrohydrodynamic" (EHD) motions. The EHD motions augment the passage of current depending upon the charge injection intensity, which in turn is determined by the applied voltage, the nature of liquid and the electrode material, Zahn [6.4], Takashima [6.9] and Atten [6.11].

Atten [6.11] explained the effect of the phenomenon of EHD motion on the conduction in liquid dielectrics under two extreme conditions, with and without unipolar injection of charge. For his experimentation, an extremely nonuniform field configuration between a knife and a plane was taken, as shown in Figure 6.15.

Consider first the case of injection of unipolar charge (ions) by the knife, Figure 6.15 (a). The field at the tip of the knife electrode is reduced because of the same polarity space charge being injected, whereas an increase in the field intensity is caused over the basic field towards the plane electrode. The space charge zone is of finite extent when a moderate voltage is applied. Under this condition, the liquid relaxation time is lower than the transit time of the ions from the knife to the plane. The total current is, therefore, higher than the ohmic current because of the enhancement of field intensity near the plane. Moreover, due to the action of the electric field on the same polarity charge close to the knife, a "stream" or "jet" of liquid towards the plane is induced in this region.

In case of pure conduction with no injection of charge, Figure 6.15 (b), an opposite polarity space charge near the knife extends further. Consequently, the space charge field at the tip increases, and is reduced with respect to the basic field

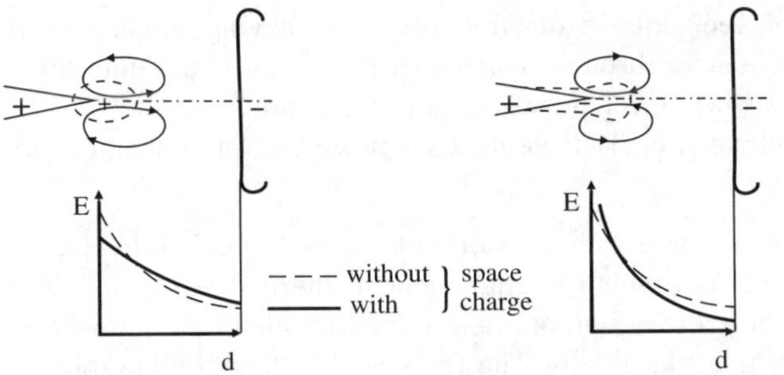

(a) injection of charge by knife

(b) constant rate dissociation with no injection

Figure 6.15 Schematic representation of space charge zone and liquid motion streams between knife-plane electrodes and the field distribution along the axis, Atten [6.11].

towards the plane. This results in lower total current than the ohmic current (sub-ohmic behavior). The opposite polarity space charge in the vicinity of the knife gives rise to a liquid motion directed towards the knife.

Taksahima et al. [6.9] also observed similar results, when they conducted experiments between a razor blade and the plane in transformer oil and liquid nitrogen. Zahn et al. [6.4] conducted their experiments with highly purified water over a temperature range of 0 to 30°C using parallel plane electrodes. An average field intensity up to 160 kV/cm was achieved on applying a positive impulse voltage. They concluded that for pulse duration times greater than 500 μs, stainless steel and copper electrodes generally injected positive charge, although under some conditions with mixed electrodes they injected negative charge. Aluminium electrodes injected only negative charge, while brass electrodes could inject either positive or negative charge. Thus by appropriate choice of electrode material combinations and voltage polarity, it is possible to have uncharged, unipolar charged negative or positive, or bipolar charged liquids.

At extremely high field intensities (1000 kV/cm and above) approaching the "intrinsic strength" of liquid dielectrics, the field emission at the electrode surfaces affects the conduction phenomenon. Under these conditions, electrons dominate the conduction current.

6.3.2 Intrinsic Breakdown Strength

The intrinsic breakdown of liquid and solid dielectrics is defined as the highest value of breakdown strength obtained after eliminating all known secondary effects, which may influence the breakdown. The concept of intrinsic breakdown is ideal. It is, however, extremely difficult to ascertain whether an observed breakdown was intrinsic or not. Intrinsic strength can be suitably interpreted as the field intensity at which the particles in the material are accelerated without any limit by the direct electrostatic action of the field.

The mean free path of an electron in insulating oils is very short, of the order of 10^{-6} cm. Extremely high field intensity is therefore required to begin the elementary process of ionization with free electrons having a very short lifetime. The elementary processes through which energy loss into the liquid takes place, determine the breakdown mechanism. Hence these processes decide the breakdown strength of a dielectric. The elementary processes of ionization are as follows:

- the most likely elementary process in case of hydrocarbon liquids is by excitation or molecular vibration, which is equivalent to thermal vibrations. The vibrational modes of a hydrocarbon are determined by elementary vibration of the C-C and C-H molecular bonds. The frequencies of such vibrations lie in the infra-red region with corresponding energy quanta of about 0.2 to 0.4 eV ($1\,eV = 1.6 \times 10^{-19}\,J$). These may lead to local growth of vapor phase of microbubbles.
- the process of dissociation of molecules in neutral, low molecular, gaseous particles is due to severe molecular vibrations that requires energy levels in the range 1.5 to 7 eV.
- excitation of metastables that may lead to ionization in a few stages, requiring energy levels of the order of 1.5 to 10 eV.
- ionization process involving scintillation of electrons accompanied with weak luminescence, indicating energy quanta of several eV, which is greater than 10 eV in some liquids.

The elementary processes mentioned above depend upon the molecular structure of a dielectric. These may occur spatially and timely together. However, there is no evidence of ionization by electron collision in liquids as in the case of gases. The electrode process may play its usual role as described.

For a mean free path of electrons of the order of 10^{-6} cm and an ionization energy level of about 10 eV, the theoretically estimated field intensity required to cause ionization in insulating liquids is of the order of 10,000 kV/cm, which is an impractical value. On the contrary, a field intensity of about 200 to 400 kV/cm may be sufficient to cause ionization by the elementary process of excitation of molecular vibration.

Electric strengths of some highly purified liquids and liquefied gases as given by Lewis in [5.7] are given in Table 4.2. These breakdown strengths appear to have been measured on very thin liquid films and represent nearest comparative values of intrinsic strengths.

6.3.3 Practical Breakdown Strength Measurement at Near Uniform Fields

The peak *ac* value of electrical breakdown strength of commercially available purified insulating oils is about 350 kV/cm. This is highly contradictory to the measured intrinsic strengths of the order of 1000 kV/cm and above. As mentioned above, it is because of the secondary effects that influence the breakdown strength. If the condition of the liquid is very bad (impure, contaminated, etc.) one may measure an electric strength only as low as 10 kV/cm.

TABLE 6.2 Electric strength of highly purified liquids and liquefied gases, Lewis in [5.7]

Liquids	Breakdown Strength kV/cm
Good oil	1000–4000
Benzene	1100
Silicone	1000–1200
Hexane	1100–1300
Hydrogen	>1000
Oxygen	2400
Nitrogen	1600–1880

In order to compare the electric strengths of different insulating liquids or the same liquid in its different conditions, the methods of measurement of practical breakdown strength have been standardized by various standards. Simple methods are thus evolved to estimate the quality of insulating liquids. The methods laid down by IS-6792 (2003), VDE-0370 and IEC-156 are almost identical. According to these, an *ac* power frequency voltage is applied to a 2.5 mm gap between two identical electrodes, so called "calottes", of dimensions shown in Figure 6.16 (a). The IS recommends the electrodes to be made of brass having a good surface finish, whereas VDE recommends copper as electrode material. These electrodes are placed in a container of a given size and filled with about 300 cc of the sample of oil as shown in Figure 6.16 (b). The measurement is conducted at room temperature (20°C) increasing the applied sinusoidal waveform alternating voltage at a rate of 2–3 kV/s till upto breakdown. The rms value of the breakdown voltage U_b is measured. The Schwaiger factor or the degree of uniformity, η for the given gap and electrode configuration is 0.97, that is, the field can be considered to be almost uniform.

For each sample of insulating liquid, at least 10 minutes after pouring the fluid in to the cell six measurements of breakdown voltages are recommended to be performed stirring the liquid thoroughly after every breakdown and a minimum wait of 2 minutes (VDE), 1–5 minutes (IEC) or 5 minutes (IS) in between the measurements. VDE and IS recommend mean of all the six values to be taken into account. If the mean rms breakdown voltage is U_b in kV, the peak value of electric strength of the liquid is given as follows,

$$\hat{E}_b = \frac{\sqrt{2}U_b}{\eta \cdot d} = \frac{\sqrt{2}U_b}{0.97 \times 0.25} \quad \text{kV/cm}$$
$$\approx 5.8U_b \quad \text{kV/cm} \tag{6.30}$$

According to the method described above, the required values of electric strengths of various pure insulating liquids available commercially, are given in Table 6.1.

The standard test procedures are the quality control tests since the breakdown strengths thus measured are primarily determined by the contaminants in the sample.

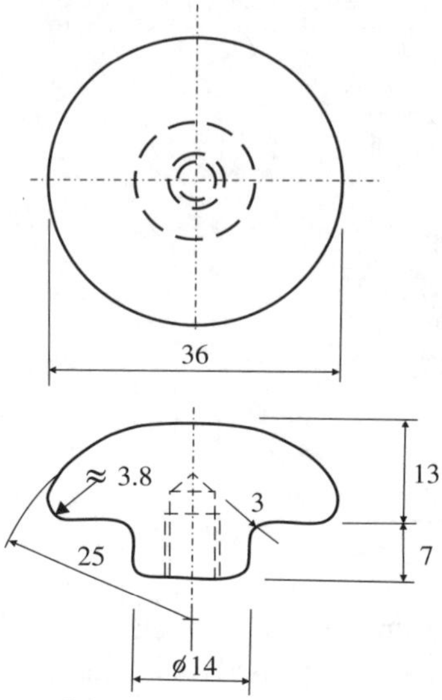

all dimensions are in mm

(a)

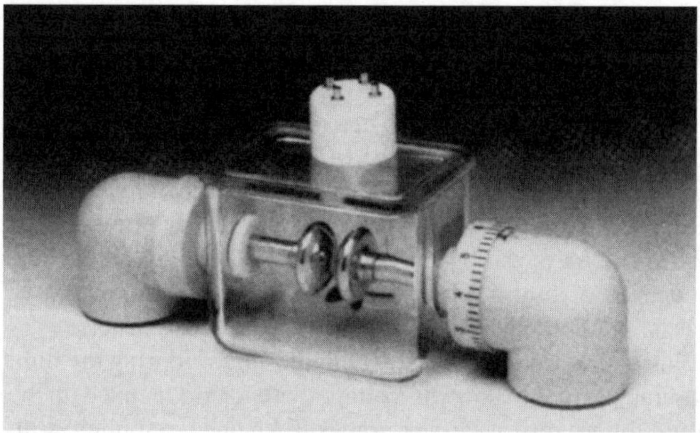

(b)

Figure 6.16 (a) Electrode design having $\eta = 0.97$ for the measurement of electric strength of commercial liquid dielectrics according to VDE-0370/IS-6792. (b) An oil test cell with PB free leads, Musil et al. [6.11].

The effects of various contaminants and the measuring conditions such as moisture, dissolved gases, types of particles, their size and quantity, temperature, sampling methods, pressure, and so on have been investigated extensively. However, Musil et al. [6.12] concluded that in spite of standard procedures, *ac* breakdown voltage of insulating liquids thus measured are valid for certain conditions of measurement. Considerable effect of the design of test apparatus and the test procedure has been found. The parameters, which influence the measurements, are:

- shape of the standardized electrodes
- electrical test circuit characteristics, e.g., trip current level setting, voltage measurement system, current limiting series resistance in the circuit etc.
- construction of the test cell, level of oil sample above the electrodes, formation of stray capacitances
- the method of circulation of liquid during the test
- liquid collection/sampling and storage practices

It is recommended to set a minimum trip current of 5 to 6 mA to avoid any false tripping. It is also desirable that the level of liquid filled in the test cell is at least 40 mm above the electrodes. A continuous, slow circulation of the liquid sample throughout the test is also desirable, Musil et al. [6.12].

Laboratory measurements may give different results for samples of the same oil collected and stored in glass and plastic bottles.

6.3.3.1 Effect of Moisture and Temperature on Breakdown Strength

Power frequency *ac* breakdown voltage/field intensity with increasing temperature measured by Holle [6.6] according to the standard method described above for transformer oil having different moisture contents are shown in Figure 6.17. These curves illustrate the effect of moisture content in oil on breakdown voltage. At very low moisture contents (less than 20 ppm) the breakdown strength is more or less independent of the temperature. For moisture contents above 20 and up to 100 ppm,

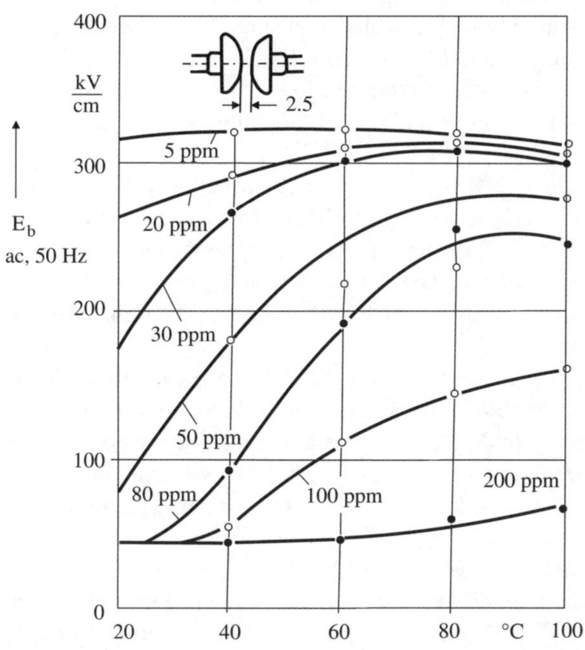

Figure 6.17 Power frequency ac breakdown field intensity (rms) with increasing temperature of a transformer oil having different water contents in ppm, Holle [6.6].

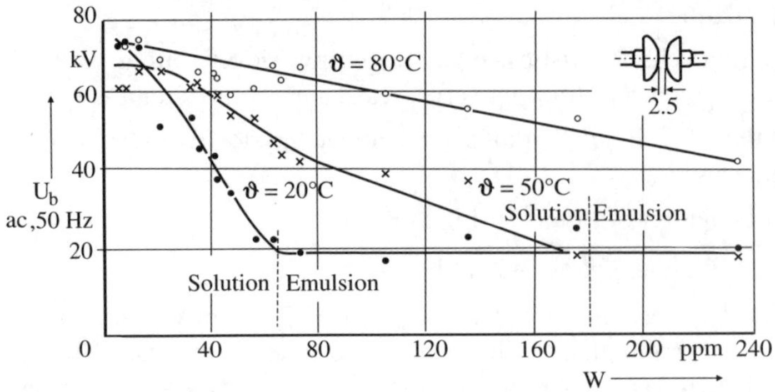

Figure 6.18 Power frequency ac breakdown voltage of a transformer oil at different constant temperature with increasing water content in ppm, Holle [6.6].

the breakdown strength is lowest at room temperature, but it increases considerably with increasing temperature. For 200 ppm again, the temperature does not appreciably affect the breakdown strength but it is quite low.

The breakdown strength is in fact strongly affected by the state of moisture in the insulating oils. As soon as the moisture content in oils acquires the emulsion state, that is tiny water droplets are unevenly distributed instead of a solution state of moisture in the oil, the lowest breakdown strengths are measured, irrespective of an increase in *ppm* water content. Holle [6.6] also measured this phenomenon, clearly illustrated in Figure 6.18, on another transformer oil sample at different constant temperatures for increasing water contents in *ppm*. When the moisture in the oil acquired emulsion state, a constant lowest breakdown voltage was measured.

In his dissertation, Haushild [6.13] made a comparison between breakdown strengths of transformer oil for different types of voltages in weakly nonuniform fields. These are illustrated in Figure 6.19, measured on a sphere-plane electrode configuration with increasing gap distance. The Schwaiger factor, or degree of uniformity of the field η, was estimated to have a value of 0.57 for the 100 mm diameter sphere and the longest gap distance of 50 mm.

Except for very small gap lengths (up to about 5 mm), the breakdown curves show linear characteristics. As in case of gaseous, in liquid dielectrics the highest mean potential gradient of the order of 360 kV/cm was also required for breakdown in the gap with *li*. The mean potential gradient in the gap for breakdown with *si* was measured lower (185 kV/cm) and it was lowest (50 kV/cm) with power frequency *ac* peak voltage.

Although in weakly nonuniform fields no stable PB takes place, still it is the propagation of an unstable streamer that leads to breakdown with an arc. In Figure 6.20 photographs taken by Fiebig [6.14] in 1968 show the breakdown in transformer oil between plane and spheres of different sizes. They represent different degrees of uniformity of the field. Because of fairly high mean field intensity at the instant of breakdown, a very dense bunch of streamer discharge is produced, as seen

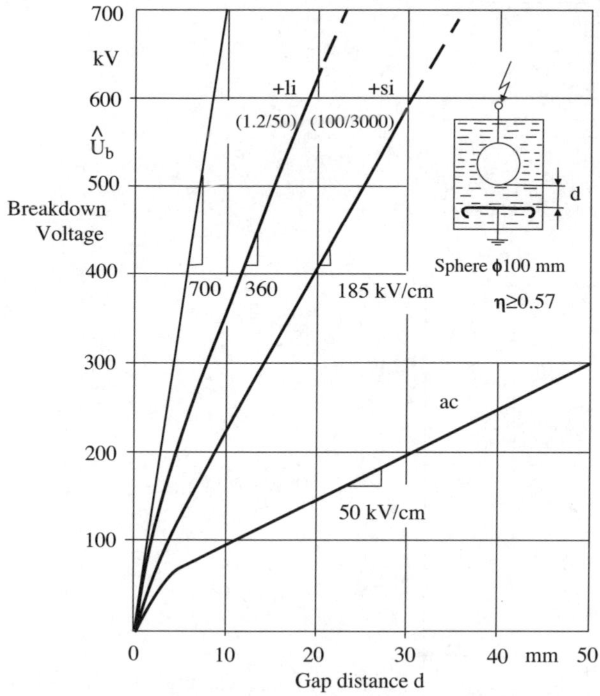

Figure 6.19 Power frequency ac, si and li breakdown voltages of transformer oil measured on sphere-plane weakly nonuniform electrode configuration with increasing gap distance, Hauschild [6.13].

in these photographs. On applying an *ac* power frequency voltage to the spheres, the breakdown always occurred at positive half cycle. Hence, a positive streamer is seen developing in all directions followed by a bright, luminous arc in the middle accomplishing the breakdown.

Figure 6.21 shows the two similar photographs taken by Torshin [6.15] in the year 2003. These Schlieren photographs, a special technique of photography, show the development of prebreakdown unstable streamer at the tip of the electrodes in weakly and extremely nonuniform fields without complete breakdown on applying positive polarity lightning impulse voltages in transformer oil. Similar work published by Top et al. [6.16] in 2002 concluded that the development and propagation of streamer in weakly and extremely nonuniform fields are different.

6.3.4 Breakdown in Extremely Nonuniform Fields and the Development of Streamer

As in gaseous dielectrics, in insulating liquids stable PB occurs in extremely nonuniform fields before the breakdown. The phenomenon of electrical breakdown in insulating liquids has been investigated extensively since the 1960s. The advent of very high speed electro-optical cameras has made it possible to study this process

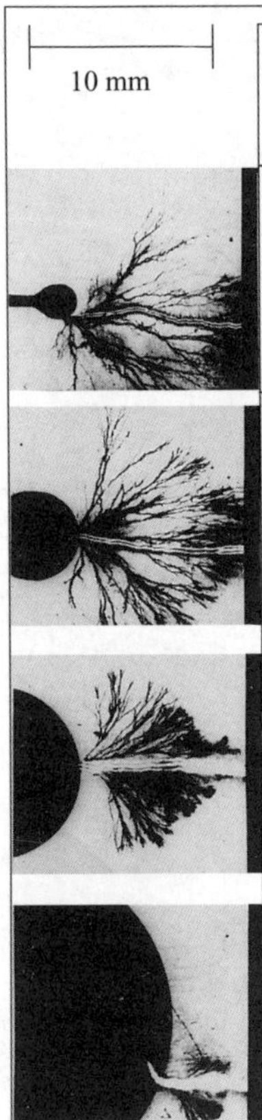

η	D (mm)	d (mm)	$\hat{U}_b$ kV/cm
0.1	2	10	52
0.2	5	10	73
0.4	10	10	120
0.7	20	5	224

Figure 6.20 Photographs of electrical breakdown of transformer oil in weakly and changing nonuniforim fields with alternating voltage for increasing degree of uniformity, Fiebig [6.14].

(a) (b)

Figure 6.21 Photographs of development of typical tree and bush type streamers at the tip of the electrodes in weakly and extremely nonuniform fields taken by Schlieren photographic technique, Torshin [6.15].

in more detail. The mechanism of breakdown in insulating liquids is more or less similar to that in gases, but the causes and the sources of charge carriers are different.

Fiebig (1968) and Hauschild (1969) [6.13 and 6.14] performed two excellent studies on the breakdown phenomenon in extremely nonuniform fields at TU Dresden in their dissertations. Similar investigation results have been published in IEEE ever since by a number of authors all over the world [6.15 to 6.22]. With the development in measurement techniques a little more detailed information could be obtained by the later work. A considerable amount of information regarding the positive and negative streamer propagation in liquids is thus available. This has helped in a better understanding of the breakdown mechanism in liquid dielectrics.

The streamer discharge propagates in liquids in a manner analogous to the propagation of discharge in gases. However, in liquids, electron avalanches contribute to the local discharge bursts, providing a means for sufficient initial charge to facilitate the propagation of streamer. Such a discharge may be caused due to particle impurity or the abrupt failure of a negatively charged oxide layer on the anode surface. In the absence of such accidental occurrence, breakdown does not take place unless the field is sufficiently high to extract electrons directly from the liquid or the electrodes.

Both the negative and the positive streamers in liquid dielectrics have been recognized to appear at low-density regions within the fluid. Thermal electrons surround them as well as the fragments produced by electron impact ionization and local/partial discharges (PB) due to enhanced electric field. Consequently, the local field dominated by the space charge intensifies the process of the growth of streamer.

Optical investigations of pre-breakdown events in insulating liquids conducted by Watson and Chadband [6.17] showed that the initial stage of the breakdown of needle-plane gaps involved growth of a vapor cavity from the needle tip. This was followed by the propagation of streamer from the cavity due to electro-hydrodynamic (EHD) instability of the vapor-fluid interface.

A schematic explanation of the process of development of positive streamer as given by Fiebig [6.14] is shown in Figure 6.22. Due to high field intensity at the tip of the needle electrode, a positive space charge is developed that expands the vapor cavity in the form of a channel. This vapor channel is described as a low density region. According to Fiebig, the avalanche process within this vapor channel is due to the production of photons (photoelectrons) by local breakdown (PB) process giving rise to light emission and current pulses. The positive space charge developed due to dissociation leaves behind a bipolar space charge. The bipolar space charge results in a field equalization effect in the region between the tip of the electrode and the extent of vapour channel, as shown in Figure 6.22.

However, within the vapour channel, displacement of the positive and negative charge carriers takes place due to the interface effect. This causes an enhancement of field intensity at the tip giving rise to new avalanche processes. Although impractical, field intensities of the order 5–50 MV/cm have been estimated at the streamer tip by Felici [6.20]. Felici has also given a similar explanation for the development of negative streamer in liquids.

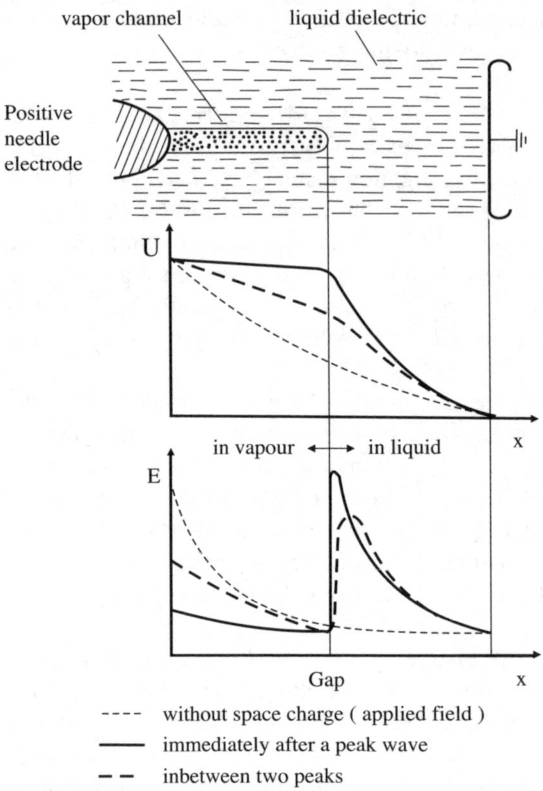

Figure 6.22 A schematic showing the development of positive streamer in liquid dielectrics, Fiebig [6.14].

Lewis [6.23] made investigations on breakdown initiating mechanisms at electrode interfaces in 2003 and also concluded that the cavity production in the fluids depends strongly upon the field intensity and the local surface/interface compositions. The cavities are likely to be filled with vapor at pressures at which impact ionization would readily take place.

The dynamics of the development of pre-breakdown cavities/channels have been widely studied using photographic techniques of pulsed Schlieren and shadowgraph, [6.17, 6.20]. The inception of prebreakdown current is associated with light emission, which precedes the first appearance of a streamer. Both bush and tree type streamer developments, along with light emission have been observed in various liquids. A tree type streamer development is generally observed for positive streamer but the negative streamer appears invariably to be a bush type. In Figure 6.23, photographs of positive and negative pre-breakdown streamers taken by Fiebig [6.14] in transformer oil at a needle electrode are shown. The typical tree and bush type appearances of streamer are easily distinguishable in these photographs. Similar results were measured by Top et al. [6.16].

In organic liquids Felici [6.20] also identified two distinct forms of streamers. A bush like slow propagating streamer at subsonic speed occurs mostly at the negative polarity electrodes. The other, observed typically at positive polarity electrodes, is filamentary, elongated and propagates faster at supersonic speeds. Since it has

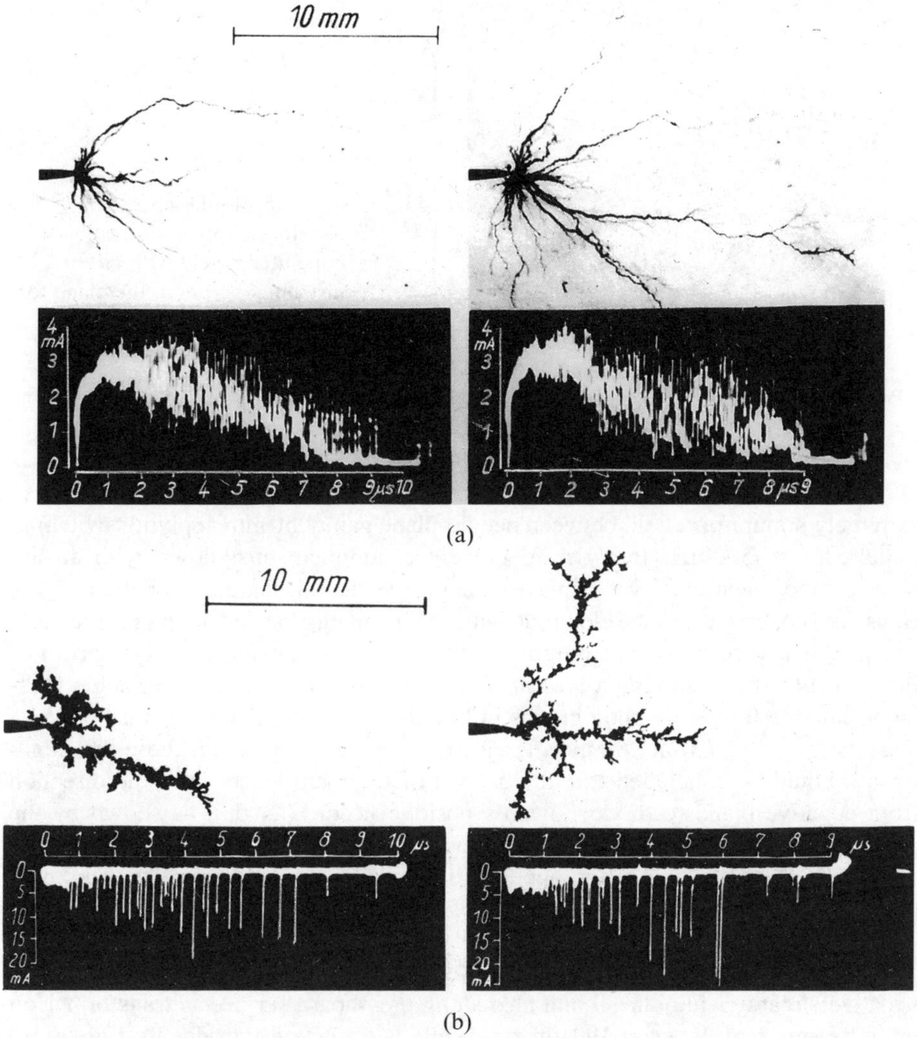

Figure 6.23 Development of streamer in transformer oil between needle-plane gaps, Fiebig [6.14]. (a) Tree type positive streamer in a gap length of 50 mm at 60 kV ac (peak). (b) Bush type negative streamer in a gap length of 50 mm at 60 kV ac (peak).

branches like a tree, it is described as tree type streamer by some authors. Felici further discovered that the fast and filamentary ones replace the slow negative streamers when an electron scavenger is added to the liquid. It has been concluded in general that streamers in liquid dielectrics travel slowly and at a very broad range of speed (0.05 to 20 mm/μs). Further, the light emitted by streamers has been correlated with the instantaneous current. Fiebig and Hauschild had also earlier discovered the above-mentioned properties of streamer development.

The ultimate breakdown is achieved when an unstable leader bridges the gap. Hauschild [6.13] investigated the breakdown mechanism in insulating oils in

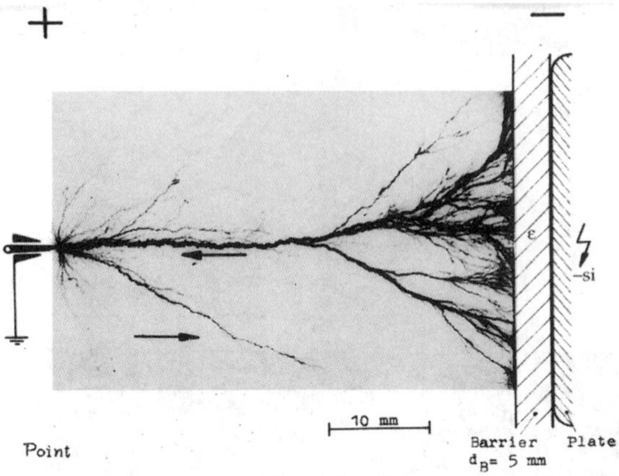

Figure 6.24 Discharge in transformer oil in extremely nonuniform field with an insulating barrier at the plane for si of negative polarity, showing stable leader developing in backward direction, Hauschild [6.13].

extremely nonuniform field between needle-plane gap of 50 mm applying switching impulse of 50/5000 μs. In order to prevent a complete breakdown with an arc between the electrodes, yet achieve a stable leader discharge through the gap, Hauschild covered the plane electrode with an "insulating barrier" having a thickness of 5 mm. The discharge extending up to the barrier in transformer oil is shown in Figure 6.24. This photograph was also taken by the Schlieren photographic technique suitable for taking photographs in liquids. The plane electrode was applied a negative polarity *si* while the needle electrode was earthed. From these investigations it could be concluded that the growth of the main leader is in the direction from negative plane (cathode) towards positive needle (anode), as shown by the arrow on the photograph. He proposed the name "backward leader" for this kind of discharge, as it develops in the opposite direction of the field intensity. This phenomenon is similar to the propagation of "negative leader" described in Section 3.3.3.2.

Experiments were performed to investigate the development of positive and negative streamers in bulk oil and also along pressboard surface in transformer oil on a 150 mm gap between 190 μm tip needle and plane electrodes by Lundgaard et al. [6.24]. Figure 6.25 shows schematic and actual tracks of positive streamer development on pressboard surface between needle and plane. It could be observed that the velocity of propagation of initial streamer development was faster (about twice) with pressboard than without the board. In this case the velocity of streamer propagation of the order of 800 mm/μs was recorded. In the presence of pressboard, faster propagation could be attributed to the surface resistivity of the dielectric, which is always lower than the volume resistivity. As observed in gaseous dielectrics, in liquid dielectrics the stable PB streamer channels are also highly conductive. Potential gradient of the order of 2 kV/mm has been estimated along the positive streamer channels in transformer oils, Lundgaard [6.24]. It is approximately four times greater than the estimated potential gradient required for positive streamer in atmospheric air, Section 3.3.2.1. This is an expected phenomenon since the dielectric strength of oils is much higher.

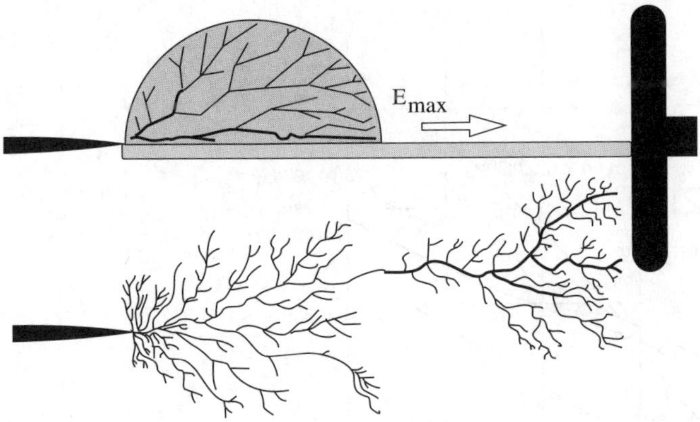

Figure 6.25 Development of positive streamer along a pressboard surface in oil. Upper: The positive streamer schematic, Lower: Actual track observed on the pressboard surface, Lundgaard [6.24].

A comparison between breakdown voltage characteristics of a sample of commercial transformer oil in extremely nonuniform field between needle and plane with *ac* power frequency, positive polarity *si* and *li* voltages is shown in Figure 6.26, also measured by Hauschild [6.13]. Up to a gap distance of about 100 mm, the *li* breakdown voltage is measured to be lower than *si*. However, above a gap distance of 100 mm the *li* breakdown voltage is much higher. A similar trend is seen in the breakdown characteristics with all the three types of voltages at a gap distance between 25 and 40 mm. As in case of gaseous dielectrics, in transformer oils too, it represents stable PB activity above a certain gap distance between given electrode systems. Compared to air, the PB inception voltages are much higher in oils.

6.4 AGING IN MINERAL INSULATING OILS

Aging of mineral oils is a continuous process in oil filled equipment. Ingress of moisture and oxygen from the atmosphere may take place in storage, handling and in service. In the presence of metallic parts in apparatus, for example, copper, lead, aluminium and others, the catalytic effects of metal ions accelerate the oxidation reactions in oils at high working temperatures. As a result, this gives rise to various oxidative products, such as peroxides, alcohols, ketones, esters and acids. These products may ultimately combine together to form higher molecular-weight compounds, both soluble and insoluble in oils. Insoluble oxidative products not only darken the oil color but in the advanced stage of oxidation lead to sludge formation. The degradation of physical as well as electrical properties of oils with time in electrical equipment is an unavoidable process.

Aging of oils leads to deterioration of their insulating properties. The electric strength of oils as well as their loss tangents and specific resistances are greatly affected by the presence of water content, contaminants and oil-soluble compounds. Resulting from oxidation, the oil soluble compounds are also known as polar

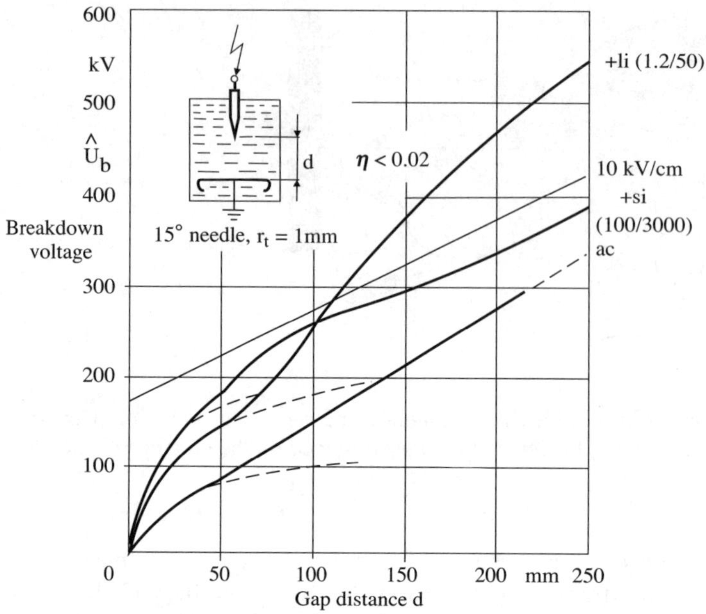

Figure 6.26 Power frequency ac, positive polarity si and li breakdown voltages of transformer oil in extremely nonuniform field between needle-plane with increasing gap distances, Hauschild [6.13].

substances [6.25]. It has been observed that the interfacial tension and viscosity of oils also change fairly rapidly during the early stages of aging. However, the rate of their change decreases with time. An interesting correlation is mentioned in [6.25] that the formation of sludge in the mineral insulating oils is possible if the interfacial tension value falls below 0.018 N/m. Deterioration in viscosity and interfacial tension of oil may also affect the cooling action in the equipment.

Fournie et al. [6.26] studied the aging process on fresh mineral insulating oils, subjecting them to photochemical, electrical and thermal stresses. During their investigations they found the evidence of increase in corrosive sulphur-index of the mineral oils when subjected to sunlight. The action of light on insulating oils caused the formation of abundant sludge, containing sulphur compounds. Infrared spectra of the sludge revealed that these were various stages of sulphur oxidations. These photo-oxidized oils showed a very poor electric strength property.

Studies made by Fournie et al. further revealed that in darkness and at room temperature, application of an electric field on fresh oils caused smaller quantity of sludge formation. However, the combined action of light and the electric field led to the formation of sludge that settled down on the conducting parts and the solid dielectrics.

The thermal aging alone does not result in oxidation of sulphur compounds, although sometimes it leads to abundant sludge formation. The degradation of electrical properties of oils is found to be much slower by thermal aging than under

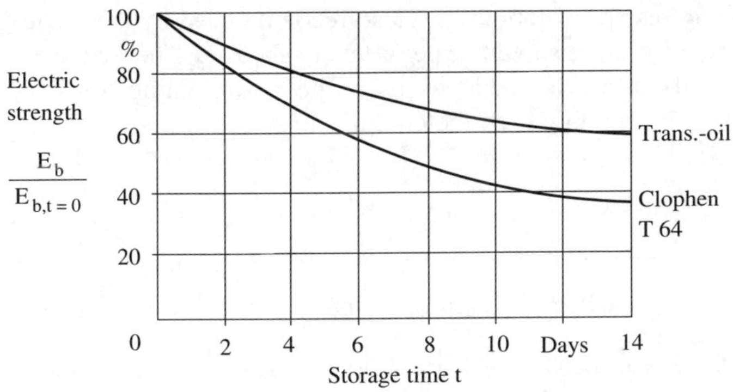

Figure 6.27 Degradation of electric strength of liquid dielectrics due to aging.

ultraviolet radiation. In worst cases, there are evidences of the presence of solid sludge in transformers and circuit breakers with sulphur contents of the order of 4 to 6%.

Degradation in electric strength of transformer oil and Clophen T-64, with respect to the storage time in air at 20°C and 63% relative humidity, are shown in Figure 6.27. Just in two weeks, the electric strength almost reduced to half when the oil was exposed to the atmosphere.

Investigation of radiation-induced chemical degradation phenomena in mineral and synthetic electrical insulating oils by Yasufuku [6.27] revealed that γ-irradiation accelerates the corrosive action of the oils in the presence of copper strips. The sulphur compounds in oil change rapidly to corrosive sulphur under irradiation and accelerate corrosion on the copper surface. Similar investigations by Krishnamoorthy et al. [6.28] revealed that the copper/oil ratio is the most significant factor in accelerating the thermo-aging process. Whereas copper has negligible effect on the rate of oxidation at 70°C, it increases with the increase in copper/oil ratio above 85°C.

The general principles and maintenance rules that can be applied for handling, reconditioning and replacement of insulating oils are found in various standards for a given type of equipment. The physical means used for removing solid particles and insoluble contaminants in suspension from the oils and to decrease their water contents to acceptable level include several types of filters, centrifuges and vacuum dehydrators. However, these cannot be effectively removed when the foreign substances are in dissolved or in colloidal form. The treatment devices do not remove air; on the contrary, they tend to aerate the oil. The water removing ability of a filter is dependent upon the dryness of the filter media. A centrifuge can handle much greater concentration of contamination than a conventional filter but cannot remove some of the contaminants as effectively as a filter. Consequently, a centrifuge is generally used for rough and bulk cleaning where a large amount of contaminated oil is to be handled. The output of the centrifuge is put through a filter for the final clean up. Apart from removing water, the vacuum dehydrators de-gas the oil and remove more volatile acids. For more details, the reader may refer to IS-1866 (2000) [6.25].

It is suggested in various specifications and also recommended by the manufacturers that the quality of oils in use should be periodically checked. The frequency of such examination of oils in service depends upon the power rating, loading, construction and other service conditions of the equipment.

REFERENCES

[6.1] Rouse, T.O., "Mineral Insulating oil in Transformers", IEEE Electrical Insulation Magazine, Vol. 14, No. 3, May/June 1998, pp. 6–16.

[6.2] Brinkmann, C., "Insulating Materials for Electrical Engineering", Springer-Verlag, Heidelberg (1975) (in German).

[6.3] Bogorodizki, N.P., Pasynkow, W.W., Tarejew, B.M., "Materials for Electrical Engineering", VEB Verlag Technik, Berlin (1955) (in German).

[6.4] Zahn, M., Ohki, Y., Rhoads, K., LaGasse, M., Matsuzawa, H., "Electro-optic Charge Injection and Transport Measurements in Highly Purified Water and Water/ethylene Glycol Mixtures", IEEE Trans. EI-20 (April 1985) 2, pp. 199–211.

[6.5] Bayer, M., Boeck, W., Möller, K., Zaengl, W., "High Voltage Engineering", (Theoretical and Practical Fundamentals for Application), Springer-Verlag, Berlin (1986) (in German).

[6.6] Holle, Karl-Heinz, "On Electrical Properties of Insulating Oils in particular the Effect of Water on their Behavior at Different Temperatures", Dissertation, TH Braunschweig (FRG), 1967 (in German).

[6.7] Umemura, T., Akiyama, K., Kawasaki, T., Kashiwazaki, T., "Electrical Conduction in Synthetic Insulating Liquid", IEEE Trans. EI-17 (Dec. 1982) 6, pp. 533–538.

[6.8] Schmidt, Werner F., "Electrons is Non-poler Dielectric Liquids", IEEE Transactions on Electrical Insulation., Vol. 26, No. 4, August, 1991, pp. 560–567.

[6.9] Takashima, T., Hunaoka, R., Ishibashi, R., and Ohtsubo, A., "I-V Characteristics and Liquid Motion in Needle-to-plane and Razor Blade-to-plane Configurations in Transformer Oil and Liquid Nitrogen", IEEE Trans. EI-23 (1988) 4, pp. 645–658.

[6.10] Nelson, Keith J., "Dielectric Fluids in Motion", IEEE Electrical Insulation Magazine, May/June, 1994, Vol. 10, No. 3, pp. 16–27.

[6.11] Atten, P., Haidara, M., "Electrical Conduction and EHD Motion in Dielectric Liquids in a Knife-plane Electrode Assembly", IEEE Trans. EI-20 (April 1985) 2, pp. 187–198.

[6.12] Musil, R. Baur, M. and Pfister, W., "Testing Practices for the ac Breakdown Voltage Testing of Insulation Liquids", IEEE Electrical Insulation Magazine, January/February 1995, Vol. 11, No. 1.

[6.13] Hauschild, Wolfgang, "Breakdown in Oil in Nonuniform field with Switching Impulse", Dissertation (1969), TU Dresden (in German).

[6.14] Fiebig, Reiner, "Discharge Appearance and Mechanism in Insulating Liquids with Alternating Voltage in Nonuniform Fields", Dissertation (1968), TU Dresden (in German).

[6.15] Torshin, Yuri V., "Prediction of Breakdown Voltage of Transformer Oil from Predischarge Phenomena", IEEE Transactions on Dielectrics and Electrical Insulation, Vol. 10, No. 6, December 2003, pp. 933–941.

[6.16] Top T.V., Massala, G., and Lesaint, O., "Streamer Propagation in Mineral Oil in Semi-Uniform Geometry", IEEE Transactions on Dielectrics and Electrical Insulation, Vol. 9, No. 1, February 2002, pp. 76–83.

[6.17] Chadband, W.G., "The Ubiquitous Positive Streamer", IEEE Trans. EI-23 (Aug. 1988) 4, pp. 697–706.

[6.18] Yamashita, H., Amano, H., "Prebreakdown Phenomena in Hydrocarbon Liquids", IEEE Trans. EI-23 (Aug. 1988) 4, pp. 739–750.

[6.19] Watson, P. Keith, Chadband, W.G., "The Dynamics of Prebreakdown Cavities in Viscous Silicone Fluids in Negative Point Plane Gaps", IEEE Trans EI-23 (Aug. 1988) 4, pp. 729–738.

[6.20] Felici, N.J., "Blazing a Fiery Trail with the Hounds", IEEE Trans. EI-23 (Aug. 1988) 4, pp. 497–503.

[6.21] Forster, E.O., "Progress in the Filed of Electrical Breakdown in Dielectric Fluids", IEEE Trans. EI-20 (Dec. 1985) 6, pp. 905–912.

[6.22] Forster, E.O., "Critical Assessment of the Electrical Breakdown Process in Dielectric Fluids", IEEE Trans. EI-20 (Oct. 1985) 5, pp. 891–869.

[6.23] Lewis, T.J., "Breakdown Initiating Mechanisms at Electrode Interfaces in Liquids," IEEE Transactions on Dielectrics and Electrical Insulation, Vol. 10, No. 6, December 2003, pp. 948–955.

[6.24] Lundgaard, Lars; Linhjell, Dag; Berg, Gunnar and Sigmond, Svein, "Propagation of Positive and Negative Streamers in Oil with and without Pressboard Interfaces", IEEE Transactions on Dielectrics and Electrical Insulation, Vol. 5, No. 3, June 1998, pp. 388–395.

[6.25] Indian Standard 1866 (2000), "Code of Practice for Maintenance and Supervision of Mineral Insulating Oil in Equipment".

[6.26] Fournie, R., Gall, Y. Le Perret, J., Recoupe, P., "Contribution to the Study of the Ageing of Mineral Oils", 3rd ISH, Milan (Aug. 1979) paper 23.05.

[6.27] Yasufuku Sachio, Ise Junichi, Kobayashi Shiegeo, "Radiation Induced Degradation Phenomena in Electrical Insulating Oils", IEEE Trans. EI-13, No. 1 (Feb. 1978) pp. 45–50.

[6.28] Krishnamoorthy, P.R., Krishnaswamy, K.R., Vijayakumari, S., and Kumar, K., "Ageing of Mineral Oils, A Diagnostic Study", Proceedings of the 3rd International Conference on Properties and Applications of Dielectric Materials, July 1991, Tokyo, Japan.

CHAPTER 7

SOLID DIELECTRICS, THEIR SOURCES, PROPERTIES, AND BEHAVIOR IN ELECTRIC FIELDS

A vast number of solid insulating materials are used in electrical power engineering. With the invention of modern insulating materials, many earlier conventional materials have been discarded, especially in case of electrical machines, capacitors and power cables. From their application point of view, the solid insulating materials can be divided into the following three broad categories:

Moulding Materials: These are used for mechanically rigid forms of insulation, for example, insulators, bushings and so on. Such materials are required to have good electrical insulating properties besides having high mechanical strength. These are usually made out of ceramics, glass (toughened glass), fiberglass, reinforced plastics and epoxy-resins.

Jacketing Materials: Polymers have been found suitable for providing extruded insulating jackets to the conductors. For example, polyethylene (PE), polyvinylchloride (PVC), natural and synthetic (ethylene propylene) rubber are extruded on the conductor in power cables. Polypropylene and paper are used in capacitors and transformers. Mica and fiberglass based polypropylene tapes are used in electrical machines.

Filling Materials: Beside oils, wax-based draining and non-draining impregnating compounds of different types are used to impregnate paper used in power cables, transformers, capacitors, and instrument transformers.

Insulating Mechanical Support: In the form of plates, pipes and ledges, insulating supports are required in transformers, circuit breakers and isolators. The products, such as pressboards, hard paper (thin paper laminates), wood (yellow teak) are used in transformers. Ebonite, also known as vulcanite, is a form of cross-linked rubber containing large proportion of sulphur. Bakelite, a hard synthetic material, and acrylic resin plates are used for providing insulating supports in other equipment.

High Voltage and Electrical Insulation Engineering, First Edition. Ravindra Arora, Wolfgang Mosch.
© 2011 Institute of Electrical and Electronics Engineers. Published 2011 by John Wiley & Sons, Inc.

Unlike gaseous and liquid dielectrics, any damage caused to solid dielectrics due to excessive electrical, thermal or mechanical stresses is often irreversible. In practical application of solid dielectrics, their thermal and mechanical properties play a very sensitive role since these considerably affect the electrical properties. The period of time through which a solid dielectric may continue to provide satisfactory insulation without failure under combined thermal, mechanical and electrical stresses is known as "life of a dielectric". This is a very important aspect to be taken into consideration for the application of solid dielectrics in power system. If exposed to atmosphere, hazards of weather (rain, storm, hail, ice deposits etc.), ultra violet radiation from the sun and pollution (dust, salts etc.) withstand capability are other aspects that must also be considered. Irrespective of the cause, if once failure of a solid dielectric takes place, it may cause not only big damage but also a discontinuation of the power supply untill alternating arrangements or tedious repairs are undertaken.

It would not be possible to describe the vast variety of solid insulating materials available, which may have found some application in the industry. However, a few most widely used insulating materials are briefly described below.

7.1 CLASSIFICATION OF SOLID INSULATING MATERIALS

The most common method of classifying the solid insulating materials is based upon their chemical compositions, distinguished mainly between inorganic, organic and the composites of the two materials.

7.1.1 Inorganic Insulating Materials

Among the inorganic materials; ceramics, glass, fiberglass, enamel, mica and asbestos are important and have found their wide application as dielectrics in the order of the mentioned sequence. They distinguish themselves in their unique ability to withstand high temperatures in addition to their being highly chemical resistant. As such, there is practically negligible sign of aging in these materials. However, it is difficult to machine or process them. These materials are inherently not homogeneous, either microscopically or macroscopically.

7.1.1.1 Ceramic Insulating Materials The word "ceramic" has its origin from Greek. Ceramic substances are those produced from clay, containing aluminiumoxide and other inorganic materials. Their thick paste is given the desired shape and form at room temperature and then baked at a high temperature (about 1400°C) to provide a solid, inelastic final structure.

Ceramics, also known as "porcelain" in one of its forms, has a history of over a century of service as insulating materials. Porcelain is widely used for insulators and bushings in the power system because of its high mechanical strength. Ceramics constitute 40 to 50% of clay, 30 to 20% of aluminium oxide and 30% of feldspar, that is, ($3Al_2O_3{\cdot}2SiO_2$) and SiO_2. Ceramics with higher mechanical strength and lower dielectric losses also contain steatites (soapstone) and talc (MgO, SiO_2).

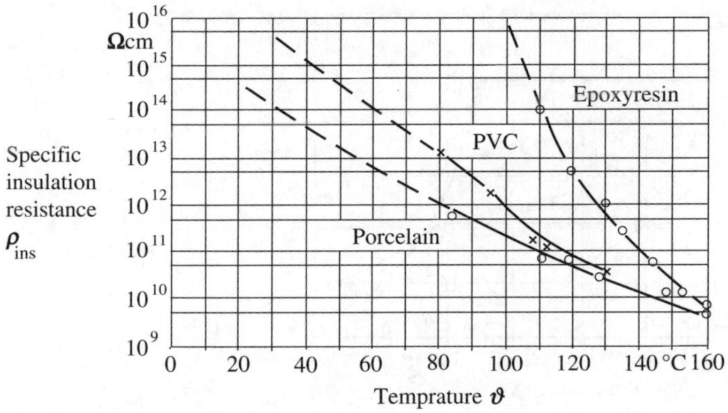

Figure 7.1 Specific insulation resistance variation with temperature of solid insulating materials.

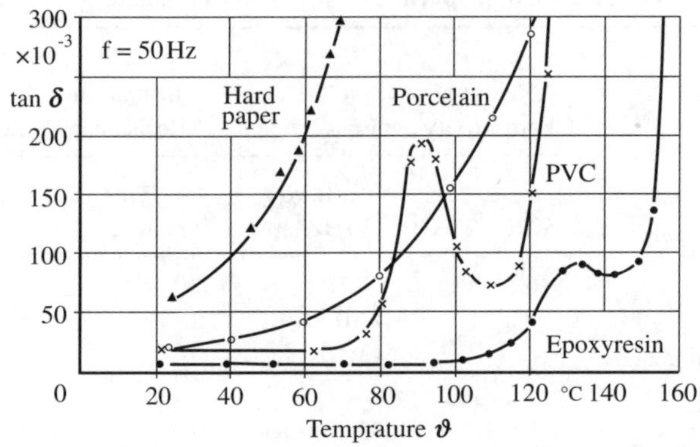

Figure 7.2 Variation of loss tangent "tan δ" with temperature of solid dielectrics at 50 Hz ac voltage.

Porcelain suitable for high frequency applications is provided with barium compounds ($BaOAl_2O_3.2SiO_2$), Bogorodizki et al. [7.1].

The specific insulation resistance of ceramics is comparatively low. Its variation with temperature is shown in Figure 7.1. The loss tangent of porcelain is high and it increases considerably with temperature as shown in Figure 7.2, thereby increasing the dielectric losses. The breakdown strength of porcelain compared to other solid insulating materials is low but it is unaffected by temperature for a wide range, as shown in Figure 7.3. In Table 7.1, standard values of some electrical properties of ceramics at power frequency and normal temperature are brought together.

The very dense nature of ceramics makes the porcelain bodies very heavy. The large size and heavy weight of porcelain products make them difficult to handle, which often require cranes as well as expensive and large structural supports. Ceramics are very brittle and thus break easily during handling, transit and installation. These are also prone to mechanical damage due to vandalism in service.

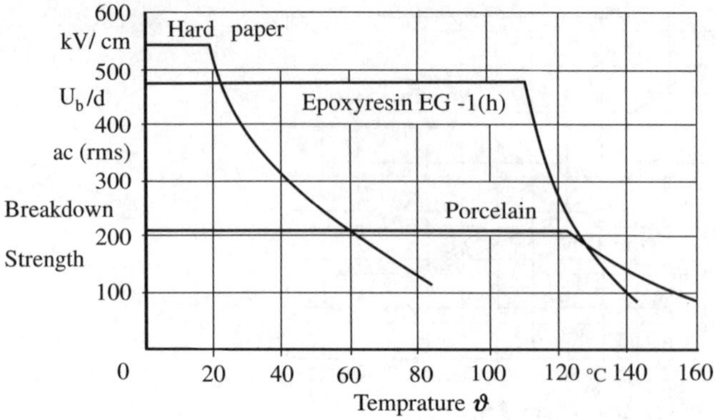

Figure 7.3 Practical breakdown strengths (ac, rms) of some solid insulating materials in uniform field as a function of temperature.

TABLE 7.1 Standard values of electrical properties of some solid insulating materials

Materials and their Classification		Relative Permittivity ε_r (50 Hz, 20°C)	Loss tangent tan δ (50 Hz, 20°C)	Specific Insulation Resistance ρ Ω.cm (20°C)	Specific Thermal Resistance* σ °C cm/W
Ceramics	Porcelain	5–6.5	20.10^{-3}	10^{11}–10^{12}	70–125
	Steatite		2.10^{-3}		40–50
Glass	Nonalkaline and E-Glass	3.8–10	$<1.10^{-3}$	10^{13}	85–135
Thermoplastic Polymers	Polyvinyl-Chloride (PVC) (Cable insulation compound)	5.0–5.3	$30–100.10^{-3}$	10^{16}	600–700
	Polyethylene (PE)	2.3	$0.1–0.2.10^{-3}$	10^{17}	350
Thermoset Polymers	Bisphenol-A	Pure 3.5	$5.5.10^{-3}$	$>>10^{17}$	—
	Epoxyresin	with filler 5.8	33.10^{-3}	—	—
	Silicon Rubber (SiR)	2.8–6.0	$5–10.10^{-3}$	10^{13}–10^{15}	500
Composite Dielectrics	Oil impregnated Cable paper	3.5–3.9	$(2.6–3.0)10^{-3}$	10^{15}	550

* inverse of Thermal conductivity "λ" in $\frac{W \cdot cm}{cm^2 \cdot °C}$.

Under polluted conditions, the hydrophilic surface of porcelain does not perform well. The water dissolves part of the pollution, thereby lowering the overall surface resistance due to conductive electrolytes. Such a situation develops especially under fog, dew or light rain when water deposits on the polluted surface of the insulators. Surface discharge or tracking process incepts at much lower voltages under this condition. Although the chemical stability of porcelain resists aging, it allows the surface to easily get wet under contamination layer. Such a condition may lead to flashovers at fairly low voltages and cause heavy damage.

7.1.1.2 Glass as an Insulating Material The main constituent of glass is silicondioxide SiO_2, available in nature in the form of quartz. A number of metal oxides—for example, lead, aluminium, sodium, barium, potassium and so on—provide a possibility of producing more than 500 types of glass. However, for the use in electrical engineering only "nonalkaline glass", or glass having alkaline content less than 0.8% are suitable. The conductivity, hence the losses in such glass are low. The so-called E-Glass, having the same properties as nonalkaline glass, is used for producing glass fiber. The "fiberglass" is used for reinforcing plastic materials to obtain high mechanical strength. These "fiberglass reinforced plastics" are utilized as tapes in electrical machines and as moulds for insulated containers, rods and so on. Some electrical properties of nonalkaline glasses are given in Table 7.1.

7.1.2 Polymeric Organic Materials

Solid organic materials used in electrical engineering are paper, wood, wax, leather, as well as a number of natural and synthetic resins, rubbers and plastics, also known as polymers. Wood being so hygroscopic is rarely used as such, but wooden poles supporting overhead cables and distribution lines are widely used in North America.

The polymeric organic insulating materials used in electrical engineering have a very high molecular weight and consist of two or more polymeric compounds of several structural units normally bound together by covalent bonds. The individual structural units may consist of single atoms or may be molecular in nature, which repeat in a regular order.

A polymer can be obtained by the reaction of compounds having at least two reactive functional groups, which can react under suitable conditions. These chemical compounds are known as "monomers". There can be one or a mixture of reacting monomers and in the latter case, the resulting material is called "copolymer", Kumar [7.2].

The early polymer insulating materials did not provide the expected service life of at least forty years. It was primarily because of inadequate UV, tear or hydrolytic resistance. However, with the development in material science, the use of high performance polymeric materials for outdoor insulation applications in the power industry has been steadily increasing over time in the last half century. Common applications include cable terminations, surge arresters, insulators, bus bar insulation and bushings, Mackevich and Shah [7.3].

The density of polymer materials is much lower than ceramics. Hence, the polymer products are significantly lighter and easier to handle and install. The reduced weight also permits the use of lighter and less costly structures. The polymeric materials resist wetting, being inherently hydrophobic. Water on the surface of such materials may form water beads. Thus the conductive contamination dissolved within the water beads becomes discontinuous. Consequently lower surface creepage current is able to flow and the probability of dry band formation increases. It appreciably prevents reduction in flashover voltages under polluted conditions.

There are several ways of classifying the polymers. One of them is by their response to heat. Accordingly, the polymers are divided into two groups of materials as follows:

7.1.2.1 Thermoplastic Polymers Thermoplastic polymers soften and become supple on heating and "solidify" back on cooling. The heating and cooling cycle within certain temperature limits can be applied to these materials several times without affecting their properties. Polymers, which generally have a linear structure, fall in this category. The synthetic thermoplastic polymeric materials used in electrical engineering for insulation purposes are polyethylene (PE), polyvinylchloride (PVC), polypropylene (PP) and polyamide (PA). The extent of their application in the industry is in the order of their above-mentioned sequence. Such materials have relatively poor thermal resistance and their properties deteriorate rapidly at higher temperatures.

7.1.2.2 Thermoset Polymers The polymers, which soften when heated for the first time resulting into cross-linking reaction (network formation), are known as thermoset polymers. This reaction, leading to the formation of network structure, is also known as curing or setting of the polymer. All monomers having functionality equal to two would give rise to linear polymers and, therefore, thermoplastic materials. However, when they have functionality more than two, the resulting polymer has a network structure. The polymers used for electrical insulation purpose are desired to retain their rubbery (flexible) properties. Hence, these are defined as "lightly cross-linked polymers".

The chemical reaction leading to cross-linking is achieved with the help of an additive, known as "agent". The term "cross-linking agent" is very general. It refers to:

(a) molecules that bridge two polymer molecules, for example, vulcanizing agents; such as sulphur, selenium and sulphur monochloride for rubbers; and polyamines in epoxy-resins,

(b) molecules that initiate a cross-linking reaction, for example, peroxides which initiate a double bond polymerization in polyesters and dicumyl peroxide in low density polyethylene (LDPE);

(c) those that are purely catalytic in their action such as acids for phenolic resins, amines in epoxies; and

(d) active site generators, for example, peroxides that abstract protons from the polymer chains.

The thermal, mechanical and electrical properties of different cross-linked materials vary considerably from each other. However, in general, after cross-linking, a material does not soften on reheating, instead it becomes thermally more stable compared to thermoplastic materials. Cross-linked polymer resins are, for example, polyester-resin, phenol-resin, silicon-resin and the most widely used in electrical engineering, the epoxy resins. Among rubbers, typical examples used for electrical insulation are natural rubber, also known as "India rubber", and a large number of synthetic rubbers such as silicon rubber (SiR), ethylene-propylene rubber (EPR), and so on. Cross-linked polyethylene (XLPE) is now being used extensively for power cables upto the highest voltages.

7.1.2.3 Polymer Compounds The bulk properties of a polymer can be suitably altered by the incorporation of a number of additives. Variations in the choice of additives can produce widely differing products. This is particularly true with PVC. In terms of functions, there are a large number of groups of additives, of which the following are the most important commonly used in preparing polymer compounds for electrical insulation purposes:

1. Fillers: usually applied to modify physical properties, mainly mechanical, of a polymer.
2. Plasticisers and Softeners: to lower the melt viscosity and also to change physical properties (softness, flexibility).
3. Colorants: normally soluble colorants (dyestuffs) are used.
4. Anti-aging additives: to prevent structural degradation due to chemical reactions like oxidation, ozone attack, dehydrochlorination (especially with PVC) and ultra-violet irradiation due to exposure to the sun.
5. Flame retarders: to improve the degree of fire resistance of polymers.
6. Cross-linking agents: to achieve intermolecular combination at the chain ends.

7.1.2.4 Polyvinylchloride (PVC) PVC is a polymerization product of the monomer vinylchloride derived from ethylene, where one of the H-atoms in the molecule is replaced by Cl-atom, as shown in Figure 7.4 (a). A catalytic agent is added under pressure to polymerize.

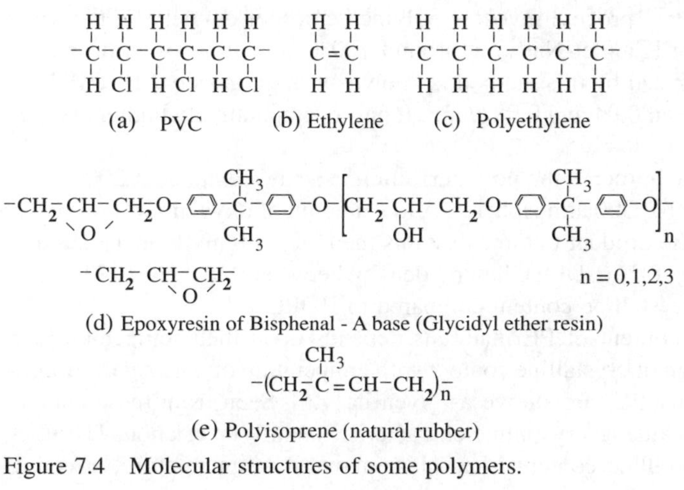

Figure 7.4 Molecular structures of some polymers.

PVC as such a polar dielectric is a very hard, brittle and thermally unstable product. It needs, therefore, to be made suitable for use as electrical insulation by adding other materials. The PVC compounds contain softening, stabilizing and filling substances, besides coloring and smoothing agents. Suitable PVC compounds are very widely used for low voltage power cables up to 1.1 kV for insulation, filler as well as outer sheathing. For providing protective outer layer it is used for up to the highest voltage power cables. A compound for these cables may contain about 68% PVC polymer; 28% plasticizer, the non-volatile solvents of PVC such as, di-iso-octyl phthalate (DIOP),di-2-ethylhexyl phthalate (DOP) etc; 3% stabilizer, lead-sulphate or lead-phthalate, and about 1% of coloring agent and other additives. Stabilizers are useful in improving the resistance of PVC to weathering, particularly against degradation by ultra-violet radiation. Whereas tribasic lead sulphate is recommended in normal compounds for electrical insulation purpose, dibasic lead phthalate is used in heat resistant insulation compounds, Brydson [7.4]. Chalk powder or china clay is used in different proportions as filler substance only for sheathing compounds. Hardwax type smoothing agents are also added in order to achieve a smoother extrusion of these compounds.

As given in Table 7.1 and illustrated in Figure 7.2, the dielectric loss tangent of PVC compounds is rather high. This has restricted its use to be suitable only for low voltage power and house wiring cable insulation. Thereby, the maximum field intensity is tried to be limited within 3 kV/mm. However, PVC compounds are very widely used for providing inner sheathing and overall covering in power cables. Variation in specific insulation resistance of electrical insulation grade PVC compound is shown in Figure 7.1 with increasing temperature.

7.1.2.5 Polyethylene (PE) Polyethylene, a polymer of ethylene, is also called polythene. It is produced by the polymerization of monomer ethylene. The molecular structures of both ethylene and polyethylene are shown in Figure 7.4 (c). Polyethylene is produced by two distinctly different processes, namely the "low pressure" and the "high pressure", classified according to the magnitude of pressure applied during manufacturing.

The "low pressure" process involves polymerization of ethylene in the presence of catalytic agent (Ziegler-Natta) at around 100°C and at a few atmospheric pressure. The PE produced by this method is known as "high density PE" (HDPE), having a density between 0.94 and 0.98 g/cm^3. It has a comparatively higher crystalline content.

In "high pressure" process the polymerization is achieved at about 200°C and a pressure between 150 to 300 atmospheres. A low amount of oxygen serves as catalyst in this process. The product obtained by this method is highly branched and is known as "low density PE" (LDPE), having density between 0.91 and 0.923 g/cm^3. It has a much lower crystalline content compared to HDPE.

The crystalline content of PE materials depends upon their temperature. In Figure 7.5, the variation of crystalline content with temperature of particular samples of high and low density PEs are shown as given in [7.5]. Seen from these curves, the high density PE retains its crystalline characteristics untill up to about 110°C. A very steep fall in crystalline content takes place above this temperature, affecting

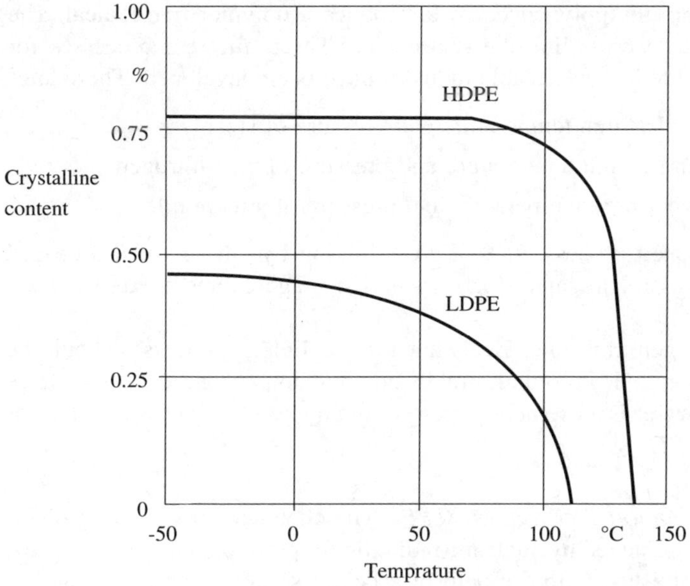

Figure 7.5 Effect of temperature on crystallization of PE, Viewag [7.5].

the electrical and mechanical properties of the material. In the case of low density PE, the decrease in crystalline content begins at –20°C and it continues up to about 115°C, which is the melting point of the material. It is because of this particular property that for an apparatus using normal HDPE, the working temperature strictly should not exceed 70°C. These characteristics show that PE as such has poor thermal properties.

The PE compound is extruded on the conductor of a cable at high temperatures. Since it shrinks by about 15%, internal mechanical stresses may develop if cooling to room temperature is not achieved gradually. Although the water absorption in PE is low, it may let through water vapour. PE is also susceptible to mechanical cracks.

In order to improve the thermal and mechanical properties of PE, chemical cross-linking of PE insulation is done. The LDPE, basically a thermoset material, is generally used for obtaining cross-linked polyethylene (XLPE) insulation. It is able to withstand a continuous temperature of 90°C and a short time fault temperature of even 250°C. The following processes can achieve cross-linking or vulcanization of PE, used for cable insulation application.

7.1.2.5.1 Chemical Process for XLPE General Electric in 1953 was the first organization to establish the application of chemical cross-linking of polyethylene for solid dielectric power cables. In this process the LDPE insulation compound is added with a cross-linking agent, such as dicumyl peroxide (DCP) to initiate a chemical reaction, accomplished under high pressure and temperature. The peroxide molecules break up, producing free radicals at elevated temperature. These abstract hydrogen from the polymer chain to produce a polymer free radical. The two radicals combine to cross-link the two chains. Three different processes for curing PE with peroxide in power cable industry have been developed. These are:

- "wet curing", under high temperature and pressure of steam
- "dry curing", under high temperature and pressure of gas (nitrogen)
- "oil curing", under high temperature and pressure of silicon oil.

The last mentioned process, that is, cross-linking by silicon oil has proved to be extremely successful in Japan for very high voltage cables rated at 500 kV and above.

Presence of oxygen molecules in chemical cross-linking processes should be avoided because they may lead to undesirable side reactions. Hence, the avoidance of oxygen and even water is extremely important in any curing process adopted for the production of XLPE.

7.1.2.5.2 Radiation Process for XLPE The chemical cross-linking of PE with DCP can also be achieved by nuclear irradiation for producing the free radicals. It is generally accomplished with the help of a cobalt source. This technique has been widely applied commercially in the United States, but only for medium voltage power cables upto 33 kV. For higher voltages, it is not a successful method as the depth of cross-linking is limited. It requires expensive equipment and extensive protective measures.

7.1.2.5.3 Silane Cross-Linked Polyethylene (SXLPE) Dow Corning in the late 1960s introduced another technology for cross-linking polyethylene using organofunctional silanes. Since the early 1970s, silane cross-linking is in application for manufacturing XLPE pipes and power cables [7.6]. This technology involves mainly three reactions; grafting of silane, $Si(OCH_3)_3$, on the PE, hydrolysis of the silane groups and condensation to form Si-O-Si links between the PE molecules, termed generally as "cross-linking". An organic peroxide initiator is used in the grafting reaction and a tin catalyst in cross-linking. In this case, the cross-linking or curing is accomplished after extrusion of the insulation simply in hot water or steam bath. It is a simpler process, hence it has much lower cost.

SXLPE, also known as Sioplast material, is hard due to the introduction of silicone. It can therefore be used for abrasion resistant cables. The SXLPE technology is so far applied only for low voltage cables (upto 33 kV) in many countries [7.6].

7.1.2.5.4 Electrical Properties of PE and XLPE PE possesses excellent electrical properties superior to those of many other dielectrics. It has a very low relative permittivity "ε_r" and being a nonpolar material, it also has very low loss tangent "$\tan\delta$", both of which are almost independent of temperature and frequency. The "specific insulation resistance" ρ, of PE is quite high. It is a function of temperature as well as the applied field intensity. The inverse of ρ, the specific dc conductivity "κ_{dc}", of LDPE measured by Krohne [7.7] with increasing field intensity at different temperatures is illustrated in Figure 7.6. Variation of loss tangent of normal LDPE and cross-linked LDPE by different processes with increasing alter-

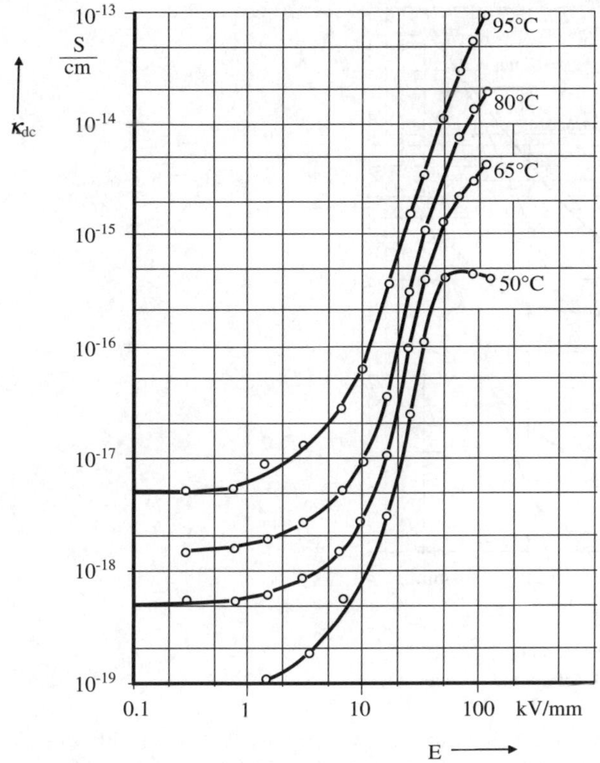

Figure 7.6 Specific dc conductivity κ_{dc} of LDPE with respect to increasing field intensity at different temperatures, Krohne [7.7].

nating 50 Hz field intensity are shown in Figure 7.7. These measurements were carried out by Sander [7.8], on test samples as shown in the figure at constant temperature of 80°C. The test cell developed contained guard-ring electrodes and it was made PB free by immersing the whole cell in SF_6 gas. Sander measured the increase in $\tan\delta$ both with increasing temperature as well as electric field. The increase is explained to be due to electrons getting activated from their donator condition by thermal and electrical stresses. According to Figure 7.7, the loss tangent of a particular PE is almost unaffected untill up to a field intensity of about 15 kV/mm. Beyond this value a sharp rise in $\tan\delta$ is measured.

As described later in Section 7.3.1, PE has a very high intrinsic breakdown strength, of the order of 700 kV/mm. Such excellent properties of PE have enabled production of extruded insulation XLPE cables up to a rated voltage of 500 kV and above in Japan. For lower voltages, XLPE cables are also very widely used all over the world since the beginning of 1970s. However, to ensure a long service life, of the order of forty years or more, the maximum electric stress in medium voltage cables (up to 33 kV) was kept within 4 kV/mm. In higher voltage cables, where the production and design demand extra care and efforts to achieve perfection, the maximum field intensities gradually increased even up to 15 kV/mm during the last two decades (after 1985).

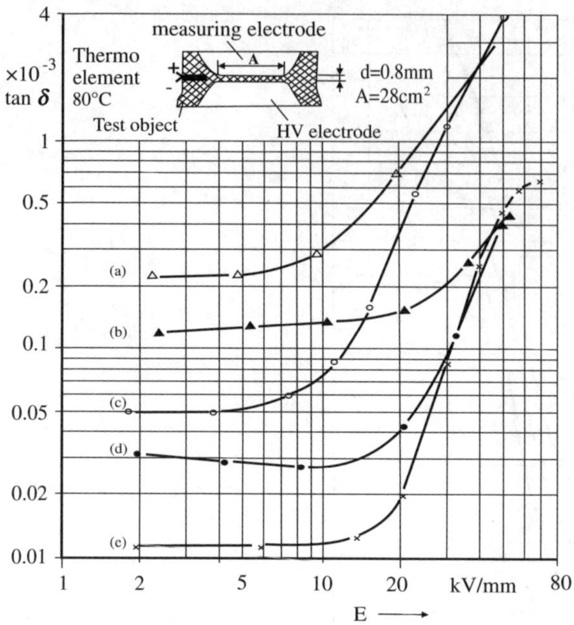

(a) XLPE (wet cured) ; (b) XLPE (wet cured , evacuated)
(c) XLPE (dry cured) ; (d) XLPE (dry cured , evacuated)
(e) LDPE

Figure 7.7 Dielectric loss tangent "tan δ" measured on different test samples of PE with increasing alternating 50 Hz field intensity at a constant temperature 80°C, Sander [7.8].

A peculiar weakness of PE and other polymeric solid dielectrics is the development of "electrical" and "electrochemical" treeing, described later in Section 7.3.4 in detail. In presence of water, the so-called "water tree" develop, which are also a type of "electrochemical tree". Treeing process may lead to a premature breakdown of the dielectric, thus reducing its life and causing hazards in power supply systems.

The problems of foreign particles and voids in extruded PE are met to a great extent by modern techniques of extrusion, using even 50 μm sieves and maintaining a high degree of cleanliness.

7.1.2.6 Epoxyresins (EP-resins) The words "epoxy" or "epoxide" refer to a compound containing more than one epoxide group per molecule, that is, an oxygen atom united with two carbon atoms already joined in some other way. Resin is any one of a class of solid or semisolid substances obtained from the exudation of many plants or by chemical processing of inorganic materials. Epoxyresin is one of the groups of resins containing at least one epoxy compound. Epoxyresins are low molecular but soluble thermosetting plastics, which exhibit sufficient hardening quality in their molecules. The EP-resins are mixed with hardening and accelerating agents besides providing filler materials, coloring agents and substances to obtain a more flexible cast. The chemical cross-linking of epoxy resins is normally carried out at room temperature either by a catalytic mechanism or by bridging across epoxy molecules through the epoxy or hydroxyl groups. Some curing process may involve both the catalytic and bridging systems. The epoxy group when catalyzed reacts with hydroxyl groups. The amines and acid anhydrides are extensively used as cross-linking (bridging) agents.

The most important EP-resin used as electrical insulating material, the so-called "Bisphenol-A Epoxyresin", prepared by reaction of acetone and phenol is shown in Figure 7.4 (d). When $n = 0$, the product is diglycidyl ether, and the molecular weight is 340. When $n = 10$, the molecular weight is about 3000, which is a high figure for resins. Hence, it can be said that in the uncured stage, the epoxy resins are polymers with low degree of polymerization. Beyer et al. [6.5] and Brydson [7.4] have reported a fairly large number of synthetic epoxy resin compounds other than normal diglycidyl ether resin being developed for their appropriate and specific applications in high voltage engineering. Such non-glycidyl ether epoxides are generally prepared by epoxidising unsaturated compounds using hydrogen peroxide or paracetic acid. Some of them are as follows:

- aromat free cyclic aliphatic expoxide resins, for example, dicyclopentadiene dioxide, which contain a ring structure as well as an epoxide group in the molecule, have a higher creepage current resistance (surface resistance), a desirable property for application in transmission lines.
- novolak-resins or hydantoin epoxyresins, which are prepared by reacting phenol with formaldehyde under acidic conditions, show better thermal properties in their final forms.
- bromised bisphenol-A epoxyresin, in which bromine compounds are used as flame retardant additive, has better fire resistance properties.

More modified compounds are also developed to acquire better flexibility and other mechanical properties.

For curing the epoxy resins, a large number of acid anhydrides are also used as hardening agents. Curing temperatures required with such hardeners are much above 100°C and the casting requires long time (up to 24 hours). On the contrary, the amine hardeners cure almost spontaneously at room temperatures. A vast variety of chemical and physical properties of the finished products are obtained by using different hardeners.

As filler materials, inorganic substances such as quartz powder (SiO_2), are used for casting application. In SF_6 gas insulated systems (GIS) having epoxy resin spacers, aluminiumoxide (Al_2O_3) and also Dolomite (Ca-Mg-Carbonate) are used as filler materials. These are found to be more compatible to the decomposed products of SF_6 by PB and arcing discharges [6.5]. In order to increase the flexibility, and in consequence the toughness of the resin, plasticizers and polymeric flexibilisers, such as polyamides, polysulphides, polyamines and polyglycol diepoxides are added.

In high voltage applications, it is very important that the cast or encapsulation should not contain voids and humidity and the material is desired to be homogeneous. Accordingly, drying and degassing of the individual components of the mixtures is advised and the casting is preferably undertaken in vacuum. The epoxyresin casts are inert to ethers, alcohols and benzol. However, most of them show solubility in mineral oils at about 70°C. Due to this, they are not suitable for application in oil filled transformers.

Epoxyresins, basically nonpolar substances, have high dc specific insulation resistance and low loss tangent compared to polar materials like PVC. However, above 100°C the specific insulation resistance begins to fall considerably and the $\tan\delta$ increases (Figures 7.1 and 7.2). The wave like variation in loss tangent curve with increasing temperature is attributed to the polarization losses. Compared to porcelain, the breakdown strength of epoxyresins is almost double at temperatures up to 100°C but falls rapidly at higher temperatures (Figure 7.3).

7.1.2.7 Natural and Synthetic Rubber Rubber is a highly flexible polymer "polyisoprene", which is produced either from natural or synthetic substance "caoutchouc". The natural caoutchouc, derived from rubber trees, is also known as "India rubber", grown mainly in India and other South East Asian countries. Isoprene is the structural unit of natural rubber shown in Figure 7.4 (e).

Rubber is a linear chain compound and is cross-linked (or vulcanized) using sulphur. Vulcanization is a heat treatment process through which a bridging bond (cross-linking) is accomplished between the added sulphur and the raw caoutchouc molecules. Choosing different base materials in varying proportions for the mixture and giving them different degrees of vulcanization can produce different products having a variety of properties. The rubber products obtained from synthetic caoutchouc possess better electrical and thermal properties as compared with the one obtained from natural caoutchouc. These synthetic products are obtained, for example, by copolymerizing butadiene (C_4H_6) with styrene (C_6H_5CH) to produce a synthetic rubber "styrene-butadiene rubber" (SBR), known by its trade name as "Buna S". By copolymerizing with acrylic, a product known as "Perbunan" is produced. The term "Buna" arose in Germany from the fact that the early polymers of butadiene were produced by sodium (Na) catalysed polymerization of butadiene (Bu). Similarly, from the isomeric hydrocarbon butylene (C_4H_8) with isoprene (C_5H_8) a product known as "butyl rubber" is obtained and from siliconcaoutchouc "silicon rubber" (SiR) and ethylene-propylene-copolymer rubber (EPR) are produced. EPR compounds contain only about 50% of the base polymer. These are complex formulations incorporating as many as ten ingredients, with clay fillers being the next largest component. These formulations vary from manufacturer to manufacturer. A typical compound usually has the following ingredients: EP rubber, filler/reinforcing agent, plasticizer, antioxidants/stabilizers, flame retardants, curing and curative agents for vulcanization, Brown [7.9]

Among the products mentioned above, EPR and SiR are more resistant to atmospheric conditions, such as light, oxygen, ozone and overall aging with time. However, these products are not oil resistant. Recent investigations revealed that EPR has an even better nuclear irradiation resistance property compared to other polymers. These rubber products have found a vast use in high voltage special-purpose power cables used in mining, shipping and nuclear power plants. A new field of application of SiR is for the shields provided on the high voltage rod shaped fiberglass reinforced plastic insulators for outdoor applications, Beyer et al. [6.5].

Rubbers as such have poorer electric strength compared to even resins. Their electric strengths are comparable to that of porcelain. Some other standard values of electrical properties of SiR are given in Table 7.1.

7.1.3 Composite Insulating System

Combining different dielectric materials has produced some very high quality solid insulating systems. Common composites of organic and inorganic materials are fiberglass reinforced plastics, mica based plastic tapes, quartz, fiber and mica mixed with synthetic resins, such as epoxyresin used in electrical machines, and so on. Oil and wax based compound impregnated insulation systems with paper and polymer tapes are commonly applied in power cables, capacitors and transformers. Recently developed uniaxially oriented polyethylene (UOPE) tape has been found to have very good compatibility with oils besides having good mechanical properties. Such composite insulating system make use of the advantages of the inherent properties of their constituent materials. Thus, it is possible to create insulating systems having better thermal, mechanical and electrical properties.

7.1.3.1 Impregnated Paper as a Composite Insulation System Impregnated paper is one of the earliest forms of composite dielectric used in high voltage engineering since the end of the nineteenth century. Paper as such has very poor dielectric properties, but when impregnated with oil or an impregnating compound, the properties of the composite dielectric considerably improve.

The paper used in electrical engineering is a produce of "cellulose" obtained mainly from pine or spruce wood sulphate pulp found on Scandinavia. Cellulose is a complex carbohydrate, that forms the chief constituent of the cell walls of plants. The cellular tissue form of paper consists of crystalline micro cells in between the layers of individual cellulose molecules. Thus, paper has an inherent micro void structure having an average capillary diameter of 10^{-1} to $10^{-2}\,\mu$m. The effective inner surface of the paper is, therefore, very large and highly hygroscopic. The paper produced for application in electrical engineering is made extra smooth. The paper used for condensers is extra thin and has a thickness of 10 to 30 μm. For transformers and cables, 80 to 130 μm thick paper is commonly applied. A desired layer of thickness comprised of thinner paper possesses higher electric strength.

From an electrical insulation point of view, paper as such can be described as a consolidate of air condensers and cellulose dielectrics. The relative permittivity and loss tangent of paper is, therefore, much smaller than that of pure cellulose ($\varepsilon_r = 6.1$ at 20°C). Variation of relative permittivity and loss tangent of paper (unimpregnated) used for electrical apparatus as a function of temperature is illustrated in Figure 7.8, measured by Jacobsen [7.10]. The strong and variable effect of temperature on the loss tangent characteristic is attributed to the physical phenomena explained as follows. In the temperature regions ranging from −270°C and above +100°C, polarization losses dominate, besides the ion conduction current increases the value of $\tan\delta$ above +50°C. An interesting result seen from these curves is that in the service temperature range of −20°C to 100°C the $\tan\delta$ values are relatively small. This is because the polarization losses in this temperature range are small at

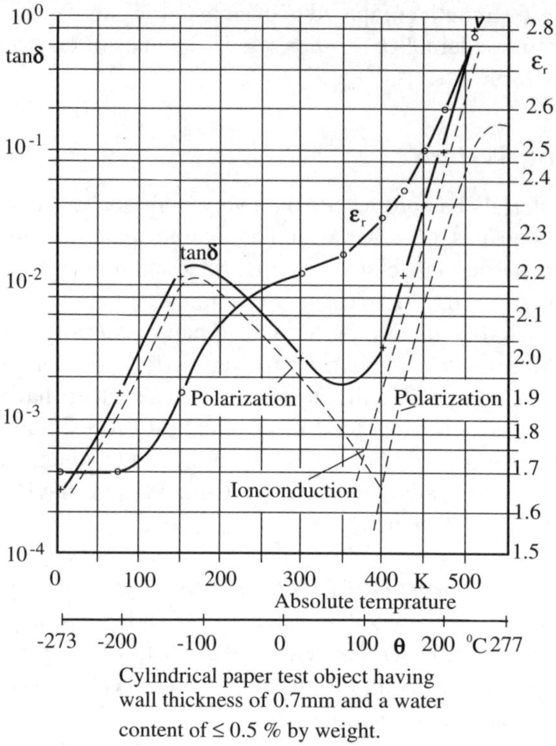

Figure 7.8 Variation of loss tangent "tan δ" and relative permittivity "ε_r" of insulating paper (unimpregnated) with temperature measured at constant rms ac voltage of 500 V, 50 Hz, Jacobsen [7.10].

the frequency of measurement (50 H_Z) and simultaneously the ion conduction current losses are also low.

Paper strips are cut from the bulk paper rolls. A number of layers of paper strips of different widths are suitably wrapped on the metal conductor to achieve the desired thickness of insulation. This is preferably accomplished under humid atmospheric conditions in order to minimize mechanical tearing of the paper. However, before impregnating the paper with liquid dielectric, it must be systematically dried by standardized methods. The drying process of paper takes place in a warm vacuum chamber. To avoid any damage to cellulose, a slow drying process is adopted with a maximum temperature of 120°C. The whole process may take a few days.

On impregnating the paper with oil or impregnating compound after drying and degassing, the voids in the paper, which were earlier filled with moist air, get filled up with the impregnating agent. It changes the values of the relative permittivity and the loss tangent of the new composite dielectric. Their values are considerably affected by the drying process, degree of impregnation, properties of the impregnating agent as well as its general condition (for example moisture content, if any). Figure 7.9 shows the variation of tan δ with voltage at constant room temperature for 20 kV impregnated paper cables. Curve 1 represents the limiting values

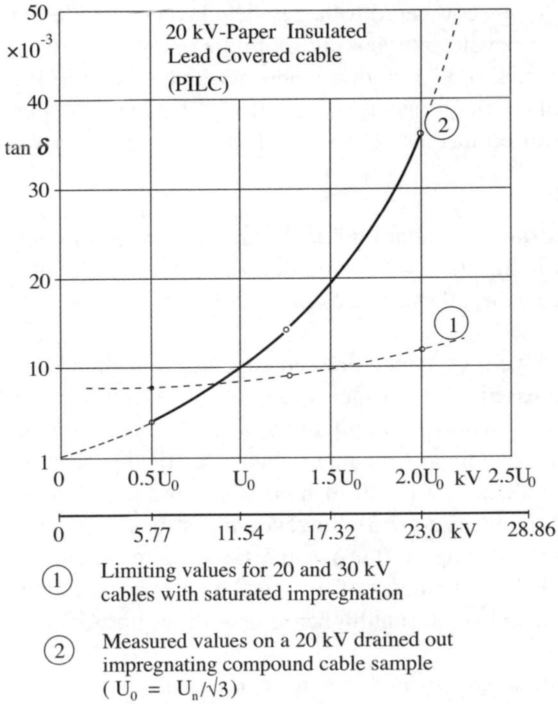

Figure 7.9 Variation of tan δ with increasing voltage for 20 kV PILC cables.

for a normal cable impregnated to saturation state according to German standard. Curve 2 represents the measured values on a drained impregnating compound cable. The paper insulation in such cables is deliberately not saturated with impregnating compound in order to make it suitable for vertical laying. The increase in $\tan\delta$ with increasing voltage reveals that the paper insulation pores and interlayer voids are filled with air instead of impregnating compound.

If the drying process is not effective or the impregnating compound or oil contains moisture, similar rise in loss tangent would be measured with increasing voltage. Moisture content above 1% by weight in impregnated paper also considerably affects its electric strength, Grossekatthöfer [7.11]. Standard values of relative permittivity, loss tangent and specific insulation resistance of paper impregnated with oil are given in Table 7.1. The maximum "service electric stress" in paper cables impregnated with wax based compounds up to 33 kV rated voltage is kept between 3 and 4 kV/mm. But in low pressure oil filled cables, produced up to 110 kV rated voltage, the maximum electric stress at the inner radius of the dielectric can be raised even up to 10 to 16 kV/mm. These cables are provided with either lead or aluminium sheath, protecting the dielectric from moisture ingress. In equipment like transformers, which are not sealed from atmosphere completely, the service stress should not exceed 2 kV/mm.

In high voltage condensers, besides paper, polypropylene (PP) is also used as part of the composite dielectric. Polypropylene possesses a better electric strength property and has lower dielectric losses compared to paper. The layers provided in between polypropylene for the movement and soaking of impregnating agents in condensers are; phenyl-xylene-ethane (PXE), mono-isopropyl-biphenyl (MIPB), Baylectrol 4900 (ditolylether) and benzyl-neocuproine (BNC) beside Clophen (PCB). Among vegetable oils, castor oil has also been found suitable especially for condensers.

7.1.3.2 Insulating Board Materials Insulating board materials are commonly required for indoor low voltage switchgears and transmission as well as distribution transformers. After giving suitable shapes, these boards provide mechanical support while insulating the live parts.

"Pressboards" are made of layers of paper pressed together with synthetic resins as binding material. Such pressed paper products are also produced in other forms, for example, plates, laminates, cylinders or any other desired suitable shape, known as "hard paper". Depending upon the resin content, their electrical properties vary from product to product. The electric strength of hard paper and pressboards is higher than that of resins or porcelain (Figure 7.3); however, it is strongly affected by temperatures above 20°C. Their loss tangent ($\tan\delta$) values also increase rapidly above this temperature (Figure 7.2). The advantage of this insulating material is that it continues to provide good mechanical support at higher temperatures upto 120°C. They are also oil resistant.

Another type of board material known as "transformer board" is produced from high quality sulphate cellulose thin paper. Since no binding material is used, a large number of thin paper having thickness of the order of 30 μm, are pressed together by a hot press method in the desired form. It is then dried, degassed and impregnated with oil. A glossy finish is provided to the surface. In the absence of binding material, the electrical properties of these boards are near that of high quality paper. The relative permittivity is low, about 2.7, and electric strength is high, of the order of 100 kV/mm when impregnated with mineral oils.

A hard, non-inflammable synthetic material available in the form of a board is known as "Bakelite", a trademark after its Belgian American inventor and chemist L.H. Baekeland. It is produced by polymerization of phenol and formaldehyde into linear chains, which are subsequently cross-linked under pressure.

7.1.3.3 Fiber Reinforced Plastics (FRP) The first fiber reinforced plastic (FRP) for electrical insulation purpose was asbestos reinforced phenolic resin compound introduced in 1915. Glass fiber reinforced plastics (GFRP) containing unsaturated polyester as the matrix resin was developed in 1943. Since then, electrical grade GFRPs have become established as structural insulating material because of their combination of high electrical insulating properties and mechanical strength. With the help of improved production techniques, these composite dielectrics are finding increased applications in the electrical power system.

The electrical grade GFRPs are composed of a combination of electrical grade glass fiber materials and thermosetting resins, such as epoxy, phenolic, silicone, vinylester, polyester and polyimide resins, as well as a number of thermoplastic resins, Yasufuku [7.12]. GFRPs are finding increased application in a wide range of products in heavy electrical apparatus, for example, GIS, transformers, vacuum switchgears, electrical machines, control devices, and so on. These are produced in the form of insulating cylinder, container, operating rod, bushing housing frame, barrier, and so on, as per requirement. Besides having light weight, high mechanical and impact strength, products with desired electrical properties, such as, PB free, tracking, flame, heat and arc resistant are commonly produced, Yasufuku [7.12].

In the above description of electrical properties of solid insulating materials, the reader might have noticed that the quantitative values of electric strength are less emphasized. The reason is simple since electric strength of solid dielectrics is affected by many internal and external factors. The quantitative values are often misleading unless exact conditions of the material and measurement methods are described. Such details are rarely available from the literature particularly for solids.

7.2 PARTIAL BREAKDOWN IN SOLID DIELECTRICS

Partial Breakdown (PB) is localized electrical breakdown within an insulation system restricted only to a certain part of the dielectric. In other words, PB is the breakdown that does not bridge the electrodes. PB occurs in gaseous, liquid, solid or in any combination of insulation system at extremely nonuniform field locations above a particular intensity of field. The field intensity, above which the PB may occur, depends upon the physical conditions of the dielectric and its properties.

The term "Partial Breakdown" includes a wide group of phenomena; corona in gaseous dielectrics, internal breakdown at voids or cavities within a liquid or solid dielectric, and surface discharge, also known as "tracking", which appear on the boundary surface of some solid and liquid dielectrics with air or any other gaseous medium.

Corona in air, their impact on power systems and atmosphere have been discussed in Chapter 3. The effects of degeneration of free air due to PB are not significant, since no irreversible damage to the dielectric is caused. Even in the case of liquid dielectrics, the damage caused by PB to the insulating properties is not of much concern. However, in the case of internal breakdown taking place within a solid dielectric, the damage is irreversible. Every breakdown event causes deterioration of the material by the energy impact of high-energy electrons or accelerated ions, inflicting electrical and chemical transformations. The actual deterioration is, however, dependent upon the material and the surrounding conditions, but with time it may lead to a permanent damage of the dielectric. Hence, the significance of PB in solid dielectrics is of great importance and will be described in detail in the following sections.

7.2.1 Internal Partial Breakdown

Partial breakdown in solid dielectrics may take place at the so-called weak points. These weak points may be voids, cavities, cuts, foreign particles or protrusions in

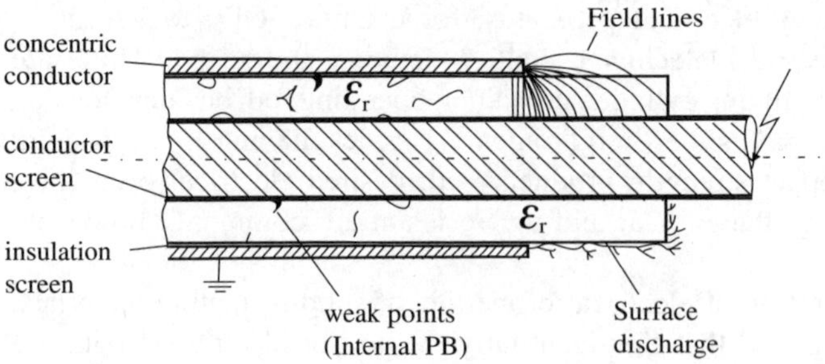

Figure 7.10 A cable cross section showing possibilities of internal and surface partial breakdown.

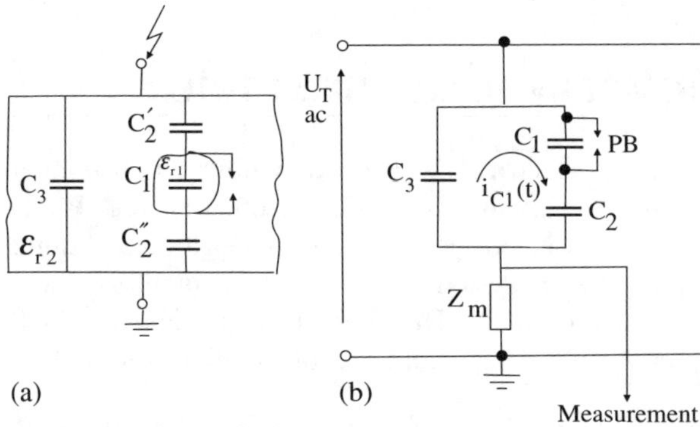

C_1 - capacitance of void

C_2' and C_2'' - capacitance between void and electrodes

C_3 - capacitance of the test objects , C_c - coupling capacitor

C_2 - is equivalent of C_2' and C_2'' in series

Z_m - measuring impedance

Figure 7.11 Simulation of internal partial breakdown, (a) schematic showing a void building a capacitance, (b) equivalent circuit diagram for measurement.

the dielectric, as shown in Figure 7.10. This figure represents a longitudinal cross sectional view of an extruded solid dielectric power cable, provided with an extruded semi conductive sheathing on the dielectric. With the development of modern technology, careful handling and maintaining extra cleanliness, the presence of voids and foreign particles in extruded, moulded and cast dielectrics have been minimized, but not ruled out. Voids and cavities, which are the most common type of weak points in solid dielectrics, may vary in size from a few *μm* to a few *mm*.

Weak points in general have lower electric strength than the dielectric itself. Still worse is the fact that they acquire higher electric fields. Such weak points can be simulated by a capacitance C_1 within the main dielectric system having capacitance C_3, as shown in Figure 7.11 (a). If the relative permittivity of a weak point, ε_{r1}, is lower than that of the dielectric ε_{r2}, the electric stress at the weak point proportionately increases. A protrusion in a dielectric acts as sharp extension of the

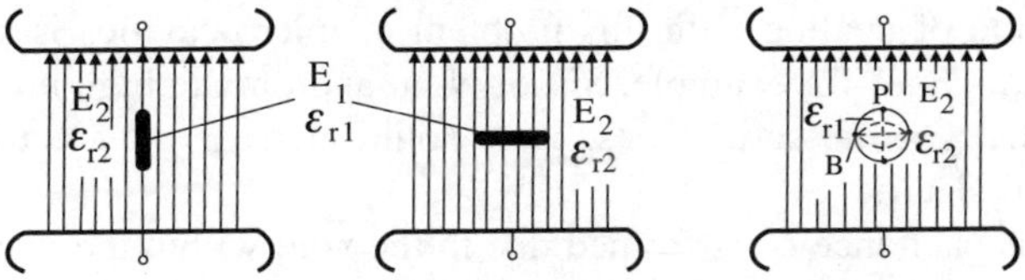

$E_1 = E_2$

1- void
2- solid dielectric.

$E_1 = \frac{\varepsilon_{r2}}{\varepsilon_{r1}} \cdot E_2$

$E_{B1} = \frac{3}{\frac{\varepsilon_{r1}}{\varepsilon_{r2}} + 2} \cdot E_2 = E_{B2}$

$E_{P2} = \frac{3}{\frac{\varepsilon_{r1}}{\varepsilon_{r2}} + 2} \cdot \frac{\varepsilon_{r1}}{\varepsilon_{r2}} \cdot E_2 = \frac{\varepsilon_{r1}}{\varepsilon_{r2}} E_{P1}$

(a) void longitudinal to field

(b) void perpendicular to field

(c) spherical void representing a combination of the fields

Figure 7.12 Fields at voids or cavities in a solid dielectric in uniform field.

electrode and causes an extreme distortion in the field locally. Thus, the electric stress within the dielectric considerably increases at the weak points.

Consider three voids of different shapes having a relative permittivity, ε_{r1}, in uniform field between two parallel plates with a dielectric having relative permittivity ε_{r2}, as shown in Figure 7.12. It is interesting to find the field enhancement in the void in these three situations. The thin flat void falling in longitudinal direction of the field, Figure 7.12 (a), represents the typical case of a longitudinal interface, described in Section 2.5.3.1. If the void dimensions are very small compared to that of the dielectric, ε_{r2}, the field in the void E_1 is equal to the applied field E_2, obeying the relation given by equation (2.52). In the second case where the void falls flat across the field, Figure 7.12 (b), it represents the case of a perpendicular interface. In this case, the field intensity E_1 in the void given by equation (2.55) is equal to $(\varepsilon_{r2}/\varepsilon_{r1} \cdot E_2)$.

The third case of a spherical void falling in the uniform field, Figure 7.12 (c), is a combination of the above two cases. The main points on the sphere are the so-called poles "P", which fall perpendicular to the direction of field E_2. The points along the horizontal belt "B" fall in the longitudinal direction of the field E_2. By solving the Laplace equation for the spherical void in uniform field, the relations for the field intensities at poles *"P"* and belt *"B"* can be obtained as given in Figure 7.12 (c).

As seen from Figure 7.12, in case of (a) i.e. a long void falling in longitudinal directional of the field, the field intensity at the void does not change and remains equal to the applied field. However, such a void can prove to be extremely dangerous, adversely affecting the electric strength of the dielectric. In case of (b) as well as (c), the field intensities at the voids depend upon the relative permittivities of the void ε_{r1} and that of the main dielectric ε_{r2}. If ε_{r1} is smaller than ε_{r2}, an overall enhancement of the field intensity at the void is expected. This is a usual case when the voids are filled with any gas. But, if the cavity is filled with a liquid dielectric, oil or an impregnating compound having high ε_r, the enhancement of the field can be checked. Another method of dealing with this problem of micro-cavities is by filling them with an insulating gas, for example, SF_6 or N_2 at appropriate high pressure, as practiced in internal gas pressure cables. This would increase the electric strength of the composite dielectric.

It is assumed that a capacitance C_1 is formed due to the void within the main dielectric between two electrodes, as shown in Figure 7.11 (a). Let C_3 be the total capacitance of the electrode system (test object), ε_{r1} and ε_{r2} the relative permittivities of the void and the main dielectric respectively. The field lines starting and ending at the void form two capacitance C_2' and C_2'' within the dielectric, C_2' and C_2'' together in series are combined to form C_2. The void capacitance C_1 is the origin of PB on applying a voltage to the test object U_T, shown in Figure 7.11 (b). Let C_c be the coupling capacitor required in the circuit to suppress the reflections from the open end during the measurement.

On applying a 50 Hz, ac test voltage U_T, the capacitance C_1 formed by the void gets charged. C_1 would experience some proportion of the applied voltage U_T, depending upon the magnitudes of different capacitances. The intensity of electric field at C_1 will depend upon the shape, size and location, and so on, of the void, as explained earlier. On further raising the applied voltage, the inception of breakdown across C_1 appears during the rising part of the half cycle, as shown in Figure 7.13, causing the void capacitance C_1 to discharge. The discharge current, $i_{c1}(t)$, which cannot be measured directly, is a very short pulse current of *ns* range of duration. The opposite polarity charge, produced by this discharge, is displaced towards the electrodes in the field direction, thus neutralizing the original electric field at the void. If the voltage is still rising at the positive or the negative slope of an *ac* cycle, the field builds up again, repeating the breakdown phenomenon *n* number of times

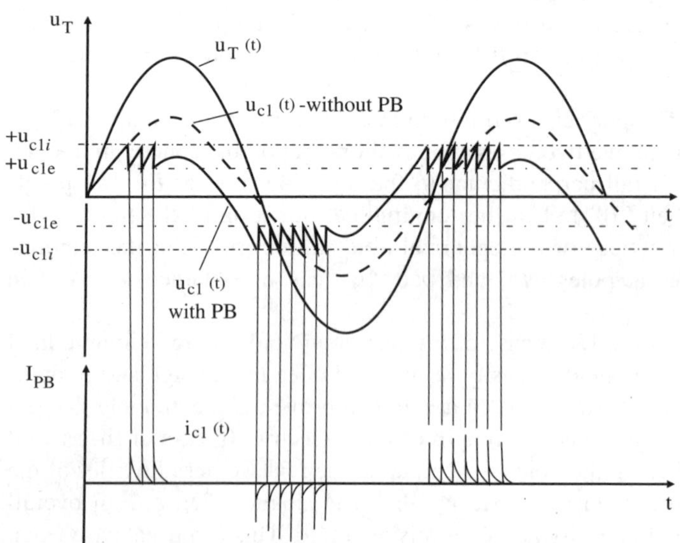

Figure 7.13 PB voltages and pulse currents at a void or cavity in the dielectric.

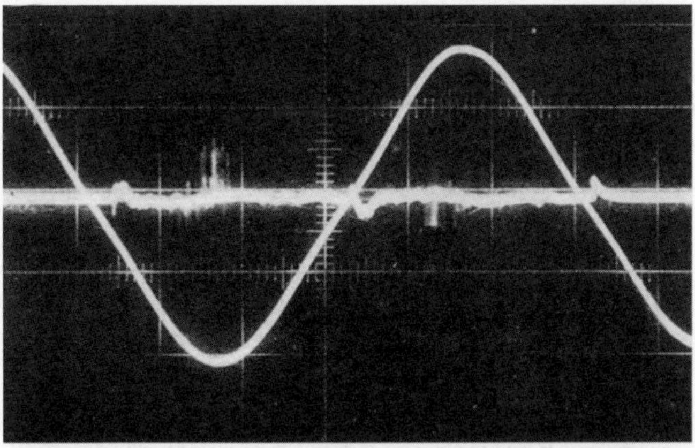

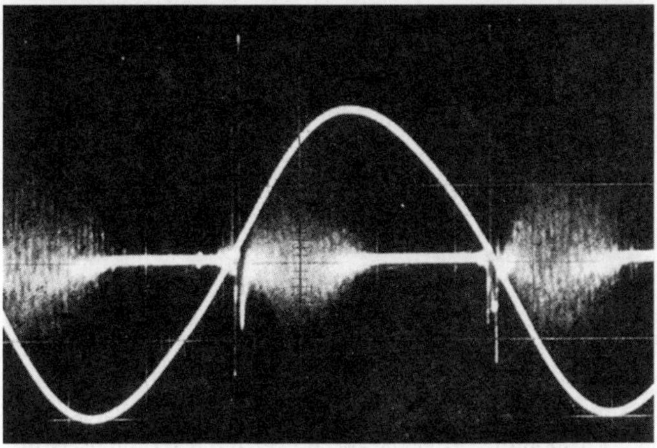

Figure 7.14 Oscillograms of internal PB in solid dielectric.

during each cycle. The PB current pulses thus produced are measured at the external circuit, giving the intensity of PB calibrated in coulombs. On further raising the applied voltage on the test object, the frequency of occurrence of breakdown at C_l and their current discharge intensity increases. In Figure 7.14, the two oscillograms show the PB inception and intensive PB taking place at both half cycles of higher applied voltage on a test object.

Consider the magnitude of the voltage at which PB at C_l incept to be U_{cli} and the voltage at which PB extinguish U_{cle}, as shown in Figure 7.13. The inception voltage U_{cli} can be given in terms of the applied voltage and capacitances shown in the equivalent circuit diagram as:

$$U_{c1i} = U_T \frac{C_2}{C_1 + C_2} \tag{7.1}$$

Due to breakdown taking place at C_1, let the drop in voltage at C_l may appear to be ΔU_{c1}. Considering no charge at C_l initially, ΔU_{c1} is given by,

$$\Delta U_{c1} = U_{c1}\frac{C_2}{C_3 + C_2}$$

If it is assumed that the voltage drop at C_l equals the corresponding drop in applied voltage at the test object, ΔU_T, it can be given by the equation,

$$\Delta U_T = U_T\frac{C_2^2}{(C_3 + C_2)(C_1 + C_2)} \tag{7.2}$$

In this process the actual charge q_{c1}, transposed to the weak point (void) by the power supply can be estimated considering C_2 and C_3 to be in series in parallel with C_1,

$$q_{c1} = \Delta U_{c1}\left[C_1 + \frac{C_2 \cdot C_3}{C_2 + C_3}\right] \tag{7.3}$$

U_{c1i} as well as q_{c1} are both not measurable quantities. However, due to the breakdown at void a corresponding event of potential drop takes place at power input terminals of the test object. From the equivalent circuit diagram given in Figure 7.11 (b), the charge thus delivered at the power input terminals can be approximated as follows,

$$q \approx \Delta U_T\left(C_3 + \frac{C_1C_2}{C_1 + C_2} + C_c\right) \tag{7.4}$$

This is a measurable quantity of charge over the measuring impedance $Z_{m.}$ Since the values of C_1 and C_2 are not known, the actual inception voltage at the void U_{c1i} and the magnitude of the charge q_{c1} transposed to the void cannot be calculated. Neither of these quantities are measurable since the actual phenomenon takes place within the dielectric. The corresponding effect at the power input terminals is the only measurable quantity. This is the reason that the measure of charge "q" at the terminals given by equation (7.4) is described by IEC-270 and also IS-6209 as "apparent charge", "q_a". It is defined as follows,

The apparent charge "q_a" of a partial breakdown is that charge which, if instantaneously injected between the terminals of the test object, would momentarily change the voltage between its terminals by the same amount as by the partial breakdown itself. The absolute value $|q_a|$ of the apparent charge is often referred to as the discharge magnitude. The apparent charge is expressed in coulombs.

The apparent charge q_a defined in this way is not equal to the amount of charge actually transferred across the cavity due to partial breakdown in the dielectric. It is implied because current measuring instruments respond to this quantity. Thus, the magnitude of apparent charge represents the intensity of an individual current pulse due to PB in a test object. Lemke [7.13] introduced a new concept in this respect in 1975. Since the apparent charge represents the single discharge of maximum intensity taking place in the test object, Lemke called it "impulse discharge" denoted by "q_i". However, in a test object, PB may simultaneously occur at several locations. Therefore, the complete phenomenon occurring in the dielectric is not taken into account by q_i. In order to evaluate all the PB taking place in a test object, Lemke introduced a new concept of "cumulative discharge", "q_s" (s-sum total). q_s takes into consideration all

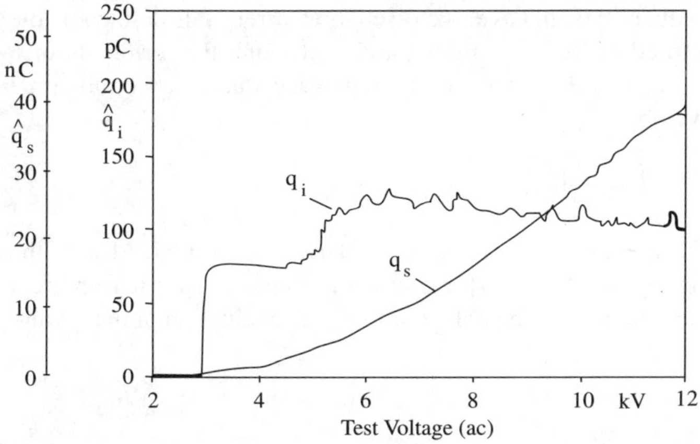

Figure 7.15 Variation of impulse charge "q_i" and cumulative charge "q_s" PB magnitudes with increasing voltage, Lemke [7.13].

the individual PB occurring in the test object within a certain time period of the applied voltage. Thus, a more comprehensive investigation of PB phenomenon in a test object is made possible by measuring q_s. In Figure 7.15, variations of measured values of q_i and q_s with increasing voltage are illustrated. These curves measured by Lemke [7.13], are for a needle-plane electrode configuration in air, where the PB (corona) at the negative half cycle of the 50 H_Z ac voltage have been recorded. Similar curves are also measured on test objects with solid dielectrics as insulation.

Beside apparent charge, the other important PB quantities for a test object are the PB inception and extinction voltages. The quantities related to individual PB are defined in IEC-270, are as follows:

PB inception voltage "U_i"

It is the lowest terminal voltage at which discharge magnitude due to partial breakdown exceeding a specified intensity is observed under specified conditions, when the voltage applied to the test object is gradually increased from a lower value at which no PB is observed.

PB extinction voltage "U_e"

It is the voltage at which PB exceeding a specified intensity cease under specified conditions when the voltage is gradually decreased from a value exceeding the inception voltage.

Repetition rate "n"

The PB pulse repetition rate "n" is the average number of current pulses produced per second.

Energy of an individual PB, "w"

The partial breakdown energy "w" is the energy dissipated by one individual breakdown. It is expressed in joules.

Among integrated quantities, the average discharge current $\bar{I}$ is defined as the sum of the rectified charge quantities passing through the terminals of a test object due to partial breakdown taking place during a certain time interval Δt, given by,

$$\bar{I} = \frac{1}{\Delta t}[|q_1| + |q_2| + |q_3| \ldots |q_n|]$$

The measurable PB quantities described above have been explained with the help of a schematic given by Lemke [7.14], as shown in Figure 7.16. Measurement of these quantities on a test object may be able to provide an evaluation of the quality

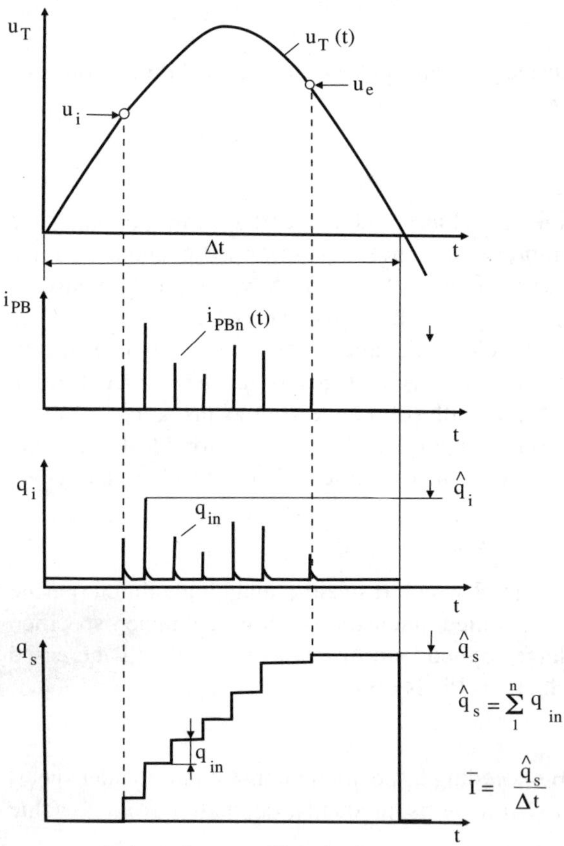

u_i - PB inception , u_e - PB extinction
Δt - time interval of measurment ,
q_i - apparent impulse discharge , q_s - apparent cumulative discharge
$\hat{q}_i$ - impulse discharge of maximum intensity ,
$\hat{q}_s$ - sum total of individual discharges in a time interval of Δt,
$\bar{I}$ - average discharge current

Figure 7.16 Measurable quantities of PB, Lemke [7.14].

of finish of the dielectric. Satisfactory performance and the life expectancy of electrical apparatus depend upon the dangerousness of the weak points in the dielectric. Limiting values of PB quantities are standardized for different products in order to ensure the quality finish of the apparatus.

7.2.2 Surface Discharge (Tracking)

PB on the surface, also known as "tracking", occurs in any gaseous medium and also in vacuum at their interfaces with solid and liquid dielectrics. Partial breakdowns on the surfaces are caused due to the presence of high tangential component of electric field along a diagonal interface between any solid or liquid and a gaseous medium. Typical examples of the presence of such interfaces in practice are in bushings, insulators, cable terminations and sites in transformers. As shown in Figure 7.10, when the concentric conductor of a coaxial cable is abruptly terminated, the field intensity at the point of termination becomes very high. An increased tangential component of the field is resulted in the gaseous medium, which stresses it with higher field intensity. Surface discharges are caused if an interruption of the creepage current path takes place across the surface of the dielectric. Under this condition the field component at the location on the surface may exceed the electric strength of the medium.

Day in [7.15] describes surface discharge as an untidy process, whose occurrence depends upon the properties of the dielectric but their inception depends upon several other factors too. Day defines tracking as the formation of conducting path on the surface of a dielectric. In most cases, the conduction results from degradation of the dielectric surface or by deposition of a contamination layer on it. Hence for tracking to occur, the following must be present:

- a conducting film across the surface of the dielectric
- a mechanism by which the creepage current through the conducting film is interrupted when the discharges occur.

The conducting film is usually a form of contamination, such as salt deposition (in coastal areas), carbonaceous dust from fuels and industrial or cellulose fiber deposits along with moisture. Interruption of moisture films is caused when a surface is dried due to the heating effect of the surface current. As the amount of contamination increases, so does the surface current, until the joule heating in the layer causes it to dry out locally, initiating partial breakdown between the receding edges of the wet film. On the surfaces of many dielectric, tracking may not occur under dry and clean conditions. An estimate revealed that tracking is likely only when the creepage currents of the order of 1 mA are able to develop on the surface.

Referring to Section 2.5.3.3 the presence of a diagonal interface gives rise to tangential and normal components of field at the surface. With this result, multiplication of charge carriers takes place right at the surface. On increasing the applied voltage, the same sequence of discharge mechanism is followed on the surface as in case of gaseous and liquid dielectrics, that is, star, streamer and leader corona on the surface. Since the charges get trapped over the surface, depending upon the discharge magnitudes track lengths with numerous branches are produced. Tracking

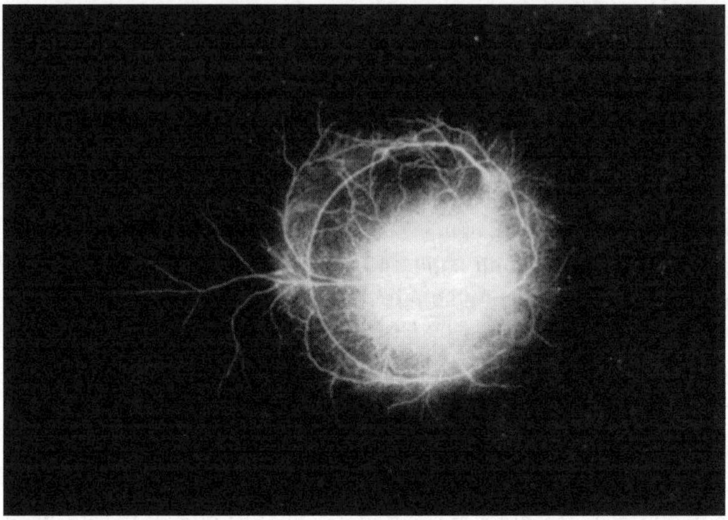

Figure 7.17 A typical autograph of surface breakdown showing stable streamer as well as leader corona, Nema [7.16].

is an undesirable property of the material. A typical autograph of surface discharges is shown in Figure 7.17 taken by Nema [7.16].

For a material to have the tendency to track the theoretical consideration is suggested in terms of chemical bond energies of the material by Parr and Scarisbrick [7.17]. Since the primary bonds are important when considering thermal stability of polymers, the tendency to track depends upon the proportion of bond that produces free carbon on pyrolysis, that is, by chemical decomposition of molecules as a consequence of exposure to a high temperature. It can be expressed as follows:

$$\frac{T_C}{T_m} = \frac{\begin{array}{c}\textit{Dissociation energy of all the bonds which on}\\ \textit{breaking produce free carbon (kcal/mole)}\end{array}}{\textit{Total bond energy of the molecules (kcal/mole)}}$$

The smaller the fraction T_c/T_m, the lesser is the likelihood of dielectrics to track. Experimental results revealed that if T_c/T_m is smaller than 0.3, the material has no tendency to track. Polymers, which do not track, are, for example, polypropylene, polyformaldehydes, and so forth. On the other hand, if this fraction is greater than or equal to about 0.5, the material may track easily. For example, polystyrene, phenol-formaldehydes and polycarbonates are among the polymeric dielectrics that track easily besides the ordinary glass being well known to track.

Since it is the free carbon produced by thermal decomposition that forms the conducting tracks on the surface, the degree of resistance to track of a material depends upon its chemical nature and the manner in which it breaks down when subjected to high temperatures. For example, the degradation products of PE are gaseous; therefore, the polymer does not normally track except when some conducting contamination is trapped on its surface during erosion. This is the reason that in order to prevent tracking, a minimum creepage distance is often specified to limit surface currents.

For the sake of obtaining a comparative knowledge of tendency to track for materials, IEC-112 and BS-3781 have introduced an index, known as "Comparative Tracking Index" (CTI), giving standard procedure of its measurements. The CTI can be used as a specification for the measurement of track-resistant quality of dielectrics. The CTI greater than 750 represents the upper stratum of materials in terms of track resistance. However, the use of a material having CTI higher than 750 may not guarantee that tracking will not occur under very bad contamination conditions.

7.2.3 Degradation of Solid Dielectrics Caused by PB

Internal partial breakdown in solid dielectrics cause their degradation and lead to the formation of solid, liquid and gaseous products due to synthesis reactions taking place within the dielectric. Organic compounds containing oxygen are readily synthesized from mixtures of CO_2, CO and H_2O undergoing local internal breakdown. Experimental investigations made by Gamez-Garcia et al. [7.18] on PE and XLPE confirmed their earlier findings that CO_2, CO, H_2O and acetophenone are the main degradation products. This was revealed by their experimental investigations of the products obtained from tests on XLPE degradation due to discharge synthesis reactions within the electrode gap space. Oxalic acid and other non-aromatic compounds result from discharge synthesis reactions involving CO and H_2O. Acetophenone, which evaporates from XLPE surface after cross-linking, plays an important role in the formation of other degradation products. Benzoic acid, benzamide, toluene and other aromatic compounds result from chemical reactions of this volatile compound during the degradation process in the presence of PB.

An important practical inference from the work performed by Gamez-Garcia et al. [7.18] could be drawn that in physical cavities, occluded within the XLPE insulation in cables, the occurrence of PB leads to the formation of oxalic acid because such cavities are not vented. As it is known that oxalic acid suppresses the PB in solid dielectrics, an appreciable amount of generation of this product may in turn lead to the disappearance of PB. Thus the formation of other acids, produced due to CO_2 and CO evolution, may be prevented. However, since the amount of H_2O available within the dielectrics is restricted, the quantity of oxalic acid formed would also be limited. The same argument is also applied regarding the formation of formic acid. Therefore, if cable insulation under consideration is to operate in a humid environment, it is probable that PB within the dielectric may extinguish with time. But, if the PB intensity is unusually high, the degradation and erosion mechanism may proceed at a sufficiently fast rate. The exact conditions determining these mechanisms are yet to be defined through more detailed investigations.

From the above, it can be concluded that PB in solid dielectrics cause physical changes through chemical synthesis reactions. The nature and extent of these physical changes depend upon the particular local condition and the time factor, which play a very important role. The limited literature available in this field of research reflects on inadequate work.

7.2.3.1 Inhibition of Partial Breakdown/Treeing in Solid Dielectrics The partial breakdown (PB) activity and the treeing process in solid dielectrics are generally correlated. Inception and development of a tree may involve different mechanisms. These are mainly governed by the extent of charge injection phenomena. Due to accelerated injection of charge, chain scissions may occur leading to the formation of micro-cavities. The inception of PB at a void thus created may be followed by the growth of treeing process in the dielectric. It is also characterized by the gas discharges in the voids. Presence of oxygen is known to significantly enhance the degradation rate in XLPE.

A new field in material research started in the mid-1970s to suppress or "inhibit" PB and hence the formation of trees by adding "peroxiradical scavengers", also known as "inhibitors" to the polymers. Phenomenological investigations of water trees made by Henkel et al. [7.19] showed that the tree formation can be inhibited by additives. A number of additives inhibiting trees have been found; barbituric acid is one among them. Stabilizers of concentrated solution of hydrogen peroxide (H_2O_2) were found to be particularly effective. Tree inhibiting effect of barbituric acid is quite obvious due to the presence of easily dissociable H-atom in the molecules. This allows them to behave as radical scavengers, thereby stopping the oxidative degradation of the polymer. In an *ac* field even regeneration of the inhibitor is possible, Henkel et al. [7.19]. Inhibitor additives act as weak electrolytes. At high field intensity ionic dissociation gives rise to a field dependent contribution to the conductivity. This moderates the local electric stress in the region of the enhanced field sites by checking the development of space charge. The voltage required to incept the PB and treeing process is, therefore, increased.

Additives may, however, act adversely if not dispersed uniformly in the dielectric. Another important aspect in the choice of additive is its retention property with the polymeric insulation during its service life. Acetophenone, though very effective in inhibiting electrical trees, is found to get easily lost over a few weeks due to its volatile property. For this reason, radical scavenger materials containing effective groups, such as polystyrene and polysulphane, have also been considered with some success [7.20].

As Patsch [7.21] puts its, in case of *ac* voltage, capacitive voltage division occurs inside the dielectrics due to low conductivity at sites. This could take place due to space charge formation resulting in very high electric stress concentration at the sites of impurities in the solid dielectrics. The local stress enhancement may lead to PB and hence inception of tree formation. During his experimental investigations it was discovered that additives, especially acetophenone, substantially increases the conductivity of the material. Possibly because of this effect, acetophenone is well known to resist development of treeing in XLPE. Acetophenone is also produced as a by-product due to decomposition of the commonly used chemical cross-linking agent dicumyl peroxide (DCP). PB inception voltage in XLPE containing various amounts of acetophenone showed a dependence on temperature and frequency. This phenomenon is expected considering the effect of different levels of capacitive conductive voltage division caused by the additives. Patsch [7.21] concluded that the occurrence of PB in XLPE, which could lead to treeing, can be reduced if the local electric stress enhancement at the so called "weak points" is brought down. One way to achieve this is by using appropriate additives that increase the field dependent conductivity within the polymer.

Another experimental investigation performed by Roseen et al. [7.22] revealed that resistance to PB in polymers is primarily determined by their chemical structure and concentration of the carbonyl. It is suggested that the carbonyl groups enhance the radical formation and hence accelerate the surface erosion process causing PB and electrical tree development. They also found out that vinylnaphthalene acts as a stabilizer in XLPE and prolongs the incubation period for oxidation.

Besides the electrical and thermal stresses, the XLPE insulation in power cables is also subjected to mechanical stresses. These could be present due to several reasons, such as residual internal stresses created during the cooling process of manufacture, external forces due to bending during installation or thermo-mechanical stresses developed by differential thermal expansion between the conductor and the polymeric material. Influence of internal mechanical stresses on electrical performance of XLPE with respect to tree development was studied by David et al. [7.23]. The time required for the inception and the extent of the growth of treeing process both were found to depend directly upon the magnitude of mechanical stresses within the dielectric. The higher the mechanical stresses, shorter inception time and longer growth of electrical trees within a specified time were measured.

7.2.4 Partial Breakdown Detection and Measurement

The phenomenon of partial breakdown (PB) is, in general, degenerative towards the quality of dielectrics causing damage over time. It involves the production of a number of typical effects due to local electric discharge in the form of repetitive current pulse generation. The main effects are electromagnetic interference (EMI), audible noise (AN), emittance of light (photon effect), and generation of gases, heat and power loss in the dielectric. Several PB detection techniques have evolved over the last century that evaluate the intensity of PB by the measurement of any one of such quantities. These techniques of detection and evaluation of PB activity in the dielectrics can be classified into the indirect and the direct methods, described in the following sections.

7.2.4.1 Indirect Methods of PB Detection The earliest detection techniques adopted measurement of PB indirectly, i.e. measurement with the help of separate tools not as a part of the electric circuit in which PB is produced. Several indirect methods described below are in practice.

EMI detection technique

It is one of the earliest techniques developed for the detection of PB activity in the power system. Depending upon the apparatus for which this detection technique is to be applied, a suitably designed antenna or a UHF sensor, also called coupler, is installed at an appropriate location to receive the electromagnetic waves radiated due to PB. The signals received at this sensor could be recorded by, for example, a spectrum analyzer for further analysis with the help of a computer. This technique is widely adopted for the detection of PB in GIS and transformers [7.24]. The bandwidth to which the EMI activity may extend is known to be more than 2 GHz [7.25].

Light-weight hand-held compact PB detectors have been developed that can capture EMI from the electrical apparatus in the power system network

Acoustic technique for detection

Acoustic noise, termed as Audible Noise (AN), is produced by PB in general or by corona in gaseous dielectrics. Sudden expansion of gas due to high temperature in the impulse current discharge channel and its movement in pockets gives rise to audible noise. Bouncing particles in GIS, which could produce PB, also produces such noise. Ultrasonic sensor or vibration transducers are commonly used in GIS for the detection and measurement of acoustic noise. A number of transducers are needed if the condition of entire GIS is required to be monitored in service continuously. These transducers operate in the frequency range of 20–100 kHz. However, extraneous noise may reduce their detection sensitivity.

This technique of PB detection is also applied in gas filled dry type of transformers. However, it has limitations for oil filled transformers in locating the exact site of PB due to significant differences in the velocities of propagation of sound in different materials, e.g. oil, metal, wood, board, and so on [7.26].

Optical detection

The light emission due to PB is measured with the help of a photomultiplier tube using photon-counting technique. The resulting pattern of signals obtained is a histogram, a frequency distribution graph, whose amplitude is related to the total number of photon count per second collected by the optical detector.

The optical method is found to have advantages over other detection techniques since it is not affected by the surrounding electromagnetic interferences. Because of its high sensitivity, the application of optical detection of PB in GIS is found suitable for monitoring their condition in service and also for the detection of very fast restrikes (section 3.4.2) in the SF_6 circuit breakers installed in GIS.

Detection by analysis of gas generated due to PB

Depending upon the dielectric medium, the PB activity gives rise to the generation of different gaseous byproducts. For example, in transformer oil PB produces hydrogen, methane, ethane and propane. Gaseous byproducts due to electrical breakdown in SF_6 gas are described in section 3.5.6. Presence of gaseous byproducts are analyzed by gas chromatography to detect the PB activity in the apparatus.

The Buchholz relay, a gas-actuated relay is also designed to detect such incipient faults, that generate gaseous byproducts. This relay is mounted on the top of the oil filled transformer tanks. The gases slowly generated due to PB are collected in this relay. A sample of this gas is periodically collected and analyzed with the help of gas chromatography to find out if any undesirable PB activity has developed within the tank housing the transformer.

7.2.4.2 Direct Methods of PB Detection and Measurement These are electrical circuit detection methods used most widely in the high voltage laboratories for both research and development and non-destructive testing. The basis of electrical detection methods is the measurement of pulse current or the magnitude of discharge due to local or partial breakdown and its analysis. There has been a continuous advancement in the techniques and technology involved with the PB detectors over the last decades. A large number of PB detectors are commercially available in the market. Started with narrow band detectors, broadband detectors are now widely applied. The detection techniques with the help of electrical circuits have been standardized by various specifications, for example, IS-6209 (2006), IEC-60270.

Since the early 1970s, measurement of apparent charge "q_a" in *p*C displaced by PB and its co-related quantities is being widely practiced. The broadband detection instruments, available commercially, can measure and analyze the PB pulses irrespective of their repetition rate and steepness. Distinction in evaluation can be made between PB taking place at a single location of highest intensity, "q_i" and sum total of discharges at all the locations in an equipment, "q_s". Measurement of PB with electrical circuit is a highly sensitive method that can detect an apparent charge, "q_a" of even less than 1.0 pC. Calibration of the circuit is accordingly performed depending upon the required sensitivity.

There are basically three electrical test circuits adopted widely over the world for the measurement of PB. These are recommended by VDE-0414 (Part 10), IEC-Publication 44-4 and 270. The circuit diagrams are shown in Figure 7.18.

The circuits constitute a PB free power source and a high voltage measurement device, preferably a capacitive voltage divider (C_1 and C_2) along with a "peak voltmeter". The test object may be connected over a suitable series impedance (or resistance) Z_s and the line is terminated with a coupling capacitor C_c to check reflections of the signals during the measurements. For connecting the PB detector and other accessories for measurement an impedance Z_m—basically an ohmic resistance—is used to connect the test object or the coupling capacitor to ground so that the PB current pulses are able to develop a voltage across it required for measurement, Figure 7.18 (a) and (b). The individual components and all the connections must be PB free up to the desired magnitude of the measurement voltage. A standard charge pulse generation device is used to calibrate the circuit to its required sensitivity. The circuit shown in Figure 7.18 (c) has special advantage that the external noise disturbances create an opposite effect on the measuring impedance Z_m and hence are neutralized, König [7.26].

7.3 BREAKDOWN AND PRE-BREAKDOWN PHENOMENA IN SOLID DIELECTRICS

The pre-breakdown phenomena in solid dielectrics leading to an irreversible rupture of the material (breakdown) are extremely complicated. In this process, the local electric field, heat transport, charge injection and accumulation, mechanical stress,

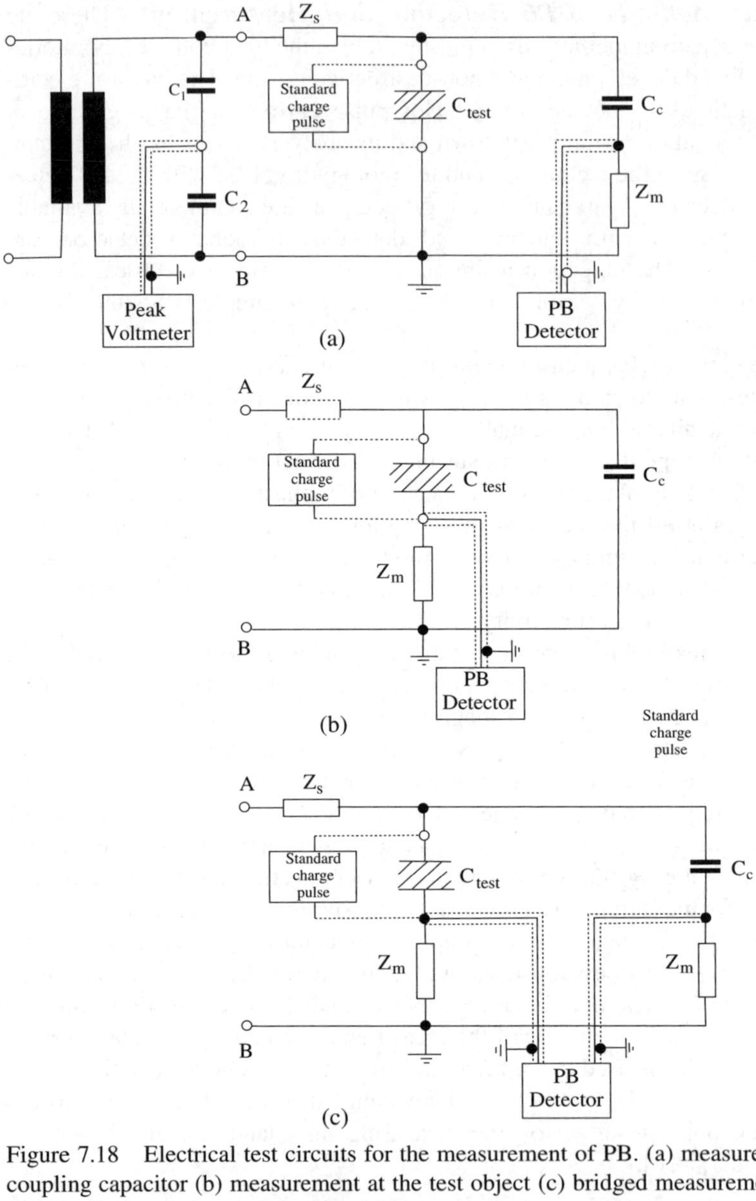

Figure 7.18 Electrical test circuits for the measurement of PB. (a) measurement at coupling capacitor (b) measurement at the test object (c) bridged measurement with both, the coupling capacitor and the test object.

chemical and mechanical stability are all strongly and often nonlinearly coupled. Above all, the time factor affecting the breakdown makes the pre-breakdown phenomena more complicated.

7.3.1 Intrinsic Breakdown Strength of Solid Dielectrics

As described in the case of liquid dielectrics, the same definition of intrinsic strength holds good also for solid dielectrics. For a homogeneous and isotropic solid dielectric, the intrinsic strength is described to be the highest value of electric strength obtainable after all known secondary effects leading to a premature breakdown (below intrinsic strength), seem to have been eliminated.

The concept remains an ideal one and its validity is always dubious, as it is very difficult to know whether an observed breakdown was or was not intrinsic. One may conclude, therefore, that the more observed results are likely to be intrinsic, the more they are independent of parameters such as temperature, thickness, and time. It is also extremely important that the test sample is prevented from self-heating and mechanical distortions caused by the field.

In order to measure intrinsic strength, preparing the large number of identical test samples is sensitive involving a considerable amount of precision techniques. As described by Garton in [7.15], only two methods of measurements are seriously considered, the "recessed specimen" and the "McKeown's technique". Lawson [7.27] measured the intrinsic strength of PE samples with the help of above-mentioned specimens at different temperatures described in the following paragraphs.

Recessed Specimen—A spherical depression is pressed, machined or ground on a circa 1.5 mm thick plane disc sample of the material. It nearly penetrates the dielectric, leaving only a thickness of approximately 50 μm, suitable for breakdown by a reasonable magnitude of voltage, as shown in Figure 7.19 (a). The radius of the spherical depression should be such that a uniform field is achieved. The surfaces of the recessed and of the opposite plane sides are then made conductive by a technique that provides an extremely good contact with the dielectric, for example, an evaporated film of aluminium, or a graphite coating applied by repeated spraying, then polished to a continuous film. Liquid electrodes have also been tried. However, limitations such as inadequate cooling of the stressed area and no mechanical support for the thin apex of the specimen remain with the provision of such electrodes.

McKeown Technique—Two ball bearings used as electrodes in this specimen are located within a cylindrical hole in a disc made of suitable insulating material such as Perspex. The lower ball is cemented in position with an epoxyresin, and the hole filled with the same resin in liquid form. Then it is gently heated in partial vacuum until degassed. As shown in Figure 7.19(b), the polyethylene test sample in the form of a thin film disc is placed in the hole on the top of the ball and the liquid is degassed again. Before encapsulation, the polyethylene film sample were treated chemically by immersion in concentrated chromic acid for five minutes to oxidize their surfaces and make a bond with the epoxide resin. The top ball is then inserted and the whole specimen cured until solid conditions are achieved. Measuring the total thickness

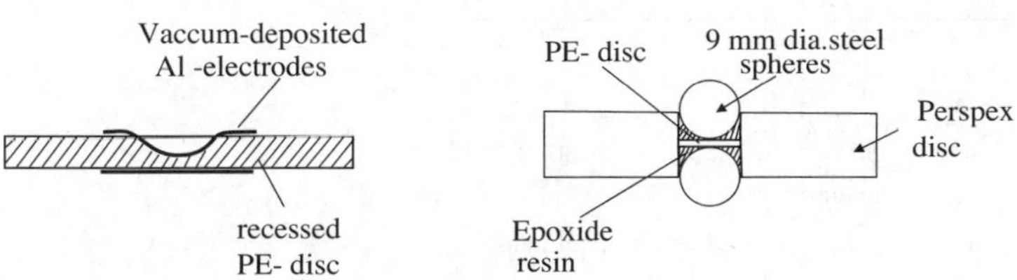

(a) Recessed specimen (b) McKeown specimen

Figure 7.19 Specimens for the measurement of intrinsic strength, Lawson [7.27].

outside the balls and subtracting twice the diameter of the ball gives the thickness of the assembled test sample. A proper adhesion of the test sample material with the epoxyresin must be achieved to avoid any partial breakdown or tracking at the interfaces. The specimen thus prepared may further require a chemical treatment with a solution of sodium in liquid ammonia in order to prevent corona or PB on the surfaces.

As reported by Garton, both of these techniques have proved extremely satisfactory at room temperatures for measuring the intrinsic strength of low-loss polymers. However, the results obtained by McKeown specimens have been consistently higher than those obtained by the "recessed" specimens.

Lawson [7.27] used McKeown techniques to measure the intrinsic strength of PE with increasing temperature between 20–85°C, as shown in Figure 7.20. In his experiments, the breakdown tests were carried out under transformer oil to prevent flashover due to surface discharge. The *dc* applied voltage, raised linearly with time, was obtained from a smoothed half-wave rectifier source. In these experiments, the breakdown occurred in about 15–20 s. In all cases, the number of specimens tested were from six to eight, and 95% confidence limit of the mean was considered.

Compared to the results obtained from recessed specimens, the fall of intrinsic strength with increasing temperature was greatly reduced by using McKeown specimens. Altogether higher values of electric strength, of the order of 800 kV/mm, were obtained, as shown in Figure 7.20.

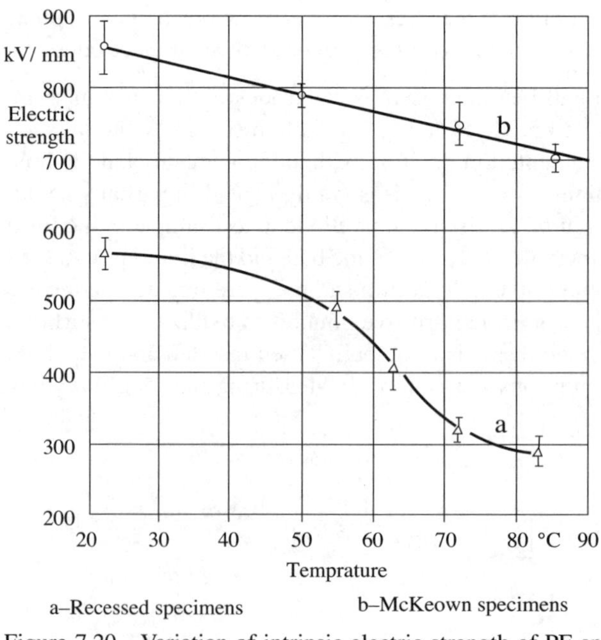

Figure 7.20 Variation of intrinsic electric strength of PE specimens measured with dc for increasing temperature.

The theories of intrinsic breakdown estimate a condition for irreversible instability that is electronic in nature. These simply assume that in case of intrinsic breakdown the irreversible electronic catastrophe produces conditions in which the lattice is destroyed. Fröhlich's amorphous or high temperature theory of intrinsic breakdown, which is applicable to PE, predicts that the electric strength "E_b" should vary with temperature above a critical temperature T_{cr} according to the relation:

$$E_b = C\exp\left[\frac{\Delta V}{2kT_o}\right] \tag{7.5}$$

where C is a constant, ΔV- spread in excited energy levels of the defect states below the conduction band, k- Boltzmann's constant $= 8.622\times10^{-5}\ eV/°\text{C}$ and T_o- lattice temperature in K.

According to Fröhlich's theory, the breakdown of amorphous solids is largely determined by the properties of slow thermal electrons, for example, on its mean free path. The straight-line characteristic of breakdown strength with McKeown specimens obtained by Lawson (Figure 7.20) agrees with Fröhlich's amorphous or high temperature theory of intrinsic breakdown. The straight line shown is drawn to be the most appropriate, conforming to the results obtained from experiments. The constants of Fröhlich's equation, calculated for PE from this characteristic by Lawson [7.27], are as follows:

$$\Delta V = 0.06\ eV$$

and

$$C = 2.66\ \text{MV/cm}$$

The McKeown techniques have also been successfully applied to maintain high but subintrinsic stresses on PE specimens for very long durations. Such PE specimens could hold 170 kV/mm even for 15,000 hr without failure [7.15].

The general aim of the measurements of intrinsic electric strength of the dielectrics has been to reveal the complexity of the breakdown mechanism at extremely high electric fields. However, opinions have often been expressed that a breakdown always occurs through some secondary mechanism or an external cause. Therefore, considering a breakdown intrinsic is mostly dubious. Measured breakdown strengths of the order of 800 kV/mm for PE by Lawson are the most creditable achievement.

7.3.2 Thermal Breakdown

Since it develops very rapidly, the intrinsic breakdown is assumed to have electronic origin. The simple theoretical approach for intrinsic breakdown has been to derive a theory for an electronic instability in the dielectric subjected to a uniform field, O'Dwyer [7.28]. While measuring the intrinsic strength, if there exists a cause which leads to other than electronic instability, the value of electric strength measured may be anywhere below intrinsic strength. One such cause is "thermal instability", particularly in the dielectrics having sufficient conductivity. The conductivity is found to increase with electric stress, but the rate of increase may not be linear, Lawson [7.27]. The "hot electron" theory of breakdown in solid dielectrics developed by

Fröhlich and Paranjape in [7.27] also predicted a stress-enhanced conductivity, that may cause thermal failure before a true instability (intrinsic failure) occurs. This theory involves lattice temperature rise processes and is derived for intrinsic or near intrinsic breakdowns.

The experimental concept of "thermal breakdown" is different as it is based upon self-heating of the dielectric due to power losses. Because of power loss due to conductivity, polarization and other forms of dielectric losses, heat is produced continuously in electrically stressed dielectrics. Depending upon the magnitude of the applied voltage, its period of application and the conduction of heat, the dielectric temperature rises. If the heat generated within a dielectric system equals the dissipation of heat to the surroundings by thermal conduction, the temperature rises to an equilibrium value, and a thermally balanced and stable operation of the insulation system takes place. But in practice, the power losses in a dielectric are also functions of temperature. Since the ionic conductivity increases with increasing temperature, the power losses increase with temperature. If the dissipation of heat by cooling processes is not adequate, it is possible that an unstable state may arise. The temperature could increase without limit, causing a breakdown.

Consider a solid insulation system in uniform field with electrode area "A" and gap distance, "d", having *dc* conductivity, "κ_{dc}". Neglecting polarization and other forms of losses, the active or real power loss due to *dc* conductivity in the dielectric can be given according to equation (6.3) as:

$$P_{dc} = U^2 \cdot A/d \cdot \kappa_{dc} \tag{7.6}$$

Since κ_{dc} strongly depends upon the temperature in an exponential function, equation (6.29) can be rewritten as:

$$\kappa_{dc}(\theta) = \kappa_{odc} \cdot e^{\beta(\theta-\theta_o)} \tag{7.7}$$

where κ_{odc} is the conductivity at 20°C, β the temperature coefficient of κ_o in per K and θ_o the ambient temperature. From the above two equations the following equation can be derived,

$$P_{dc}(\theta) = U^2 \cdot A/d \cdot \kappa_{odc} \cdot e^{\beta(\theta-\theta_o)} \tag{7.8}$$

If the dissipation of heat to the surroundings is considered to be linearly proportional to the temperature, then,

$$P_{diss}(\theta) = K(\theta - \theta_o) \tag{7.9}$$

where K is a constant.

A schematic of graphical solution for the condition $P_{dc}(\theta) = P_{diss}(\theta)$ is shown in Figure 7.21. For a given voltage U_1 two values of temperature θ_1 and θ_2 are obtained, whereas at a particular voltage U_2 higher than U_1 the two curves meet at only one point θ_c. At all three points the real power losses in the dielectric are equal to the power dissipated to the surroundings. Under such a condition, the following equation holds true,

$$\frac{dP_{dc}}{d\theta} = \frac{dP_{diss}}{d\theta}$$

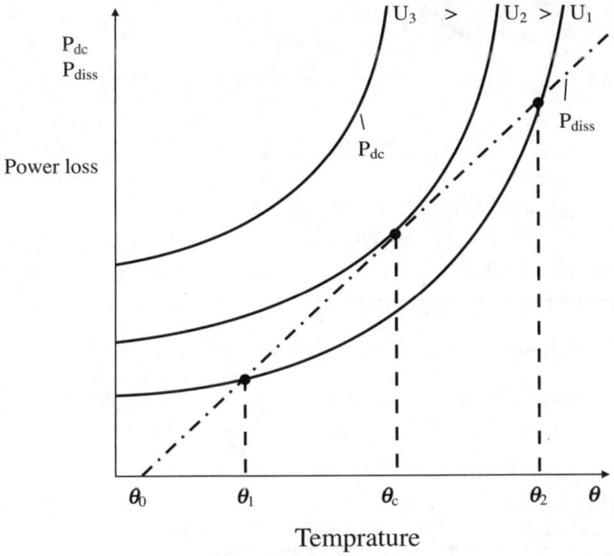

Figure 7.21 Schematic of power loss and power dissipated in a dielectric system causing a rise in temperature when stressed with different magnitudes of applied dc voltages.

θ_c can be described to be the critical temperature. At voltages higher than U_2 and at greater temperature than θ_c at U_2, the real power loss in the dielectric is always more than the power dissipated to the surroundings. Under such conditions the dielectric temperature may rise without limit. Hence, it can be concluded that at voltages U_2 and above, thermal breakdown is inevitable. Since the total energy loss in a dielectric depends upon the period of application of the voltage, the experimental concept of thermal breakdown is coupled with time as well as the electrode geometry of the insulation system.

On applying an alternating voltage, the real power loss in a volume of dielectric V in uniform field E is given by equation (6.27) as,

$$P_{ac} = \varepsilon_o \cdot \omega \cdot E^2 \cdot V \cdot \varepsilon_r \tan\delta \tag{7.10}$$

For a given volume of dielectric and frequency of the applied voltage, this equation can be written in terms of a constant K as,

$$P_{ac} = K \cdot (\varepsilon_r \tan\delta) \tag{7.11}$$

The quantity (ε_r tan δ), defined as "loss index", strongly depends upon the temperature of the dielectrics, besides the frequency and magnitude of the applied voltage. In order to estimate the amount of energy produced, the duration of application of the voltage must also be taken into account. A schematic showing the pre-breakdown temperature rise of a dielectric with respect to time on applying different magnitudes of voltage is given in Figure 7.22. Such an illustration is demonstrative of experimental measurement on an insulation system for thermal breakdown.

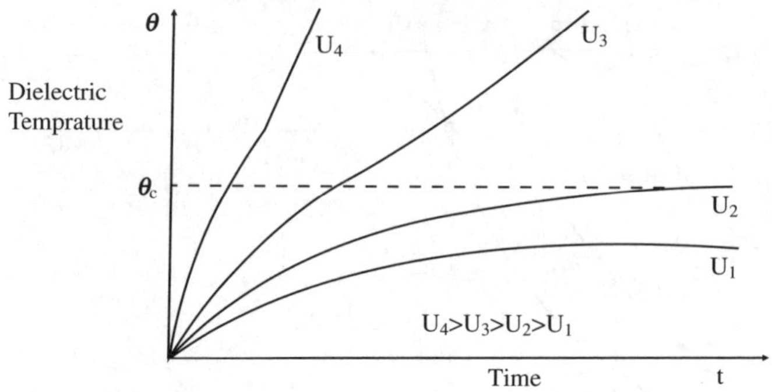

Figure 7.22 Development of temperature within a dielectric with time on applying different magnitudes of voltages.

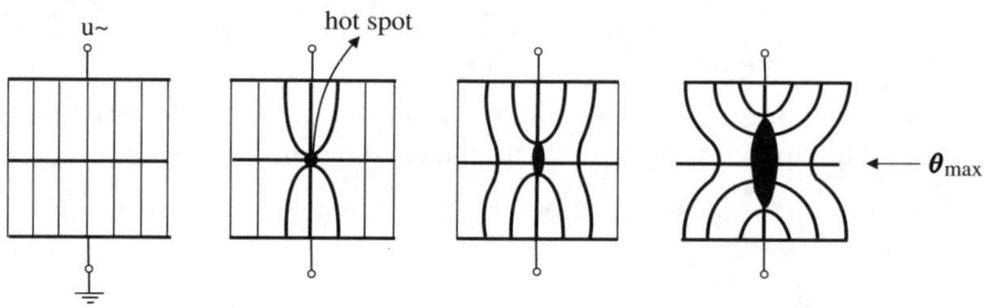

Figure 7.23 Schematic of the development of "hot spot" leading to thermal breakdown.

At a comparatively lower magnitude of voltage U_1, temperature acquired by the dielectric remains below the critical temperature even after a very long time of application. A balanced condition between heat produced and heat dissipated prevails and no thermal breakdown takes place. On applying a voltage, U_2, higher than, U_1, the critical temperature that may lead to thermal breakdown is acquired after a considerably long time. At still higher voltage, U_3, the critical temperature in the dielectric is reached within a short time and a thermal breakdown is most likely to occur. If a very high impulse form of voltage U_4 is applied, an impulse thermal breakdown may take place, since the heat dissipation within such extremely short duration is not possible and the losses proportional to the voltage magnitude are very high.

On applying voltage to a block of a dielectric, the temperature rises due to heat produced from losses within the dielectric. It is presumed that first a sort of temperature gradient field is formed within the dielectric. The field pattern of heat is determined by the isothermals along the electrode surface (Figure 7.23). The highest temperature may be acquired somewhere in the middle of the dielectric. If the heat dissipation is inadequate, a so called "hot spot" develops in this region due to thermal heterogeneity. Since the electrical conductivity in the high temperature region increases, there is a possibility of higher field intensity due to space charge entrapment. A breakdown channel develops from the hot spot towards the electrodes.

Most of the electrical power equipments such as cables, transformers and generators operate at considerably high service temperatures. In such equipment, the main cause of high working temperature is the power loss (I^2R) in the conductor or windings. The heat produced due to such losses is dissipated through the dielectric. The dielectric power losses only marginally augment the main power losses. Adequate arrangements for forced cooling are often essential to maintain the maximum permissible working temperature of the dielectrics within limits. Such arrangements are, for example, hydrogen and water cooling in generators, cooling by the circulation of oil in high voltage power cables and transformers, and so on. The problem of thermal breakdown is faced more by equipment, that has to operate at extremely high working temperatures. There is a need for the development of heat conducting solid dielectrics.

7.3.3 Mechanism of Breakdown in Extremely Nonuniform Fields

While designing electrode configurations in practice, the basic approach is to avoid extremely nonuniform field configurations, especially for solid dielectrics. Thus, the solid dielectrics are normally not stressed with extremely high nonuniform fields. But, the pre-breakdown process generally begins at an extreme intensity or distortion of local fields that may arise due to protrusions, foreign particles or voids. Under the action of this high local electric field, a solid dielectric may either globally or locally loose its electrical insulating property and mechanical stability, leading to its complete or partial disintegration. A review of the breakdown and prebreakdown phenomena in solid dielectric made by Zeller [7.29] has brought out certain important conclusions described in the following paragraphs.

At high field intensities closer to breakdown, massive space charge injection from the electrodes is inevitable. The local electric stress, dominated by the space charge, may increase considerably further. The local electromechanical action due to high electric force causes dielectric instabilities in solids, giving rise to initiation and growth of electrical trees, as described in the next section.

The electric field exerts a force on the space charge distribution, developing space charge currents within the dielectric. The local current density due to space charge concentration not only depends upon the applied voltage, but also upon the voltage rise time, as long as PB has not set in. These currents within the dielectric produce heat locally. Depending upon the heat flow conditions, the local increase in temperature may lead to thermal instabilities resulting in breakdown.

Zeller described further that qualitatively, the space charge current decays into the dielectric filaments. The discharge current produced at a weak point due to PB proceeds along the easiest course determined by the material inhomogeneities, not necessarily in the direction of applied field. Numerical simulation also confirmed that material inhomogeneities lead to a considerable meandering of the filament currents. Adverse conditions of PB may give rise to treeing in the dielectric, leading to complete breakdown at comparatively much lower voltage.

7.3.4 "Treeing" a Pre-Breakdown Phenomenon in Polymeric Dielectrics

Treeing is an electrical prebreakdown phenomenon. The name is given to the type of damage caused to the solid dielectrics, progressing through parts under electric stress in extremely nonuniform fields. The path of its growth appears like a miniaturized tree, hence the name. Treeing process normally begins at a very sharp, pointed electrode tip due to internal partial breakdown. Extreme field distortions at protrusions, foreign particles and voids in the dielectrics may also give rise to the inception of treeing process.

Treeing may occur and develop slowly due to PB. In the presence of moisture, it may develop slowly even without any measurable PB quantity. However, treeing develops rapidly when an impulse voltage is applied. While treeing is generally associated with *ac* and impulse voltages, it also takes place under high *dc* stresses. These is evidence that the treeing process is aggravated by the presence of water/moisture, chemical environment, voids and contaminants such as foreign particles. Treeing may or may not be followed by complete electrical breakdown of the part of the dielectric where it occurs. However, in solid polymeric dielectrics it is one of the most common causes of electrical failure. The whole process may or may not take a long time, for it strongly depends upon local conditions.

7.3.4.1 Forms of Treeing Patterns The great variety of patterns that appear like stems and branches comprising an electrical tree are described in the literature with large number of corresponding names of the figures such as dendrite, fan, plume, delta, bush broccoli, string and bow tie, and so on. Eichhorn [7.30]. The reader should therefore not get confused as they all represent the same phenomenon of treeing under different local conditions. In Figure 7.24 (a), a tree, and in 7.24 (b) a bush having tree-like branches are shown. These were produced in Piacryl, a transparent solid dielectric, in extremely nonuniform field between needle and plane by Pilling [7.31].

Besides depending upon the inhomogeneity of the dielectric material, the pattern of growth of treeing also depends upon the radius of the tip of needle, diameter of needle, gap length "*d*" as well as shape and size of the plane electrode, Löffelmacher [7.32].

In Figure 7.25, a "bow tie" type tree is shown, Eichhorn [7.30]. Such a tree developed at a void on the surface and grew up along it on both sides.

7.3.4.2 Classification of Treeing Process Three types of main categories of treeing processes are commonly mentioned in the literature. These are electrical, electrochemical and water treeing. However, water treeing in its physical sense is a type of electrochemical treeing process. This reduces to basically two major categories of treeing, electrical (ET) and electrochemical treeing (ECT), Vogel [7.33].

The growth of an "electrical tree" is controlled by the dielectric instability arisen from, for example, local field enhancement due to space charge injection. If the local stress exceeds the stable condition, it may lead to the formation of a void at electrode-dielectric interface. This void then becomes the nucleation point

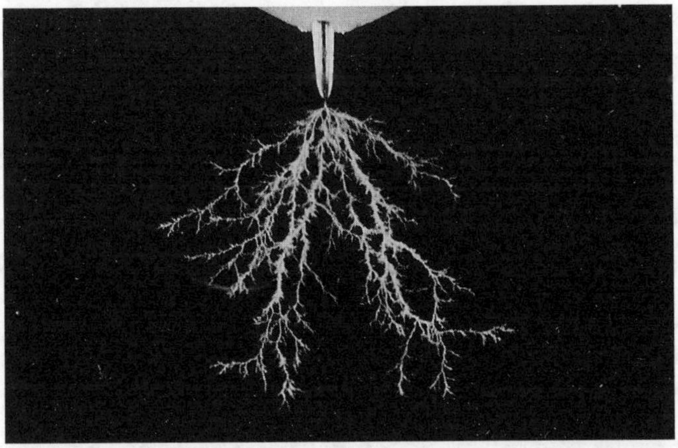

(a)

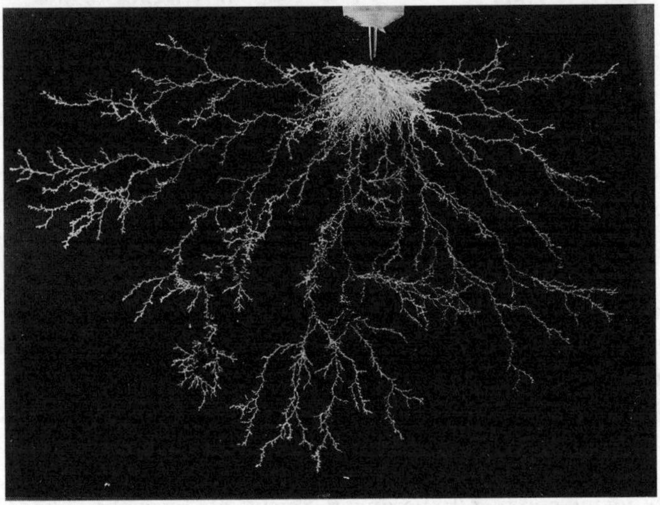

(b)

Figure 7.24 Treeing in Piacryl, Pilling [7.31] (a) A Tree, (b) A Bush.

for the tree formation having a preferential growth in the mechanically easy direction due to electric forces. Depending upon the local conditions, this instability may spread in its own form. Electro-erosion is undoubtedly a process that takes place under the influence of high electric stress. A familiar hypothesis is that treeing proceeds due to electro-erosion induced by gas discharges in the conductive channels thus formed.

As Zeller [7.29] reports, carefully controlled experiments by Baumann et al. have revealed that prolonged electrical stress at fields lower than "critical field intensity" never leads to PB inception. Below this field level neither hot electron damage nor other local instabilities occur in a mechanically stable material.

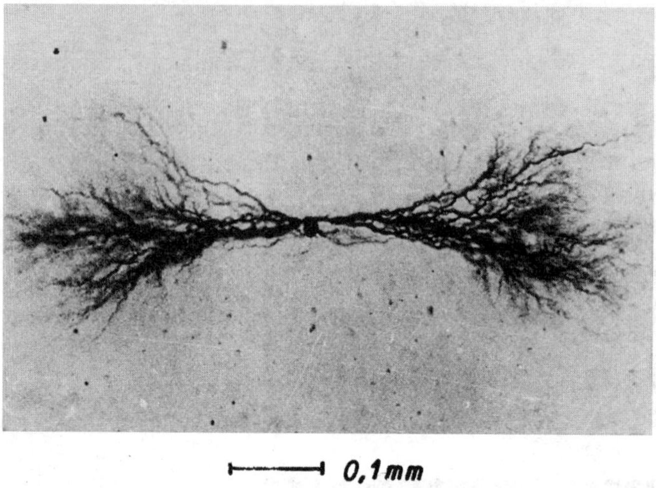

Figure 7.25 A bow tie type tree, [7.30].

Once a tree begins to grow, the situation becomes rather complex. The growth is then controlled by three criteria; first, the local PB occurrence conditions, second, the solid dielectric instability at the end point, and third, the spatial extension of the gas discharge inside the tree channels.

"Water treeing", which is basically an "electrochemical treeing" process, occurs in polymeric materials subjected to electric stress in the presence of water/moisture. Water treeing has received tremendous attention since the beginning of 1970s, as it has been one of the main cause of polymeric insulation degradation, leading to premature breakdown in HV power cables.

The space charge, injected from the electrodes, sets in across the polymer-water interface. Electrochemical oxidation and free radical formation were observed by Zeller [7.29]. Once the free radicals are formed, oxidation proceeds by an autocatalytic process, penetrating the polymer. Experimental results of Given et al. [7.34] have shown that the free radical formation, besides the properties of individual polymer radicals, alter the growth rate of water trees. The initial penetration of free radicals into the undisturbed region of the polymer distorts its structure, causing polymerization at the surface. This reaction causes enhancement in the polymerization rate at which water molecules could migrate through the polymer, leading to the disruption of the structure of polymer and increasing the penetration of water to the mechanism of treeing process.

As described above, an important factor in the growth of electrochemical trees is their access to the free air and water surroundings. Trees, which start their growth at surfaces with an unlimited supply of air and water, grow through the dielectric well, bridging the electrodes completely in comparatively shorter time. These are called as "vented trees". One such growth of tree is shown in Figure 7.25 on the surface of the dielectric. Trees that start at an internal void or a foreign inclusion are known as "nonvented trees". The nonvented trees rarely grow as large as the vented trees.

7.3.5 Requirement of Time for Breakdown

Consider a solid dielectric in extremely nonuniform field between needle and plane applied an alternating field, as shown in Figure 7.26. The field intensity at the needle tip in the dielectric will be maximum in this electrode configuration if there is no other cause of field distortion/enhancement present elsewhere in the dielectric. In other words, a local field enhancement will be caused at the tip of the electrode, leading to a favourable condition for the production of free electrons and their acceleration. On increasing the applied voltage when this process acquires certain minimum intensity, the development of a gas filled channel may begin after certain time lags. Let the time required for the first appearance of such a local breakdown channel be called "channel appearance" time "$t_{ch.a}$". This is the beginning of the degradation process by treeing in the dielectric. The development of treeing process has been described in the previous section. The time required for the development of a tree up to the opposite electrode "$t_{ch.d}$" may vary considerably in different dielectrics from location to location, depending upon the favorable (or unfavorable for a user) local conditions. As soon as the foremost channel grows up to the opposite electrode, the breakdown takes place. Thus, the time required for breakdown "t_b"

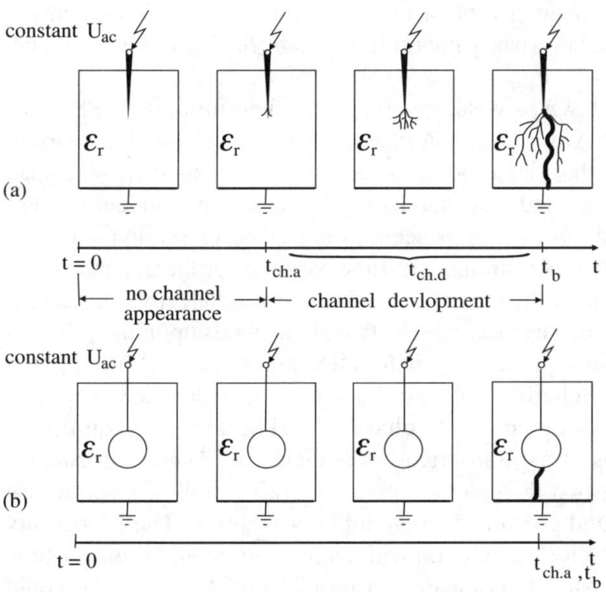

$t_{ch.a}$ - time required for channel appearance ,

$t_{ch.d}$ - time required for channel devlopment

t_b - time required for breakdown

(a) breakdown in extremely nonuniform fields

(b) breakdown in weakly nonuniform fields

Figure 7.26 Schematic of breakdown process in solid dielectrics.

can be given as the sum of time required for the appearance of the first channel and its development up to the opposite electrode,

$$t_b = t_{ch.a} + t_{ch.d} \tag{7.12}$$

Practical experience has revealed that in case of extremely nonuniform fields, the time required for the appearance of the first channel "$t_{ch.a}$" is generally very small compared to the time required for the development of a tree up to the opposite electrode "$t_{ch.d}$". In other words, it can be expressed as,

$$t_{ch.d} \gg t_{ch.a} \text{ (in extremely nonuniform fields)}$$

However, in case of weakly nonuniform fields existing between the electrodes, the time-wise development of complete breakdown is quite different. As shown in the schematic in Figure 7.26 (b), on applying alternating field between sphere and plane in a dielectric, the time required for the appearance of first partial breakdown channel "$t_{ch.a}$" is normally very long compared to the time required for the development of final jump of the breakdown. Almost an abrupt development of breakdown channel takes place in this case, mostly without any PB or distinct treeing process. Hence,

$$t_{ch.d} \ll t_{ch.a} \text{ (in weakly nonuniform fields)}$$

This phenomenon is explained by the fact that because of higher field intensity, which can be generally applied in weakly nonuniform fields, the breakdown channel extends itself much faster.

Photograph of a breakdown in weakly nonuniform field in a 20 kV PE cable is shown in Figure 7.27 (a), Arora [7.35]. A clean breakdown channel appears to have developed abruptly in the dielectric between two semiconductive extruded screens. In Figure 7.27 (b), a breakdown channel in PE due to high tangential component of field at a cable end termination is seen, as described in Section 7.2.2.

Statistical analyses of the measurement of time required for the appearance of first channel of treeing "$t_{ch.a}$" as well as breakdown through the dielectric "t_b", established the probability of their occurrence, which strongly depends upon the physical conditions of the dielectric. Such probability distribution curves for "$t_{ch.a}$" on PE and XLPE samples measured by Scharff [7.36] are illustrated in Figure 7.28. The time "$t_{ch.a}$" were measured on applying constant voltage of 10 kV, ac (peak), on needle electrodes having a tip radius of 5 μm inserted in the dielectric. These experiments were performed maintaining two different temperatures (20 and 90°C) on the test samples. The XLPE compound contained tree-inhibiting additive. The probability curve plots clearly show that XLPE insulation with antioxidant required longer time for the appearance of first channels compared to normal PE. A conclusion could therefore be drawn, that XLPE with additive is more tree resistant and may have longer life expectancy. Further, it was revealed that the appearance of first channel due to PB also depends upon the temperature.

The time required for breakdown in solid dielectrics strongly depends upon the applied field intensity or voltage. In certain cases, the breakdown process may even take decades to develop. This is described as "electrical aging" process. Accordingly, depending upon the magnitude of the applied electric stresses, a certain life period of a dielectric may be expected during which trouble-free service is

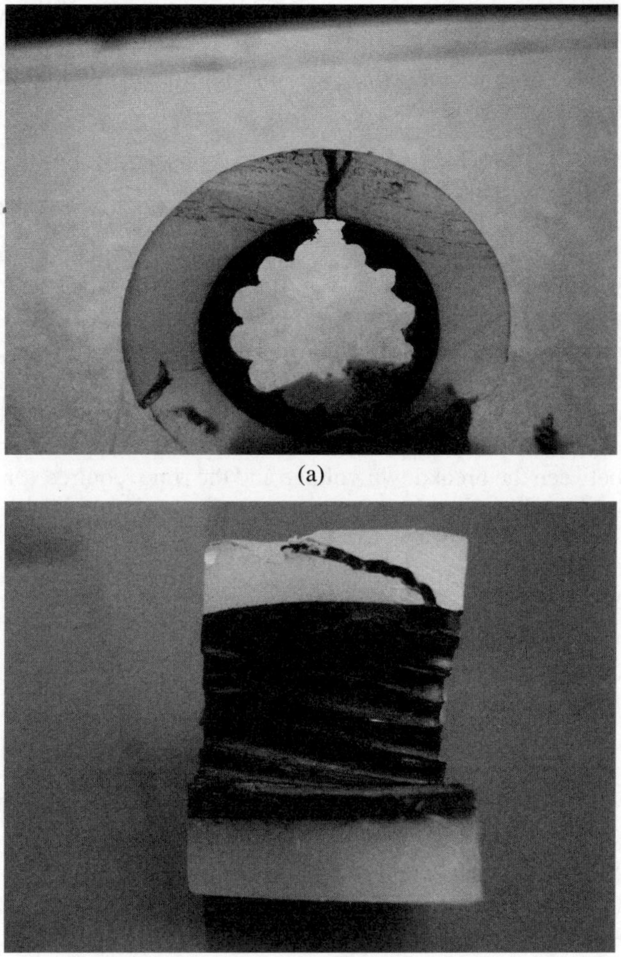

(a)

(b)

Figure 7.27 Breakdown channels in coaxial PE Cables, Arora [7.35].

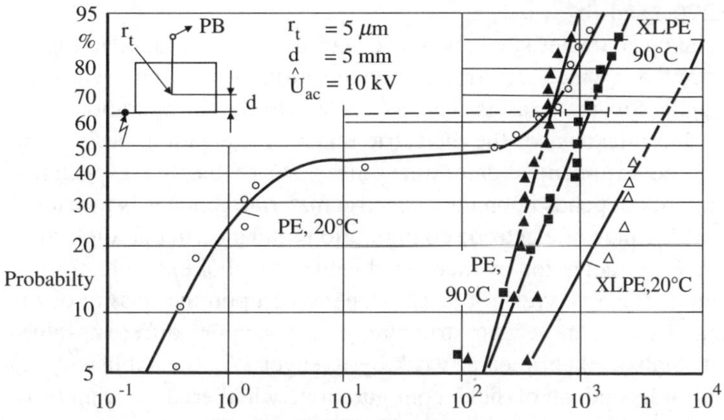

Figure 7.28 Probability distribution of time required for the appearance of first channel at the tip of a needle electrode in PE and XLPE with antioxidant dielectrics at different temperatures, Scharff [7.36].

rendered. Practical estimation of such trouble-free life of a dielectric is described in the following section.

7.3.6 Estimation of Life Expectancy Characteristics

In the previous section it has been established that the formation of PB channels in solid dielectrics is not only a time dependent process, but it also depends upon the magnitude of applied voltage and the temperature of the material. Besides the time required for the first appearance of the channel, $t_{ch.a}$, the time required for the development of the PB channel, $t_{ch.d}$, as well as the time required for breakdown t_b, also depend upon the magnitude of the applied voltage or in other words, on the intensity of applied electric field. The so-called life expectancy characteristics are plotted between the breakdown voltage and the rime required for breakdown t_b, as explained in the following.

$$\hat{U}_b = f(t_b) \tag{7.13}$$

or

$$\hat{U}_b = k_b \cdot t_b^{-1/n_t} \tag{7.14}$$

or

$$\log \hat{U}_b = -\frac{1}{n_t}\log t_b + \log k_b \tag{7.15}$$

where k_b is a constant and n_t the life expectancy characteristics exponent.

Combining with the Weibull-function, the distribution function of the breakdown voltage for constant breakdown time, as shown in Figure 7.29 (c) is given as follows:

$$F(\hat{U}_b) = 1 - \exp\left[-\left(\frac{\hat{U}_b}{\hat{U}_{b63\%}}\right)^{C_u}\right] \tag{7.16}$$

where the Weibull exponent $C_u = n_t \cdot C_t$ and n_t represent the slopes of the curves as shown in Figure 7.29(a) and (b).

The life expectancy characteristic, given by equation (7.15) is plotted in Figure 7.28 (b). The constant k_b, which determines the position of this characteristic, depends upon the dielectric material as well as the electrode configuration. It is affected by the area and thickness of the dielectric under investigation, besides the electrode geometry. The exponent n_t determines the slope of the life expectancy characteristics and mainly depends upon the dielectric material. Since n_t is measured to be constant for a wide range of electrode configurations and a particular dielectric, different dielectrics are characterized by their fixed value of n_t. For example, for PE, n_t is equal to 9 whereas for epoxyresin it is 12. Depending upon the quality of the dielectric and its macroscopic molecular structure, n_t may acquire different values for the same dielectric also. Experimental work by Kreuger [7.37] established and confirmed that n_t is independent of field configuration, whether it is weakly or extremely nonuniform.

In the solid dielectrics it is possible that the aging process may completely change at different magnitude of voltage, or after a certain time of application of

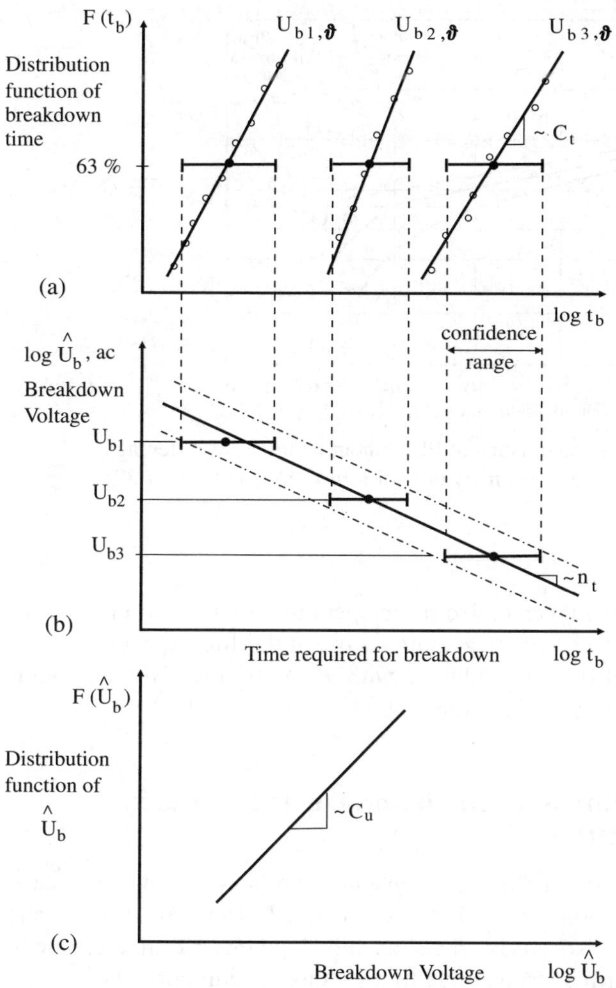

Figure 7.29 Schematic of empirical estimation of life expectancy curves of solid dielectrics.

the same voltage level. It may also depend upon the rate of change of the applied voltage. This could result in sharp bends in the measured life expectancy characteristics. Therefore, the extrapolation of these curves is often not advised, as it may lead to completely wrong characteristics. There are enormous varieties of aging mechanisms that may take place in solid dielectrics depending upon complexity of their molecular structure, behavior and physical parameters. All these factors individually as well as collectively affect the life expectancy of the dielectric.

Life expectancy characteristics of ordinary LDPE measured by Li [7.38], at different temperatures are shown in Figure 7.30. These measurements were conducted on actual insulated conductors (30 cm long) by impinging a needle having tip radius of 50 μm in the dielectric, as shown in the figure. A sharp bend in the

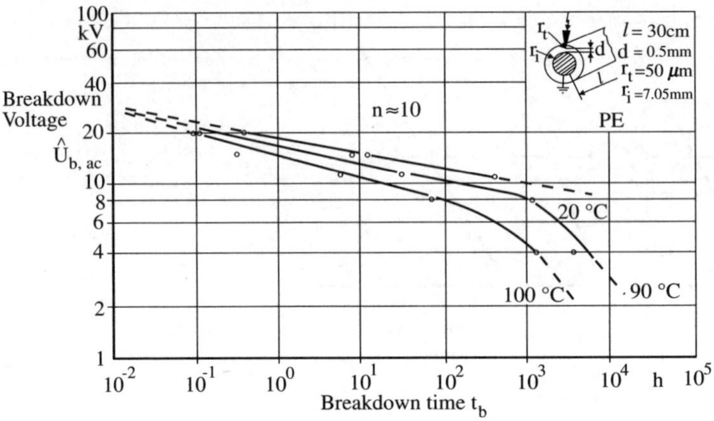

Figure 7.30 Life expectancy characteristics of PE without additives at different temperatures with artificially created extremely nonunifrom weak point, Li. [7.38].

characteristic is measured at higher dielectric temperature after a certain level of time required for breakdown. A very large number of similar life expectancy characteristics measured under different conditions with a variety of electrode systems and parameters are available in the literature.

7.3.7 Practical Breakdown Strength and Electric Stress in Service of Solid Dielectrics

The average life expectancy of an electrical apparatus is at least twenty-five years. During this time, the insulation must withstand the rated electrical, thermal and mechanical stresses. It must also survive the short time high electric stresses due to over voltages as well as thermal stresses due to high short circuit currents.

As explained in this chapter, the breakdown strength of solid dielectrics depends upon a vast number of external and internal factors. It is, therefore, not possible to standardize the breakdown strengths of different solid dielectrics. However, some practical values are given in Table 7.2 for their comparative evaluation. The unpredictable behavior of solid dielectrics has lead to the utilization of only about 2% of their practical breakdown strength in order to have a high factor of safety and reliability. In Table 7.2, practical values of electric strengths of some solid dielectrics as well as their common values of electric stresses in service are shown, as given by Philippow [7.39]. As seen from this table, the service electric stresses for insulations impregnated with oil are much higher compared to other solid dielectrics. This is because the reliability of insulation systems impregnated with oils has established itself to be higher compared to other solid dielectrics. However, with the development in chemical technology and manufacturing techniques, the maximum in service electric stress in high voltage XLPE insulated cables produced for rated voltages of 500 kV and above has been increased to 15 kV/mm and above.

TABLE 7.2 Practical breakdown strengths and service stresses of some solid dielectrics

Insulating material	Apparatus	$\hat{E}_{bp}$ kV/mm	$\hat{E}_s$ kV/mm	$\frac{\hat{E}_s}{\hat{E}_{bp}}\times 100 = \%$ $\hat{E}_s$ as % of $\hat{E}_{bp}$
Polyethylene, PE	Cable	140	3.... .10 (15)	2 . . . . . . 7 (11)
Paper—oil	Cable	55....80	10 12	≈ 16
Paper—oil	Condenser	180	15 28	≈ 12
Porcelain	Insulators	125	1... .2.5	0.8....2.0
Epoxyresin—Mica	Generators	75	1... .2.5	1.3....3.3
Epoxyresin	Bus bars	125	1.5 . . . 3	1.2....2.4

E_{bp}—practical breakdown strength, E_s—in service electric stress.

REFERENCES

[7.1] Bogorodiski, N.P., Pasynkov, V.V., Tareev, B.M. "Electrical Engineering Materials", Verlag Technik, Berlin (1955) (in German) Mir Publishers, Moscow (1979)(in English).

[7.2] Kumar, Anil and Gupta, Santosh K. "Fundamentals of Polymer Science and Engineering", Tata McGraw- Hill, New Delhi (1978).

[7.3] Mackevich, Jeffry and Shah, Minesh, "Polymer Outdoor Insulating Materials Part I:Comparison of Porcelain and Polymer Electrical Insulation", IEEE Electrical Insulation Magazine, Vol. 13, No. 3, May/June 1997.

[7.4] Brydson, J.A., "Plastic Materials" Btterworths, (1989), fifth edition.

[7.5] Vieweg, Richard (Editor), "Handbook on Synthetic Materials", Volume IV Polyoelfins, Hanser, Munchen (1969) (in German).

[7.6] Yutao Zhu, Ho Gyu Yoon and Kwang S. Sub, "Electrical Properties of Silane Crosslinked Polyethylene in Comparison with DCP Crosslinked Polyethylene", IEEE Transactions on Dielectrics and Electrical Insulation, Vol. 6 No. 2, April 1999.

[7.7] Krohne, Rolf, "About the developed of space charge in PE at high dc, ac and combined voltage", Dissertation, TU Hannover (1978) (Faculty of Mechanical Engineering).

[7.8] Sander, Gerhard, "The effect of structure on conduction mechanism in PE at high alternating fields", Dissertation, TU Hannover (1977) (in German).

[7.9] Brown, Morton, "Compounding of Ethylene-Propylene Polymers for Electrical Applications", IEEE Electrical Insulation Magazine, Vol. 10, No. 1, January/February 1994.

[7.10] Jacobsen, Peter, "Influencing factors on drying process of paper insulated high voltage cables", Dissertation, TU Hannover (1977) (in German).

[7.11] Grossekatthöfer, H.D., "The drying process of paper insulation and the effect of remaining moisture on the electrical properties of impregnated dielectrics", Dissertation, TU Hannover (1973) (in German).

[7.12] Yasufuku, Sachio, "Application of Glass Fiber-Reinforced Plastics to Electrical and Electronic Apparatus in Japan", IEEE Electrical Insulation Magazine, Vol. 10, No. 1, January/February 1994.

[7.13] Lemke, Eberhardt, "A contribution to the electrical measurement of PD with special consideration to the broad band measurement techniques of cumulative discharge", Dissertation, TU Dresden (1975) (in German).

[7.14] Lemke, Eberhardt, "A new PD measuring procedure based on the evaluation of cumulative charge", Proceedings of the "Conference on PD in Electrical Insulation", Institution of Engineers (India), Bangalore (1976).

[7.15] Alston, L.L., "High Voltage Technology", Oxford University Press (1968).

[7.16] Nema, R.S., Shivkumar, B., "Breakdown of thin polythene films by surface discharges", Indian Journal of Pure & Applied Physics, Vol. No. 2 (1974) pp. 163–165.

[7.17] Parr, D., Scarisbrick, R.M., "Performance of synthetic insulation materials under polluted condition", Proc. IEE 112 (1965) pp. 1625–32.

[7.18] Gamez-Garcia, M., Bartnikas, R., Wertheimer, M.R., "Synthesis reactions involving XLPE subjected to practical discharge", IEEE Trans. EI-22 (April 1987)2, pp. 199–205.

[7.19] Henkel, H.J., Müller, N., Nordmann, J., Rogler, W., Rose, W, "Relationship between the chemical structure and the effectiveness of additives in inhibiting water trees", *IEEE Trans.* EI-22 (April 1987) 2, pp. 157–161.

[7.20] Electrical Degradation and Breakdon in Polymers by Len A. Dissodo and John C. Fothergill, Peter Peregrims Ltd., IET (1992), ISBN 0863411967.

[7.21] Patsch, Rainer, "On Tree-Inhibition In Polyethylene", IEEE Transactions on Electrical Insulation, Vol. 14 No. 4, August 1979.

[7.22] Roseen, P.A., Reitbeger, T., Gubanski, S.M. and Gedde, U.W., "PD Resistance of Thermally Aged Polyethylene and Carbonyl Containing Model Polymers", IEEE Transaction on Dielectrics and Electrical Insulation, Vol. 6 No.2, April 1999.

[7.23] David, E., Parpal, J.L. and Crine, J.P., "Influence of Internal Mechanical Stress and Strain on Electrical Performance of Polyethylene Electrical Treeing Resistance", IEEE Transaction on Dielectrics and Electrical Insulation, Vol. 3 No.2, April 1996.

[7.24] Hampton, B.F., Meats, R.J., "Diagnostic measurement at UHF in gas insulated substations", IEE Proceedings, Vol. 135, Pt.C.No.2, March 1988.

[7.25] Garg Sachin, "Frequency domain analysis using antenna and spectrum analyzer for the measurement of partial breakdown activity produced by ac and dc voltages", M.Tech. thesis, Department of Electrical Engineering, IIT Kanpur, June 2008.

[7.26] König, D., Rao, N., "Partial Discharges in Electrical Power Apparatus", Vde-Verlag gmbh, Berlin (1993) Germany.

[7.27] Lawson, W.G., "Effect of temperature and techniques of measurement on the intrinsic electric strength of polyethylene", Proc. IEE 113 (1966)1, pp. 197–202.

[7.28] O' Dwyer, J.J., "The theory of dielectric breakdown of solids", Oxford (1964).

[7.29] Zeller, H.R., "Breakdown and prebreakdown phenomena in solid dielectrics", IEEE Trans. EI-22 (April 1987) 2, pp. 115–122.

[7.30] Eichhorn, R.M., "Treeing in solid extruded electrical insulation", IEEE Trans. EI-12 (Feb. 1977) 1, pp. 2–18.

[7.31] Pilling, Jurgen, "A contribution to the interpretation of life curves and dielectric measurements and investigation of solid polymer dielectrics", Dissertation B, TU Dresden (1976) (in German).

[7.32] Löffelmacher, G., "On the physical and chemical processes in the development of discharge channels in PE and epoxyresin in nonunuform field with ac", Dissertation, TU Hannover (1976) (in German).

[7.33] Vogel, Ulrich, "A contribution towards clarification of the development and effects of electrochemical treeing in polyethylene", Dissertation, TU Hannover (1982) (in German).

[7.34] Given, M. J., Fouracre, R.A., Crichton, B.H, "The role of ions in the mechanism of water tree growth," *IEEE Trans.* EI-22 (April 1987) 2, pp. 151–156.

[7.35] Arora, Ravindra "A contribution to the estimation of dynamic short circuit stresses in plastic insulated cables of medium voltage range", Dissertation, TU Dresden (1973) (in German).

[7.36] Scharff, C, "Effect of cross-linking and increase in temperature on the life expectancy of PE insulation under alternating voltage", Dissertation A, TU Dresden 1988 (in German).

[7.37] Kreuger, F.H., "Long duration testing of PE insulated cables, their methods and criterion", CIGRE-Report No. 21-02 (1968).

[7.38] Li, Chol Su, "Life expectancy of PE-Cable insulation with simultaneous thermal and electrical stresses with alternating voltage", Dissertation, TU Dresden (1989) (in German).

[7.39] Philippow, E, "*Handbook of Electrical Engineering*", Volume 6, VEB Verlag Technik Berlin (1988).

Index

High Voltage and Electrical Insulation Engineering, First Edition. Ravindra Arora, Wolfgang Mosch.
© 2011 Institute of Electrical and Electronics Engineers. Published 2011 by John Wiley & Sons, Inc.

IEEE Press Series on Power Engineering

1. *Principles of Electric Machines with Power Electronic Applications, Second Edition*
   M. E. El-Hawary
2. *Pulse Width Modulation for Power Converters: Principles and Practice*
   D. Grahame Holmes and Thomas Lipo
3. *Analysis of Electric Machinery and Drive Systems, Second Edition*
   Paul C. Krause, Oleg Wasynczuk, and Scott D. Sudhoff
4. *Risk Assessment of Power Systems: Models, Methods, and Applications*
   Wenyuan Li
5. *Optimization Principles: Practical Applications to the Operations of Markets of the Electric Power Industry*
   Narayan S. Rau
6. *Electric Economics: Regulation and Deregulation*
   Geoffrey Rothwell and Tomas Gomez
7. *Electric Power Systems: Analysis and Control*
   Fabio Saccomanno
8. *Electrical Insulation for Rotating Machines: Design, Evaluation, Aging, Testing, and Repair*
   Greg Stone, Edward A. Boulter, Ian Culbert, and Hussein Dhirani
9. *Signal Processing of Power Quality Disturbances*
   Math H. J. Bollen and Irene Y. H. Gu
10. *Instantaneous Power Theory and Applications to Power Conditioning*
    Hirofumi Akagi, Edson H. Watanabe, and Mauricio Aredes
11. *Maintaining Mission Critical Systems in a 24/7 Environment, Second Edition*
    Peter M. Curtis
12. *Elements of Tidal-Electric Engineering*
    Robert H. Clark
13. *Handbook of Large Turbo-Generator Operation Maintenance, Second Edition*
    Geoff Klempner and Isidor Kerszenbaum
14. *Introduction to Electrical Power Systems*
    Mohamed E. El-Hawary
15. *Modeling and Control of Fuel Cells: Disturbed Generation Applications*
    M. Hashem Nehrir and Caisheng Wang
16. *Power Distribution System Reliability: Practical Methods and Applications*
    Ali A Chowdhury and Don O. Koval

$103.95

17. *Introduction to FACTS Controllers: Theory, Modeling, and Applications*
Kalyan K. Sen and Mey Ling Sen
18. *Economic Market Design and Planning for Electric Power Systems*
James Momoh and Lamine Mili
19. *Operation and Control of Electric Energy Processing Systems*
James Momoh and Lamine Mili
20. *Restructured Electric Power Systems. Analysis of Electricity Markets with Equilibrium Models*
Xiao-Ping Zhang
21. *An Introduction to Wavelet Modulated Inverters*
S.A. Saleh and M. Azizur Rahman
22. *Probabilistic Transmission System Planning*
Wenyuan Li
23. *Control of Electric Machine Drive Systems*
Seung-Ki Sui
24. *High Voltage and Electrical Insulation Engineering*
Ravindra Arora and Wolfgang Mosch
25. *Practical Lighting Design with LEDs*
Ron Lenk and Carol Lenk
26. *Electricity Power Generation: The Changing Dimensions*
Digambar M. Tagare
27. *Electric Distribution Systems*
Abdelhay A. Sallam and Om P. Malik
28. *Maintaining Mission Critical Systems in a 24/7 Environment, Second Edition*
Peter M. Curtis
29. *Power Conversion and Control of Wind Energy Systems*
Bin Wu, Yongqiang Lang, Navid Zargan, and Samir Kouro
30. *Integration of Distributed Generation in the Power System*
Math Bollen and Fainan Hassan

Forthcoming Titles

Doubly Fed Induction Machine: Modeling and Control for Wind Energy Generation Applications
Gonzalo Abaci, Jesus Lopez, Miguel Rodriguez, Luis Marroyo, and Grzegorz Iwanski

High Voltage Protection for Telecommunications
Steven W. Blume